Cell Press Reviews:
Cancer Therapeutics

Cell Press Reviews:

Cancer Therapeutics

Curated by

Rebecca Alvania; Scientific Editor, Cell Press
Ann Cheung; Scientific Editor, Cell Press

**Original Articles Edited by the Following
Cell Press Scientific Editors**

Karen Carniol
Ann Cheung
Michaeleen Doucleff
Laurie Gay
Rhiannon Macrae
Fabiola Rivas
Joanna Schaffhausen

ELSEVIER

AMSTERDAM • BOSTON • HEIDELBERG • LONDON
NEW YORK • OXFORD • PARIS • SAN DIEGO
SAN FRANCISCO • SINGAPORE • SYDNEY • TOKYO
AP Cell is an imprint of Elsevier

CellPress

Emilie Marcus, CEO, Editor-in-Chief
Joanne Tracy, Vice President of Business
Development
Keith Wollman, Vice President of Operations
Peter Lee, Publishing Director
Deborah Sweet, Publishing Director
Katja Brose, Editorial Director, Reviews
Strategy
Elena Porro, Editorial Director, Content
Development
Meredith Adinolfi, Director of Production
Jonathan Atkinson, Director of Marketing

Science and Technology Books

Suzanne BeDell, Managing Director
Laura Colantoni, Vice President & Publisher
Amorette Pedersen, Vice President, Channel
Management & Marketing Operations
Tommy Doyle, Senior Vice President,
Strategy, Business Development & Continuity
Publishing
Dave Cella, Publishing Director, Life Sciences
Janice Audet, Publisher
Elizabeth Gibson, Editorial Project Manager
Julia Haynes, Production Manager
Ofelia Chernock, Portfolio Marketing
Manager
Melissa Fulkerson, Senior Channel Manager
Cory Polonetsky, Director, Channel Strategy &
Pricing

AP Cell is an imprint of Elsevier
32 Jamestown Road, London NW1 7BY, UK
225 Wyman Street, Waltham, MA 02451, USA
525 B Street, Suite 1800, San Diego, CA 92101-4495, USA

British Library Cataloguing-in-Publication Data
A catalogue record for this book is available from the British Library

Library of Congress Cataloging-in-Publication Data
A catalog record for this book is available from the Library of Congress

ISBN: 978-0-12-420192-7

For information on all AP Cell publications visit our
website at www.store.elsevier.com

Typeset by TNQ Books and Journals

This book has been manufactured using Print On Demand technology.
Each copy is produced to order.
14 15 16 17 18   10 9 8 7 6 5 4 3 2 1

Contents

About Cell Press

Cell Press is a leading publisher in the biological sciences and is committed to improving scientific communication through the publication of exciting research and reviews. Cell Press publishes 30 journals, including the Trends reviews series, spanning the breadth of the biological sciences. Research titles published by Cell Press include *Cell*, *Cancer Cell*, *Cell Stem Cell*, *Cell Host & Microbe*, *Cell Metabolism*, *Chemistry & Biology*, *Current Biology*, *Developmental Cell*, *Immunity*, *Molecular Cell*, *Neuron*, *Structure*, and the open-access journal *Cell Reports*. In addition to publishing high-impact findings, Cell Press research journals publish a wide variety of peer-reviewed review and opinion articles, essays from leaders in the field, graphical SnapShots, science news articles, and much more. Cell Press is also the publisher of three society journals: *Biophysical Journal*, *American Journal of Human Genetics*, as well as the open-access journal *Stem Cell Reports*.

The *Trends* reviews journals are also part of the Cell Press family and consist of 14 monthly review titles that publish in a range of areas across the biological sciences. Peer-reviewed and thoroughly edited review and opinion articles cover the most recent developments in relevant fields in an authoritative, succinct and broadly accessible manner. Together with a range of additional shorter formats, *Trends* journals collectively provide a forum for hypothesis and debate.

As part of its mission to be a leader in scientific communication, Cell Press also organizes scientific meetings across a wide range of topics, hosts online webinars to bring leading scientists to the broadest international audience, and is committed to promoting innovation in scientific publishing.

Contributors

Corresponding authors and affiliations

David J. Adams
Department of Medicine, Duke University Health System, Duke Box # 2638, Research Drive, Durham, NC 27710, USA

C. Anthony Blau
Department of Medicine/Hematology, Institute for Stem Cell and Regenerative Medicine, University of Washington, Seattle, WA 98109, USA; Partners in Personal Oncology, Seattle, WA 98102, USA

Alberto Chiarugi
Department of Preclinical and Clinical Pharmacology, University of Florence, Viale Pieraccini 6, 50139 Firenze, Italy

Michele De Palma
The Swiss Institute for Experimental Cancer Research (ISREC), School of Life Sciences, Swiss Federal Institute of Technology Lausanne (EPFL), CH-1015 Lausanne, Switzerland

Hugues de Thé
University Paris Diderot, Sorbonne Paris Cité, Hôpital St Louis 1, Avenue Claude Vellefaux, 75475 Paris, Cedex 10, France; INSERM UMR 944, Equipe labellisée par la Ligue Nationale contre le Cancer, Institut Universitaire d'Hématologie, Hôpital St Louis 1, Avenue Claude Vellefaux, 75475 Paris, Cedex 10, France; CNRS UMR 7212, Hôpital St Louis 1, Avenue Claude Vellefaux, 75475 Paris, Cedex 10, France; Pole Sino-Francais des Sciences du Vivant et de Génomique de l'Hôpital Rui Jin, Rui-Jin Hospital affiliated with Jiao Tong University, 197 Rui Jin Road, Shanghai 200025, China; AP-HP, Service de Biochimie, Hôpital St Louis 1, Avenue Claude Vellefaux, 75475 Paris, Cedex 10, France

Haian Fu
Department of Pharmacology, Emory University, Atlanta, GA 30322, USA; Department of Hematology and Medical Oncology, Emory University, Atlanta, GA 30322, USA; Emory Chemical Biology Discovery Center, Emory University, Atlanta, GA 30322, USA

Michael S. Glickman
Infectious Diseases Service and Immunology Program, Memorial Sloan Kettering Cancer Center, 1275 York Avenue, New York, NY 10065, USA

Daniel A. Haber
Massachusetts General Hospital Cancer Center, Harvard Medical School, Charlestown, MA 02129, USA; Howard Hughes Medical Institute, Chevy Chase, MD 20815, USA

Thomas J. Hudson
Ontario Institute for Cancer Research, Toronto, ON M5G 0A3, Canada; Department of Medical Biophysics, University of Toronto, Toronto, ON M5S 1A1, Canada; Department of Molecular Genetics, University of Toronto, Toronto, ON M5S 1A1, Canada

Tony Kouzarides
Gurdon Institute and Department of Pathology, University of Cambridge, Tennis Court Road, Cambridge CB2 1QN, UK

Claire E. Lewis
Department of Oncology, Sheffield Cancer Research Centre, University of Sheffield Medical School, Sheffield, S10 2RX, UK

Martin A. Nowak
Program for Evolutionary Dynamics, Harvard University, Cambridge, MA 02138, USA; Department of Mathematics, Harvard University, Cambridge, MA 02138, USA; Department of Organismic and Evolutionary Biology, Harvard University, Cambridge, MA 02138, USA

Charles L. Sawyers
Human Oncology and Pathogenesis Program, Memorial Sloan Kettering Cancer Center, 1275 York Avenue, New York, NY 10065, USA; Howard Hughes Medical Institute

Matthias Schwab
Dr Margarete Fischer-Bosch Institute of Clinical Pharmacology, Stuttgart, Germany; Department of Clinical Pharmacology, Institute of Experimental and Clinical Pharmacology and Toxicology, University Hospital, Tübingen, Germany

Leonard W. Seymour
Department of Oncology, University of Oxford, Oxford, OX3 7DQ, UK

Hensin Tsao
Wellman Center for Photomedicine, Massachusetts General Hospital, 55 Fruit Street, Boston, MA 02114, USA; Department of Dermatology, Harvard Medical School, Edwards 211, 55 Fruit Street, Boston, MA 02114, USA; MGH Cancer Center, Massachusetts General Hospital, 55 Fruit Street, Boston, MA 02114, USA

Louis M. Weiner
Department of Oncology, Lombardi Comprehensive Cancer Center, Georgetown University Medical Center, 3800 Reservoir Road NW, Washington, DC 20007, USA

Alan Wells
Department of Pathology, University of Pittsburgh and Pittsburgh VAHS, Pittsburgh, PA 15213, USA

Preface

We are very pleased to present *Cell Press Reviews: Cancer Therapeutics*, which brings together review articles from Cell Press journals in order to offer readers a comprehensive and accessible entry point into the rapidly advancing field of cancer therapy. Articles were selected by the editorial staff at Cell Press with an eye toward providing readers an introduction to timely and cutting-edge research written by leaders in the field. Although *Cell Press Reviews: Cancer Therapeutics* is not an exhaustive overview of current cancer therapies, our aim is to give readers insight into some of the most exciting recent developments and the challenges that remain. A wide range of topics are covered within this publication, from the influence of genetics on treatment decisions to the role of the immune system in cancer therapy and the current state of drug discovery.

We are pleased to be able to include contributions from Daniel A. Haber, Director of Massachusetts General Hospital Cancer Center and Professor at Harvard Medical School; Tony Kouzarides, Professor at the University of Cambridge, Deputy Director of the Wellcome Trust/Cancer Research UK Gurdon Institute, and a founder of the cancer drug discovery company Chroma Therapeutics; Charles L. Sawyers, Chair of the Human Oncology and Pathogenesis Program at Memorial Sloan Kettering Cancer Center, President of the American Association for Cancer Research, member of the presidentially appointed National Cancer Advisory Board, and recipient of the 2013 Breakthrough Prize in Life Sciences; and many other prominent researchers in the field. Their insights will offer readers, both experts and those new to the field, a fascinating perspective into this critically important and evolving area of research.

Cell Press Reviews: Cancer Therapeutics is one in a series of books being published as part of an exciting new collaboration between Cell Press and Elsevier Science and Technology Books. Each book in this series is focused on a highly timely topic in the biological sciences. Editors at Cell Press carefully select recently published review articles in order to provide a comprehensive overview of the topic. With the wide range of journals within the

Cell Press family, including research journals such as *Cell* and *Cancer Cell* as well as review journals like *Trends in Molecular Medicine* and *Trends in Pharmacological Sciences*, these compilations provide a diverse and accessible assortment of articles appropriate for a wide variety of readers. You can find additional titles in this series at http://www.store.elsevier.com/CellPressReviews. We are happy to be able to offer this series to such a wide audience via the collaboration with Elsevier Science and Technology Books, and we welcome all feedback from readers on how we might continue to improve the series.

CHAPTER 1

Cell

The Evolving War on Cancer

Daniel A. Haber[1,3,*], Nathanael S. Gray[2], Jose Baselga[1]

[1]Massachusetts General Hospital Cancer Center, Harvard Medical School, Charlestown, MA 02129, USA, [2]Department of Cancer Biology, Dana Farber Cancer Institute and Department of Biological Chemistry and Molecular Pharmacology, Harvard Medical School, Boston, MA 02115, USA, [3]Howard Hughes Medical Institute, Chevy Chase, MD 20815, USA

*Correspondence: haber@helix.mgh.harvard.edu

Cell, Vol. 145, No. 1, April 1, 2011 © 2011 Elsevier Inc.
http://dx.doi.org/10.1016/j.cell.2011.03.026

SUMMARY

Building on years of basic scientific discovery, recent advances in the fields of cancer genetics and medicinal chemistry are now converging to revolutionize the treatment of cancer. Starting with serendipitous observations in rare subsets of cancer, a paradigm shift in clinical research is poised to ensure that new molecular insights are rapidly applied to shape emerging cancer therapies. Could this mark a turning point in the "War on Cancer"?

INTRODUCTION

In the past year, a startling series of clinical studies has brought molecularly targeted therapies to the treatment of diverse cancers. An inhibitor that blocks a specific mutant of the serine/threonine kinase B-RAF (V600E-B-RAF) demonstrated dramatic efficacy in treating metastatic melanoma, a cancer that was long thought to be among the most refractory (Flaherty et al., 2010). An inhibitor of the anaplastic lymphoma kinase (ALK) proved highly effective against a subset of nonsmall cell lung cancers that had been prescreened for an oncogenic EML4-ALK translocation (Kwak et al., 2010). A Janus kinase 2 (JAK2) inhibitor proved beneficial against a form of preleukemia, myelodysplastic syndrome, that harbors an activating JAK2 mutation (Verstovsek et al., 2010). In B cell hematological malignancies, which selectively express the class I phosphatidylinositol 3-kinase (PI3K) δ isoform, a PI3K δ-specific inhibitor has shown remarkable activity in refractory tumors (Furman et al., 2010). Even the immunology front, after years of setbacks, made significant strides when antibodies abrogating anticytotoxic T lymphocyte antigen (CTLA)-4 improved

1

CellPress

survival in patients with advanced melanoma (O'Day et al., 2010). In addition, we anticipate promising results from approaches that combine known treatments, such as using both mitogen-activated protein kinase kinase (MEK) and PI3K inhibitors to treat cancers with mutations in K-RAS (Engelman et al., 2008) and combining antibodies and kinase inhibitors to treat breast cancers with the *HER2* gene amplified (Baselga et al., 2010).

Most impressive, however, is the sense that the timeline of translating molecular studies into the clinic is accelerating. For example, it took 6–8 years to demonstrate the efficacy of B-RAF and epidermal growth factor receptor (EGFR) inhibitors in appropriate clinical studies (Davies et al., 2002; Flaherty et al., 2010; Lynch et al., 2004; Mok et al., 2009; Paez et al., 2004). In contrast, it took just more than 1 year for researchers to demonstrate clinical responses to ALK inhibition in a cohort of genotyped patients, after the initial discovery of the EML4-ALK translocations in a subset of lung cancers (Soda et al., 2007; Kwak et al., 2010).

In this Essay, we explore the important advances in cancer genetics, medicinal chemistry, and clinical strategies contributing to these recent successes and enhanced pace. Then, we discuss the key challenges and objectives for continual success. Clearly, the future of targeted therapies depends upon a detailed understanding of the molecular genetic abnormalities that drive different subsets of cancer, a rich pipeline of promising compounds targeting such lesions, and the appropriate use of companion diagnostics, both to prescreen patients that are likely to respond and to detect early signs of either drug response or acquired resistance.

BREAKING THE CANCER WAR STALEMATE

The "War on Cancer," now 40 years old, has been declared a failure, both in the medical literature (Bailar and Gornik, 1997) and in the general press (Leaf, 2004). In contrast to the marked success of cholesterol-lowering drugs and antihypertensives in reducing the risk of cardiovascular disease, the decline in cancer mortality has been relatively modest. This decrease has been attributed primarily to screening for breast and colon cancer and preventive measures, such as a reduction in smoking and the declining use of postmenopausal estrogen replacement. Could the new forms of cancer treatment provide hope for reduced mortality? And what are the critical components that are required to achieve a sustained impact on the disease?

It is perhaps ironic that the bill signed by President Richard M. Nixon in 1971, which was ultimately labeled the War on Cancer, triggered a massive investment in basic science. The research was undertaken with the assumption that unbiased fundamental research would hold the key to unlocking the secrets of cancer cells, and indeed, our current knowledge about cellular biology and molecular genetics owes much to this national investment.

Although the anticipated timeline for clinical applications of basic research may have been optimistic, recent successes in targeted therapies are based on the accumulated knowledge generated by years of fundamental research in genetics, signaling, cellular, and molecular biology. In particular, we have learned that the genesis of cancer mirrors the process of mutation and selection underlying our own evolution, and the signals that drive or suppress proliferation, apoptosis, differentiation, and quiescence in cancer cells mimic those of normal development.

We are finally capable of manipulating these signals in some cancers to achieve a profound initial response in the tumor. However, the acquisition of resistance by cancer cells prompts the need for even greater understanding of cellular mechanisms underlying cancer cells' plasticity and adaptation. Single driving genetic lesions may provide clear therapeutic targets in some cancers, but a more profound understanding of interconnected signaling pathways may be required to tackle the majority of tumors.

DRAWING FROM SOMATIC CANCER GENETICS TO FIND THERAPEUTIC TARGETS

Early cancer genetic studies focused on inherited cancer syndromes in which a single germline mutation is responsible for cancer susceptibility within a family. Such studies provide clear and compelling evidence for the one lesion, typically in a tumor suppressor gene, that is capable of initiating tumorigenesis, as opposed to a vast number of somatic aberrations accumulating during cancer progression. Directed genotyping studies uncovered additional somatic mutations, chromosome translocations, and loss of heterozygosity patterns, pointing to recurrent (and hence probably significant) lesions in many cancers. Functional studies of introduced oncogenes and tumor suppressor genes in cells and model organisms readily demonstrated the powerful impact of mutating a single gene in triggering malignancy.

However, it was the decision to undertake whole-scale sequencing of cancer genomes, first by the Sanger Center and then by US and international consortia, that provided a comprehensive view of somatic mutational landscapes in cancer and potential therapeutic targets. Important successes of genome-scale sequencing included the discovery of highly recurrent and specific mutations in BRAF and PI3K (Davies et al., 2002; Samuels et al., 2004). However, most mutations identified in such analyses appear to be relatively rare and are not shared across multiple cancers (Beroukhim et al.,

2010). Although such mutations may be grouped within large functional pathways, predominant therapeutic targets have yet to emerge for the majority of epithelial cancers, which constitute ~85% of all cancers. Thus, although large cancer genome sequencing projects are underway for many different cancer types, the identification of promising drug targets may ultimately require detailed biological insights to complement mutational analyses.

Given that numerous mutations in human cancers occur at low frequency, it is essential to distinguish the mutations that constitute essential "drivers" of tumorigenesis from those that are "passengers" without functional significance (for more on driver and passenger mutations, see Ashworth et al., 2011 in this issue). Of course, the response of cancers to a targeted inhibitor is ultimately the most compelling evidence for the relevance of a given target. For example, the dramatic clinical response to EGFR inhibitors by patients whose lung cancers harbor an activating EGFR mutation clearly demonstrates that these EGFR mutations are not simply bystanders of mutational load, but rather drivers of malignant proliferation (Lynch et al., 2004; Paez et al., 2004). In fact, the EGFR inhibitors trigger massive tumor apoptosis in these lung cancers, suggesting that these cells are "addicted" to the EGFR signaling pathway for survival.

The concept of "oncogene addiction" emphasizes the critical role played by the "wiring" of signaling pathways in a cancer cell; specifically, a cancer cell's signaling networks may depend on a single mutationally activated driving pathway that, when disrupted, triggers cell death (in effect, its "Achilles heel") (Weinstein, 2002). However, identifying such extraordinary drug targets and translating these into effective therapy is not always straightforward, as illustrated by the development of B-RAF inhibitors against melanoma. The failure of sorafenib, an inhibitor of B-RAF, in clinical trials of metastatic melanoma challenged the validity of V600E-B-RAF as a good drug target until PLX4032, a more potent and specific inhibitor against V600E B-RAF, showed dramatic efficacy (Eisen et al., 2006; Flaherty et al., 2010). It is in this setting that preclinical modeling of oncogene addiction, using large panels of cancer-derived cell lines, is emerging as an effective approach to validating both target and inhibitor. Individual cell lines have limited predictive value. However, when they are studied in large aggregates, they recapitulate much of the genetic heterogeneity of human cancers, as well as their specific hypersensitivity profiles to inhibitors of EGFR, ALK, MET, FGFR, and B-RAF (McDermott et al., 2007).

Preclinical efforts at target identification and validation have also relied on strategies that suppress gene expression using interference RNA (RNAi), although clinically validated targets have not yet emerged from such genome-wide screens. Instead of matching a single compound against the entire genetic heterogeneity of human cancers, RNAi screens have typically focused on screening the entire kinome against a selected cell type. Important applications of this strategy include: searching for new targets for which RNAi-induced knockdown may reverse acquired resistance to first-line targeted inhibitors, and screens for "synthetic lethality," in which a target only becomes essential within a specific genetic context (Berns et al., 2004). For instance, the clinical effectiveness of poly-ADP ribose polymerase (PARP) inhibitors in breast and ovarian cancers with BRCA gene mutations is attributable to their inhibition of a parallel DNA repair pathway that becomes critical for cell viability only in the setting of BRCA gene inactivation. Finally, both RNAi and drug screens are essential to model effective combinations of agents required to block interdependent cellular pathways implicated in intrinsic and acquired drug resistance. Drug resistance has been attributed to numerous mechanisms, including the appearance of additional or "second site" mutations within drug-binding sites (Kobayashi et al., 2005; Pao et al., 2005; Talpaz et al., 2006; Zhou et al., 2009), the activation of parallel signaling pathways (Engelman et al., 2007), or even the induction of drug resistance quiescence states (Sharma et al., 2010). To counter all these mechanisms requires an even better understanding of the wiring diagram of cancer cells.

CHALLENGING MEDICINAL CHEMISTRY TO DESIGN NEW CLASSES OF INHIBITORS

The history of targeted cancer therapy is now intricately linked with the success of imatinib (Gleevec) to treat chronic myeloid leukemia (CML). The prototype kinase inhibitor imatinib blocks the ABL kinase, which is activated in the BCR-ABL chimeric fusion. From its first identification as the Philadelphia chromosome (Nowell and Hungerford, 1960) to its molecular characterization as the sole genetic driver of CML, BCR-ABL is the perfect drug target (Druker, 2004). The development of imatinib was the result of a structure-activity relationship-guided optimization of a phenylaminopyrimidine lead compound that was originally identified as an inhibitor of protein kinase C (Capdeville et al., 2002). Indeed, imatinib's eventual application as a BCR-ABL inhibitor owes much to serendipity, intuition, and the dedication of individual scientists.

The success of imatinib resolved numerous concerns about using ATP-mimics as kinase inhibitors. It demonstrated that these inhibitors can compete effectively for binding with the abundant pool of cellular ATP and that they can achieve selectivity despite closely related catalytic pockets in other kinases. In addition, it proved that a class of signaling molecules with broad expression patterns in multiple normal tissues could indeed be successful drug targets, given the exceptional sensitivity of genetically defined subsets of cancer. Imatinib was initially aimed at suppressing PDGFR signaling that is implicated in the proliferation of coronary endothelial cells; its fortuitous "off-target" effects on the ABL and c-KIT kinases launched a revolution in cancer therapy, and its anti-PDGFR effects ultimately found their place in the treatment of rare leukemias with PDGFR-dependent translocations (Druker, 2004). The admittedly serendipitous saga of imatinib spawned a broad and systematic effort throughout the pharmaceutical industry that is now poised to radically change the essential tools for cancer treatment.

Tremendous investment in medicinal chemistry, primarily from the pharmaceutical sector, has resulted in the development of numerous efficient strategies for designing potent and selective kinase inhibitors. For example, PLX4720 was developed by an innovative approach called "fragment-based screening," in which small molecular fragments with low affinity for V600E B-RAF were identified and then optimized using rational structure-guided drug design (Tsai et al., 2008). In addition, structural biology of kinases has become an integral part of drug optimization, with large-scale initiatives such as the Structural Genomics Consortium providing an increasingly large fraction of new depositions (Marsden and Knapp, 2008).

Nonetheless, serendipity still plays a major part in successful drug discovery. For example, crizotinib was originally designed as a c-MET inhibitor, but its fortuitous off-target activity against ALK drove its efficacy in EML4-ALK-dependent lung cancer (Kwak et al., 2010). Significant developmental hurdles still exist for kinase inhibitors, including the high interspecies variation in their toxic side effects and the difficulty in deciphering which of these are on-target versus off-target effects of the inhibitor. Despite their large investment, commercial enterprises have been reluctant to develop inhibitors against kinases that have not already been subject to intense biological investigation. This has resulted in an abundance of drugs against a relatively small number of well-validated targets but a dearth of inhibitors with sufficient selectivity to validate the vast majority of the kinome pharmacologically (Fedorov et al., 2010). Although there are remarkable examples of how broadly active kinase inhibitors, such as sunitinib, may be well tolerated, recent trends have focused more on inhibition of specific mutant targets. These include targeting the specific B-RAF mutant V600E in melanoma, mutationally activated forms of EGFR in lung cancer, and imatinib-resistant T315I-BCR-ABL in CML (Flaherty et al., 2010; Zhou et al., 2009).

Although many potential cancer therapies remain in the pool of untargeted kinases, new classes of cancer drugs on the horizon are reaching outside of the kinome to target modulators of protein turnover and folding, phosphatases, chromatin-modifying enzymes, and regulators of cellular metabolism.

Transcription factors and adaptor proteins with relatively large interaction surfaces remain a major challenge for drug design. From a biological standpoint, tumor suppressor genes, which commonly sustain loss-of-function mutations in cancer, are only druggable through pathway components that display synthetic lethality. Moreover, the most common villains of cancer cells, mutations in the oncogene K-RAS and the tumor suppressor p53, have remained recalcitrant to direct targeting approaches. Many of these targets will require chemists to invent entirely new classes of compounds or develop novel approaches for modulating their activity.

Targeted inhibitors stand to benefit from integrated development within the appropriate biological and genetic contexts. Preclinical screens need to test inhibition of an oncogenic target within the appropriate cancer cells, in which the targeted gene is biologically relevant. When possible, biomarkers that identify the responsive subset of cancers should be selected during early preclinical development, rather than awaiting retrospective analyses of clinical trials. As with cancer cell line studies, animal models used to optimize drug dosing and toxicity profiles are most informative if they recapitulate the genetic context of the relevant subset of human cancer. These approaches mark a departure from classical drug testing paradigms that have typically relied on testing a small number of nonannotated cancer cell lines and xenograft mouse models matched to tissue type, rather than underlying genetic lesion. However, the increasing cost of testing large numbers of new agents that may be highly selective for subsets of cancer necessitates detailed information to match the drug, disease subtype, and biomarker before entry into clinical trials.

REDESIGNING CANCER CLINICAL TRIALS: GENOTYPE FIRST, MONITOR IN REAL TIME

There is a growing consensus that traditional designs of cancer clinical trial are not well suited to address the current needs of drug development. These clinical trials typically begin with a phase one, in which the maximally tolerated dose of a new drug is defined by increasing dosage in a small number of patients (who have failed standard therapies and are willing to be treated with a new untested compound with uncertain benefits). This is followed by a phase two trial in a larger patient cohort with a specific cancer type (without regard to the cancer's genotype or biomarkers); this phase aims at defining efficacy at the appropriate drug dose. Eventually, if clinical activity has been observed in the phase two trial, a large randomized phase three trial is conducted in which the new drug, either alone or in combination with currently used drugs, is directly compared with the standard regimen. If successful, phase three leads to FDA registration for the new drug.

This strategy has resulted in a high failure rate, and even in studies with a positive outcome, the benefits are mostly incremental in nonselected patient populations. Besides time and expense, the underlying premise of such trial designs is not suited to the new world of targeted cancer therapy. In particular, defining the maximal tolerated dose for selective inhibitors in patients whose tumor does not carry biomarkers predictive of response for that inhibitor is far less relevant than preselecting tumors that are likely to respond to the inhibitor and defining the therapeutic index within these relevant cases. Traditional multi-institutional clinical trials have relied on enrolling large numbers of unselected patients, followed by retrospective analysis of tumor markers in a fraction of cases enrolled. Thus, they do not readily lend themselves to detailed pretreatment genotyping and may be underpowered to assess drug effectiveness against rare tumor subsets. Given the early and dramatic responses in genotype-selected cancers, the need and even the ethics of large clinical trials—such as the one randomizing V600E-positive melanoma patients between highly effective PLX4032 and minimally effective cytotoxic chemotherapy—is being vigorously debated. For all of these reasons, a new model is emerging, one with an extended phase one that seeks from the beginning to build upon preclinical information and enrich trial populations for the genetic marker of interest (Figure 1).

Changes in both method and expected outcome do not come easily, either for physicians or for pharmaceutical companies. The fragmentation of a market among many subtypes of cancer, each requiring a different treatment, means the end of blockbuster compounds for companies. However, it is worth noting that a highly effective drug such as imatinib, administered daily for many years to patients throughout the world, can still provide major financial returns to its manufacturer. For clinicians, targeted cancer therapy requires breaking down traditional barriers in clinical medicine. Phase one and novel drug testing teams are no longer focused exclusively on patients with advanced refractory disease, irrespective of tumor characteristics. Instead, early drug testing is being integrated into the initial care of patients, with the expressed goal of trying to match the drug under study with specific subsets of cancers, within a timeline that may derive real benefit for patients participating in such trials.

Most importantly, the pathologist is now in the critical position of going beyond the standard histopathological diagnosis of cancer, providing the oncologist with the key biomarkers that direct appropriate targeted therapy. In some cases, this may involve prescreening tumors to enable a successful clinical trial. For example, even for a common cancer, such as nonsmall cell lung cancer, genotype-drug combinations require testing a large number of cases to enrich for EGFR mutations (10% of all cases), EML4-ALK

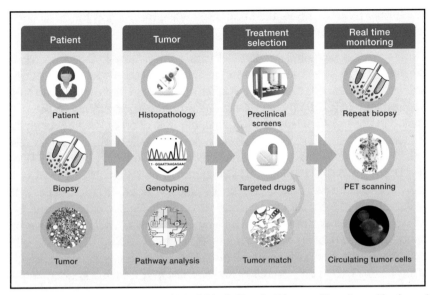

FIGURE 1 Matching Each Cancer with Individually Targeted Therapy: Steps toward Implementing a New Clinical Strategy

In genotype-directed cancer therapy, the malignant tumor specimen is subjected to detailed biomarker analysis, including immunohistochemical as well as DNA-based markers. These biomarkers are interpreted within the relevant signaling pathway to identify a potential vulnerability for the cancer. Selection of an appropriate therapy (often oral agents of limited toxicity) is dependent upon prior preclinical validation of drug/genotype pairing and detailed analysis of drug/target interactions. Early drug response and development of acquired resistance are monitored by repeat biopsy of the tumor or, noninvasively, by functional imaging or circulating tumor cell analysis.

rearrangements (4%), or MET, HER2, and B-RAF abnormalities (1%–2% each). In the landmark clinical trial of crizotinib for EML4-ALK-translocated lung cancer, 1500 patients were prescreened to identify 82 cases with the translocation. These patients were then selected for treatment with crizotinib, producing benefit in 90% of cases (57% responses and 33% stable disease) (Kwak et al., 2010).

Despite dramatic initial responses in genotype-selected clinical trials of epithelial cancers and melanoma, acquired resistance to targeted agents has emerged as a primary challenge. We are beginning only now to understand the underlying mechanisms of this resistance. Some tumors acquire second-site mutations in the target, which interferes with drug binding (Kobayashi et al., 2005; Pao et al., 2005; Talpaz et al., 2006), whereas other tumors evolve alternative signaling pathways, which compensate for the disrupted oncogenic signal (Chandarlapaty et al., 2011; O'Reilly et al., 2006; Tabernero et al., 2008). Achieving long-lasting control of malignancy in this setting may require sequential treatments while monitoring the evolution of tumor genotypes during therapy.

In some cases, signaling feedback loops immediately activate alternative pathways, which must be suppressed at the outset by using a combination of targeted therapies. For example, the initial observation that mammalian target of rapamycin (mTOR) inhibitors have limited anticancer activity in the clinic resulted in the identification of compensatory activation of insulin growth factor 1 receptor (IGF-1R) when mTOR is blocked (O'Reilly et al., 2006; Tabernero et al., 2008). This led to a phase one clinical study in which both mTOR and IGF-1R are inhibited. Indeed, this combination treatment has shown notable clinical efficacy in breast cancer (Di Cosimo et al., 2010). Similarly, the combined use of inhibitors against the two critically important PI3K and ERK pathways has shown remarkable activity in preclinical models of cells harboring K-RAS activating mutations (Engelman et al., 2008). These and other approaches that rationally combine therapies are currently being explored in clinical trials. Though very promising, these trials may elicit a number of regulatory challenges because they typically involve the combined use of investigational agents, each of which may have a separate path to regulatory approval.

Innovative clinical trial platforms are needed to address the opportunities and challenges posed by the array of targeted agents and their appropriate clinical deployment. In addition to biomarkers for preselection of responsive cancers, early markers of clinical benefit are needed to rapidly assess effectiveness and inform ongoing monitoring of clinical trials. The most promising approaches include: repeat biopsies of tumor sites to measure the impact of new therapies on the degree of target inhibition; functional tumor imaging through positron emissions tomography (PET) or functional magnetic resonance imaging (fMRI); and molecular analysis of both genotypes and signaling pathways within circulating tumor cells in the blood (Maheswaran et al., 2008). Together, these approaches may provide early warnings of acquired drug resistance and identify specific resistance pathways that may direct the choice of second line therapy (Figure 1).

Early tumor interrogation may also result in a dynamic "real-time" measurement of response or failure to the drug under study, allowing for rapid adjustments within clinical trial settings. For example, in breast cancer, the testing of new agents immediately prior to surgery (known as "neo-adjuvant therapy") allows for monitoring of tumor response at the time of surgical resection. Compared to the large number of patients and prolonged clinical follow-up required for traditional postsurgery (or "adjuvant") trials in breast cancer, neo-adjuvant trials allow for rapid testing of multiple drug combinations within small and affordable trials. These new designs rely upon surrogate markers, such as changes in proliferation or apoptosis markers, or even absence of visible tumor at the time of surgery, which will then be correlated with endpoints of clinical benefit, such as time free of disease or improved overall survival.

THE NEXT STEPS IN THE WAR ON CANCER

The War on Cancer has not been lost, nor is it won. Instead, we are now at a turning point, where fundamental knowledge gathered over the past 40 years can, for the first time, be applied directly to the care of patients with cancer. Early results in selected types of cancer are exciting in themselves and in what they forecast for the future of cancer treatment. However, the rush for translational applications of these initial breakthroughs should not be interpreted as evidence that we now know all we need to know about the pathogenesis, progression, and vulnerabilities of cancer. Far from it, basic research in cancer biology is progressing as never before. New and unpredicted discoveries continue to shed light on fundamental mechanisms, from new insights into long-studied genes like *p53*, to advances that are transforming the fields of cancer metabolism, chromatin regulation, and noncoding RNA biology. There is still much to discover, and continued support for basic research is essential to continue the progress in the War on Cancer.

Although the early successes in translating scientific discoveries into targeted treatments are full of promise, they also point to important future challenges. The development of resistance by cancer cells, even to the most dramatically effective therapies, underscores the need for a detailed understanding of these resistance pathways. In addition, we need tools to monitor the emergence of resistance pathways in "real time" and new generations of second line drugs to suppress them. For the majority of cancers, we still have not identified molecular drivers as targets for treatment, and common drivers, such as K-RAS, are currently "undruggable." Both of these facts point to the need for further analyses of genetic and epigenetic changes that drive different cancers, as well as the need for innovative approaches to drug design. Finally, "smaller and smarter" clinical trials are necessary to rapidly capitalize on credible therapeutic signals and apply these to the treatment of early cancers, for which major improvements in long-term survival may be expected (Smith et al., 2007).

Ironically, at this time of unparalleled promise in cancer biology, the biggest risk to progress may be economic. Shrinking support for basic research threatens the very foundation of the success that we are now witnessing in the clinic. Uncertainties in healthcare delivery models may emphasize more uniform and economical application of the "standard of care," prompting the pharmaceutical industry and medical community to shy away from developing and testing innovative and initially expensive new therapies. Public support and enthusiasm are not enhanced by the application of costly targeted therapies in a nontargeted setting, where the measurable clinical impact is often marginal.

Despite these immediate and serious concerns, we remain optimistic that the next 10 years will witness unprecedented progress in the fight against cancer, both in terms of our fundamental understanding and direct clinical applications. The extraordinary complexity of pathways driving malignant proliferation may be daunting, but the clinical impact of a focused strategy targeted at susceptible nodes within a cancer are compelling. This is a time of exceptional promise and a great reward for cancer researchers. It comes none too soon for patients with cancer.

ACKNOWLEDGMENTS

We are most grateful to our colleagues, Drs. Bruce Chabner, Jeff Engelman, David Fisher, Keith Flaherty, Shyamala Maheswaran, and Jeff Settleman, for helpful discussions and critical comments.

REFERENCES

Ashworth, A., Lord, C.J., and Reis-Filho, J.S. (2011). Cell *145*, this issue, 30–38.

Bailar, J.C., III, and Gornik, H.L. (1997). N. Engl. J. Med. *336*, 1569–1574.

Baselga, J., Bradbury, I., Eidtmann, H., Di Cosimo, S., Aura, C., De Azambuja, E., Gomez, H., Dinh, P., Fauria, K., Van Dooren, V., et al. (2010). San Antonio Breast Cancer Symposium. Abstract S3-3. Presented December 10, 2010.

Berns, K., Hijmans, E.M., Mullenders, J., Brummelkamp, T.R., Velds, A., Heimerikx, M., Kerkhoven, R.M., Madiredjo, M., Nijkamp, W., Weigelt, B., et al. (2004). Nature *428*, 431–437.

Beroukhim, R., Mermel, C.H., Porter, D., Wei, G., Raychaudhuri, S., Donovan, J., Barretina, J., Boehm, J.S., Dobson, J., Urashima, M., et al. (2010). Nature *463*, 899–905.

Capdeville, R., Buchdunger, E., Zimmermann, J., and Matter, A. (2002). Nat. Rev. Drug Discov. *1*, 493–502.

Chandarlapaty, S., Sawai, A., Scaltriti, M., Rodrik-Outmezguine, V., Grbovic-Huezo, O., Serra, V., Majumder, P.K., Baselga, J., and Rosen, N. (2011). Cancer Cell *19*, 58–71.

Davies, H., Bignell, G.R., Cox, C., Stephens, P., Edkins, S., Clegg, S., Teague, J., Woffendin, H., Garnett, M.J., Bottomley, W., et al. (2002). Nature *417*, 949–954.

Di Cosimo, S., Bendell, J.C., Cervantes-Ruiperez, A., Roda, D., Prudkin, L., Stein, M.N., Leighton-Swayze, A., Song, Y., Ebbinghaus, S., and Baselga, J. (2010). ASCO Meeting Abstracts *28*, 3008.

Druker, B.J. (2004). Adv. Cancer Res. *91*, 1–30.

Eisen, T., Ahmad, T., Flaherty, K.T., Gore, M., Kaye, S., Marais, R., Gibbens, I., Hackett, S., James, M., Schuchter, L.M., et al. (2006). Br. J. Cancer *95*, 581–586.

Engelman, J.A., Zejnullahu, K., Mitsudomi, T., Song, Y., Hyland, C., Park, J.O., Lindeman, N., Gale, C.M., Zhao, X., Christensen, J., et al. (2007). Science *316*, 1039–1043.

Engelman, J.A., Chen, L., Tan, X., Crosby, K., Guimaraes, A.R., Upadhyay, R., Maira, M., McNamara, K., Perera, S.A., Song, Y., et al. (2008). Nat. Med. *14*, 1351–1356.

Fedorov, O., Müller, S., and Knapp, S. (2010). Nat. Chem. Biol. *6*, 166–169.

Flaherty, K.T., Puzanov, I., Kim, K.B., Ribas, A., McArthur, G.A., Sosman, J.A., O'Dwyer, P.J., Lee, R.J., Grippo, J.F., Nolop, K., and Chapman, P.B. (2010). N. Engl. J. Med. *363*, 809–819.

Furman, R.R., Byrd, J.C., Flinn, I.W., Coutre, S.E., Benson, D.M., Jr., Brown, J.R., Kahl, B.S., Wagner-Johnston, N.D., Giese, N.A., and Yu, A.S. (2010). ASCO Meeting Abstracts *28*, 3032.

Kobayashi, S., Boggon, T.J., Dayaram, T., Jänne, P.A., Kocher, O., Meyerson, M., Johnson, B.E., Eck, M.J., Tenen, D.G., and Halmos, B. (2005). N. Engl. J. Med. *352*, 786–792.

Kwak, E.L., Bang, Y.J., Camidge, D.R., Shaw, A.T., Solomon, B., Maki, R.G., Ou, S.H., Dezube, B.J., Jänne, P.A., Costa, D.B., et al. (2010). N. Engl. J. Med. *363*, 1693–1703.

Leaf, C. (2004). Fortune Magazine. March 22 issue, 365076.

Lynch, T.J., Bell, D.W., Sordella, R., Gurubhagavatula, S., Okimoto, R.A., Brannigan, B.W., Harris, P.L., Haserlat, S.M., Supko, J.G., Haluska, F.G., et al. (2004). N. Engl. J. Med. *350*, 2129–2139.

Maheswaran, S., Sequist, L.V., Nagrath, S., Ulkus, L., Brannigan, B., Collura, C.V., Inserra, E., Diederichs, S., Iafrate, A.J., Bell, D.W., et al. (2008). N. Engl. J. Med. *359*, 366–377.

Marsden, B.D., and Knapp, S. (2008). Curr. Opin. Chem. Biol. *12*, 40–45.

McDermott, U., Sharma, S.V., Dowell, L., Greninger, P., Montagut, C., Lamb, J., Archibald, H., Raudales, R., Tam, A., Lee, D., et al. (2007). Proc. Natl. Acad. Sci. USA *104*, 19936–19941.

Mok, T.S., Wu, Y.L., Thongprasert, S., Yang, C.H., Chu, D.T., Saijo, N., Sunpaweravong, P., Han, B., Margono, B., Ichinose, Y., et al. (2009). N. Engl. J. Med. *361*, 947–957.

Nowell, P., and Hungerford, D. (1960). Science *132*, 1497.

O'Day, S., Hodi, F.S., McDermott, D.F., Weber, R.W., Sosman, J.A., Haanen, J.B., Zhu, X., Yellin, M.J., Hoos, A., and Urba, W.J. (2010). ASCO Meeting Abstracts *28*, 4.

O'Reilly, K.E., Rojo, F., She, Q.-B., Solit, D., Mills, G.B., Smith, D., Lane, H., Hofmann, F., Hicklin, D.J., Ludwig, D.L., et al. (2006). Cancer Res. *66*, 1500–1508.

Paez, J.G., Jänne, P.A., Lee, J.C., Tracy, S., Greulich, H., Gabriel, S., Herman, P., Kaye, F.J., Lindeman, N., Boggon, T.J., et al. (2004). Science *304*, 1497–1500.

Pao, W., Miller, V.A., Politi, K.A., Riely, G.J., Somwar, R., Zakowski, M.F., Kris, M.G., and Varmus, H. (2005). PLoS Med. *2*, e73.

Samuels, Y., Wang, Z., Bardelli, A., Silliman, N., Ptak, J., Szabo, S., Yan, H., Gazdar, A., Powell, S.M., Riggins, G.J., et al. (2004). Science *304*, 554.

Sharma, S.V., Lee, D.Y., Li, B., Quinlan, M.P., Takahashi, F., Maheswaran, S., McDermott, U., Azizian, N., Zou, L., Fischbach, M.A., et al. (2010). Cell *141*, 69–80.

Smith, I., Procter, M., Gelber, R.D., Guillaume, S., Feyereislova, A., Dowsett, M., Goldhirsch, A., Untch, M., Mariani, G., Baselga, J., et al. (2007). Lancet *369*, 29–36.

Soda, M., Choi, Y.L., Enomoto, M., Takada, S., Yamashita, Y., Ishikawa, S., Fujiwara, S., Watanabe, H., Kurashina, K., Hatanaka, H., et al. (2007). Nature *448*, 561–566.

Tabernero, J., Rojo, F., Calvo, E., Burris, H., Judson, I., Hazell, K., Martinelli, E., Ramon y Cajal, S., Jones, S., Vidal, L., et al. (2008). J. Clin. Oncol. *26*, 1603–1610.

Talpaz, M., Shah, N.P., Kantarjian, H., Donato, N., Nicoll, J., Paquette, R., Cortes, J., O'Brien, S., Nicaise, C., Bleickardt, E., et al. (2006). N. Engl. J. Med. *354*, 2531–2541.

Tsai, J., Lee, J.T., Wang, W., Zhang, J., Cho, H., Mamo, S., Bremer, R., Gillette, S., Kong, J., Haass, N.K., et al. (2008). Proc. Natl. Acad. Sci. USA *105*, 3041–3046.

Verstovsek, S., Kantarjian, H., Mesa, R.A., Pardanani, A.D., Cortes-Franco, J., Thomas, D.A., Estrov, Z., Fridman, J.S., Bradley, E.C., Erickson-Viitanen, S., et al. (2010). N. Engl. J. Med. *363*, 1117–1127.

Weinstein, I.B. (2002). Cancer Sci. *297*, 63–64.

Zhou, W., Ercan, D., Chen, L., Yun, C.H., Li, D., Capelletti, M., Cortot, A.B., Chirieac, L., Iacob, R.E., Padera, R., et al. (2009). Nature *462*, 1070–1074.

CHAPTER 2

Trends in Genetics

Can We Deconstruct Cancer, One Patient at a Time?

C. Anthony Blau[1,2,*], Effie Liakopoulou[1,2]

[1]Department of Medicine/Hematology, Institute for Stem Cell and Regenerative Medicine, University of Washington, Seattle, WA 98109, USA, [2]Partners in Personal Oncology, Seattle, WA 98102, USA

*Correspondence: tblau@uw.edu

Trends in Genetics, Vol. 29, No. 1, January 2013 © 2013 Elsevier Inc.
http://dx.doi.org/10.1016/j.tig.2012.09.004

SUMMARY

Patients with cancer face an ever-widening gap between the exponential rate at which technology improves and the linear rate at which these advances are translated into clinical practice. Closing this gap will require the establishment of learning loops that intimately link lab and clinic and enable the immediate transfer of knowledge, thereby engaging highly motivated patients with cancer as true partners in research. Here, we discuss the goal of creating a distributed network that aims to place world-class resources at the disposal of select patients with cancer and their oncologists, and then use these intensively monitored individual patient experiences to improve collective understanding of how cancer works.

CANCER TREATMENT IN THE AGE OF OMICS

Although genomic, transcriptomic, proteomic, and metabolomic (here collectively referred to as 'omic') technologies are dramatically improving understanding of health and disease, the goal of exploiting this information for the benefit of patients is only beginning to be met. Cancer research is especially well positioned to benefit from advances in omic technology, and next-generation sequencing (NGS) has already yielded profound new insights into cancer biology. We now know that each cancer arises from a distinct combination of genomic alterations and, thus, is unique [1]. Cancers are also remarkably heterogeneous within individual patients, and subclones expand and contract in response to selective pressure [2–5]. These discoveries pose significant challenges to the design of clinical trials in patients with

15

CellPress

cancer. Two strategies reflect diametrically opposite frameworks for advancing cancer treatment: (i) an established paradigm that compares outcomes between groups of patients randomized to different treatments and implicitly deals with cancer as a black box; and (ii) a nascent mechanistic approach that aims to understand cancer within individuals. Fully capitalizing on the opportunities that omic technologies provide requires that we acknowledge the limitations of randomized trials and develop next-generation trials that rigorously evaluate intensively monitored individual patient experiences. An essential feature of this new approach is the involvement of patients with cancer as active partners in discovery-based research.

CANCER AS A BLACK BOX

Despite their lack of molecular tools, early pioneers in cancer therapeutics were remarkably successful in developing treatments for childhood acute lymphoblastic leukemia (ALL), Hodgkin's disease, and testicular cancer. These gains were achieved through a largely empirical process culminating in the evaluation of drugs and drug combinations in randomized clinical trials. Randomized trials assign patients to two or more groups that are designed to be comparable. Patients within groups receive the same treatment, whereas patients assigned to different groups receive different treatments, and the outcomes between groups are compared. To achieve comparability, the randomization process can be adjusted to balance groups according to recognized prognostic factors, such as tissue of origin, age, clinical status, microscopic appearance (grade), and extent of disease (stage). The assumption inherent in randomized trials is that patients assigned to different treatment arms are sufficiently similar as to allow for a valid comparison between treatments. This assumption withstands scrutiny either when a large fraction of tumors responds to an empirically defined regimen (e.g., childhood ALL), or when a molecular marker strongly predicts treatment responses, as occurs with the breakpoint cluster region–c-*abl* oncogene 1 (BCR-ABL) translocation in chronic myeloid leukemia (CML), human epidermal growth factor receptor 2 (HER2) amplification in breast cancer, or point mutations in v-*kit* Hardy-Zuckerman 4 feline sarcoma viral oncogene homolog (c-KIT) or v-*raf* murine sarcoma viral oncogene homolog B1 (BRAF) that occur in gastrointestinal stromal tumor (GIST) and melanoma, respectively. A significant advantage of randomized trials is that they allow effective therapies to be developed even when the disease is not well understood. For example, randomized trials demonstrating the efficacy of lenalidomide [6] and bortezomib [7] in patients with myeloma represented significant advances even though the molecular basis for their efficacy is not well understood.

However, randomized trials can encounter difficulty in the face of increasing knowledge about cancer mechanisms and the rational development of therapeutic drugs. One well-publicized example, reported in the *New York Times*, described two cousins that both had metastatic melanoma bearing the BRAFV600E mutation, and they both enrolled in a randomized clinical trial comparing an investigational BRAF inhibitor and a toxic and largely ineffective standard therapy (DTIC) [8,9]. The cousin assigned to the BRAF inhibitor responded and served as pallbearer at the funeral of his DTIC-assigned relative, who died rapidly. The significance of this example rests not in an 'N-of-2' notion that one treatment was superior to the other, but rather in the implication that randomized trials may subjugate the best interests of individual patients with cancer to the best interests of science. This perception is especially unwelcome in view of the widespread complaint that fewer than 5% of all adult patients with cancer enroll in clinical trials [10]. One might ask why patients with cancer should be expected to participate in trials for which they are not intended beneficiaries? However, it should be noted that, fortunately, many randomized trials have provisions for crossing over to the other arm if the initial treatment fails.

Randomized trials have other limitations. They can be nonsensical in situations where interventions are so obviously effective that control arms defy reason, as with parachutes and bulletproof vests [11]. On the other end of the knowledge spectrum, randomized trials can be insensitive to detecting biologically (and clinically) relevant signals because of noise attributable to the heterogeneity of the population examined. The fact that cancer cures are notoriously more prevalent in mice than in humans is attributable in large part to the relative homogeneity of inbred mouse strains that allows for experimental repetition, in contrast to the heterogeneity of germline and tumor genomes that makes each human subject a unique object of investigation. Thus, whereas randomized trials provide the foundation for wide adoption (and often regulatory approval) of a new treatment, they are insensitive to new discoveries.

Despite their limitations, randomized trials are likely to remain the gold standard for defining the safety and efficacy of new therapies for many years to come, because even the best characterized cancer therapies fall short of the parachute standard. Nevertheless, continuous improvements in understanding of cancer, combined with increasingly sophisticated interventions, may one day make randomized trials difficult to justify.

DECONSTRUCTING CANCER, ONE PATIENT AT A TIME

Increasingly, randomized trials will be forced to share the stage with innovative trials that deeply investigate cancer within individuals. These new investigations are being enabled by advances in omic technologies, which allow for the comprehensive characterization of a patient with cancer and his or her tumor. NGS results show that cancer genomes contain small numbers of recurrent mutations accompanied by much larger numbers of mutations that are rarely or never found in the tumors of other patients [12]. Whereas recurring mutations are probably involved in disease pathogenesis, rare or unique mutations likely represent mixtures of driver mutations that collaborate to promote tumor growth and passenger mutations that are uninvolved in cancer genesis or progression [13,14]. Distilling this vast information into a cohesive framework that describes how a cancer operates within an individual, and unveiling its potential vulnerabilities, is one of the grand challenges of our time. Deciphering the general operating principles of cancer will require methods for converting the results of omic testing into candidate networks, and then monitoring how these networks evolve and respond to treatment over time. Success requires discerning biologically relevant signals from vast backgrounds of noise. A major challenge is a phenomenon known as 'overfitting' [15]. Overfitting occurs when a statistical model is confounded by random noise instead of revealing true biological relationships, and it is inevitable when the number of measured parameters outstrips the number of observations. In other words, efforts to synthesize the massive data sets that omic studies generate into a cohesive framework will invariably lead to large numbers of seemingly reasonable solutions, most of which will be wrong. In the context of single-patient studies, overfitting presents an enormous challenge, and developing approaches for meeting this challenge represents a crucial avenue of investigation.

A second challenge is that the results of omic studies provide a signal average: a summation of measurements generated from all cell types present within the tumor. Many tumors are highly heterogeneous, comprising multiple morphologically distinguishable cell types, and different types of cells can cooperate to sustain tumor growth [16,17]. Even tumors that are morphologically homogeneous comprise subclones that differ in their mutational content. Deconvoluting these relationships presents a significant challenge that will require methods that can use omic data from unfractionated tumors to infer the number and size of subpopulations [18] as well as from single-cell approaches [19].

Additional technical challenges include identifying and accounting for systematic errors that pervade omic data arising from batch effects [20], and assuring that tissue specimens are collected in a manner that safeguards the integrity of information that they contain, because factors such as ischemia time can dramatically influence results [21].

Approaches for discriminating signal from noise can be broadly classified into two categories: static and dynamic.

- Static approaches: cancer in snapshot. Several reports have described the use of omics to identify the potential vulnerabilities of a cancer.

Although strong predictions and effective interventions based on the aberrant regulation of a single gene have been described [22,23], unambiguously favorable outcomes are likely exceptions to the rule [24]. In most cases, strong evidence for dependency on a particular pathway will be lacking, and confidence levels for any given prediction will need to be assessed using a variety of approaches. For example, testing multiple independent samples from the same patient at a given time point to assess the reproducibility of key findings across specimens may increase confidence, although heterogeneity in the mutational content of tumor subclones is an important caveat. Another approach relies on developing intersecting lines of experimental evidence. Corroboration of a genomic result (e.g., upregulation of a receptor tyrosine kinase by RNA-Seq) with a proteomic result (phosphorylation of the corresponding substrates by reverse phase protein arrays) would provide valuable additional evidence. Placing the cancer under investigation into a broader context of prior knowledge using publicly available databases will also be a useful tool to identify candidates. For example, understanding where a particular cancer fits within the constellation of all other previously characterized cancers should prove increasingly valuable as omic profiling extends to all cancer clinical trials, thereby establishing associations between omic profiles and treatment responses. Similarly, the genetic diversity of at least some cancers appears to condense around a limited number of known signaling pathways [25], thus improving interpretability. Finally, correlating omic information with clinical information will be critical to interpreting these types of data. For example, in patients with metastatic cancer undergoing treatment, disease stabilization sometimes occurs at one or more sites whereas progression occurs at other sites. Comparing omic results at sites of growing versus stable disease might point to mechanisms that tumors use to resist treatment.

■ Dynamic approaches: cancer in motion. Whereas static measurements are encumbered by the enormous challenge of discriminating signal from noise, repeated assays over time can significantly improve the ability to resolve correlations between data points and discern underlying biological relations. A landmark example was recently described by Chen, Snyder, and colleagues, in which a single healthy individual (the senior author of the study) was monitored by extensive omic testing involving 20 separate assessments of peripheral blood mononuclear cells over 400 days, totaling more than 3 billion data points [26]. Results across genomic, transcriptomic, proteomic, and metabolomic platforms were normalized and integrated, revealing clusters of temporally correlated measurements that segregated into biologically plausible pathways. A serendipitous yet striking discovery involved abnormalities in the insulin regulatory pathway that arose after the second of two viral infections, and further testing led to a new diagnosis of diabetes that resolved following a strict regimen of diet and exercise. Might a similar approach be applied in patients with cancer? Deeper insights into how cancer functions will require similar systems approaches that convert omic data into presumptive regulatory networks and point to candidate vulnerabilities. Confidence in any particular model can be gained or lost by treating the patient with those drugs predicted to be most likely to disrupt key nodes within the network. Although many predictions will prove wrong, embedding mechanisms for learning into the process should, through iteration, improve accuracy. Monitoring the omic profiles of tumors before and following treatments designed to interfere with tumor growth could provide information that is valuable both to science and to patient care. It will provide insight into whether interventions designed to target specific signaling pathways worked as intended. For example, if a targeted therapy designed to interrupt a presumptive driver signaling pathway fails to interfere with tumor growth, it would be important, both scientifically and for patient care, to discern whether the targeted signaling pathway was blocked as intended. Furthermore, serial monitoring will provide a means for examining how networks respond to perturbations [27], thereby providing insight into how the networks work.

Establishing this type of learning loop is challenging to implement because of concerns that patients should not be exposed to research methods that have not been extensively validated. At center stage is a widely publicized scandal at Duke University, which involved pervasive scientific misconduct in clinical trials that used omic data to guide therapy in patients with cancer [28]. In response, the Institute of Medicine (IOM) of the National Academy of Sciences recently recommended that a 'bright line' separate testing that is done in the discovery phase of research from clinical studies in which only fully specified and validated tests with 'locked down' computational procedures are used to direct patient care [29]. The core question is: should fully informed patients with cancer, under appropriate clinical and ethical circumstances, be allowed to participate in discovery-based research? Answering 'no' places a firewall between patients with cancer and the latest scientific discoveries. Although this may help to protect patients from misconduct or less nefarious mistakes, such a firewall could also prevent them from accessing the latest scientific and technological discoveries. As a consequence, highly motivated patients with cancer with no good alternatives would be prevented from participating as true partners in advancing cancer research.

THE ROAD AHEAD

Fully exploiting the opportunities that technology creates requires a reassessment of the approach to cancer. Whereas many of the challenges described above are scientific and technical in nature, other important limitations (listed in Box 1) are self-imposed. Overcoming another major challenge to implementation (the recovery of costs) will ultimately require a restructuring of reimbursement patterns, but this issue is beyond the scope of this article.

Partners in Personal Oncology as a paradigm for advancing cancer research and treatment

One of us (CAB) co-founded Partners in Personal Oncology, a not-for-profit organization, based on the belief that a new patient-centered framework

---

BOX 1 SOME PROPOSED CHANGES IN THE CURRENT APPROACH TO CANCER

(i) Increase longitudinal monitoring. Sequential biopsies provide an important means for understanding how cancer works, but are rarely performed because until now, they have had only a limited role in clinical care. Sequential biopsies could be performed in significant subsets of patients with cancer, and might eventually be replaced by evaluating circulating tumor cells.

(ii) Improve quality control in tumor acquisition. Tumor specimens are generally not processed in accordance with the tightly controlled conditions necessary for reliably assessing RNA and protein. More attention must be paid to acquiring tumor samples in a manner that retains the integrity of the information they hold.

(iii) Improve access to investigational drugs. Regulatory approval of a new drug by the US Food and Drug Administration (FDA) requires clinical trials that begin with dose-finding (Phase I) studies and end with one or more randomized controlled trials. Outside this framework, single patient investigational new drug (IND) applications can allow patients with cancer who would not otherwise qualify for clinical trials to access investigational drugs on a 'compassionate use' basis. Given that the results of omic studies may suggest unanticipated vulnerabilities, patients need a path for rapidly accessing those drugs to which their tumors are predicted to respond, even if an established clinical trial is not available. Single patient INDs are labor intensive and require the consent of the manufacturer. Pharmaceutical companies are often reluctant to grant compassionate use access because of the risk of adverse events (which must be reported to the FDA, potentially delaying regulatory approval), and the lack of upside benefit (for the company) from anecdotal responses. Combining 'compassionate use' access with intensive longitudinal monitoring might enable drug companies to learn more about the molecular contexts in which their investigational drugs work, thereby improving the probability of success in later trials.

(iv) Allow patients with cancer to participate in discovery-based research. The rigidity of the clinical trials system risks constraining innovation. For example, the aforementioned IOM report recommends that only tests that have been extensively validated preclinically should be used in patients [29]. The uniqueness of cancer threatens to make such an approach impractical. Do we always need to know that a test is valid before applying it in a patient, or are there circumstances under which we can evaluate the validity of a test in a patient? Better ways of balancing access to innovation while protecting patients from avoidable errors need to be identified.

Both (iii) and (iv) point to a need for increased regulatory flexibility as we work to find the optimal path for translating omic information into clinical practice.

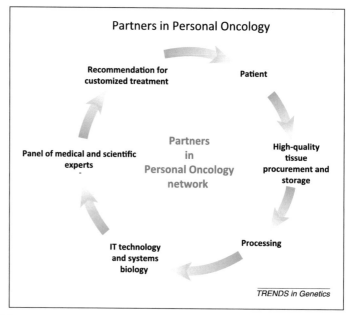

FIGURE 1

The Partners in Personal Oncology cycle.

Partners in Personal Oncology will provide: (i) optimized methods for biospecimen collection and banking; (ii) access to state-of-the-art testing using omic and other technologies; (iii) a databank that will house the results of testing annotated with the patient's clinical information; and (iv) a panel of medical and scientific experts that will examine the results of testing and convey a recommendation for treatment back to the patient and their oncologist.

is needed to optimize understanding and treatment of cancer (http://www. personaloncology.org). By building a network of linked partnerships, Partners in Personal Oncology is developing an approach that can be summarized as 'all for one, and one for all': placing world-class technology and expertise in the service of individual patients with cancer and their oncologists (all for one) and using individual patient experiences to advance our collective understanding of how cancer works for the benefit of future patients (one for all). There are several noteworthy aspects of the cycle of Partners in Personal Oncology (Figure 1). First, patients will enter the cycle repeatedly over the course of their disease, with serial biopsies and longitudinal monitoring enabling the creation of 'molecular movies' of how cancer evolves within individual patients over time. Second, the cycle of Partners in Personal Oncology will foreshorten the 'learning lag' that separates science and medicine, intertwining discovery-based research and clinical care in a more intimate and immediate manner than exists currently. Researchers will benefit by testing their hypotheses directly in patients with cancer, and patients will benefit by gaining access to world-class technology and expertise that aims to understand the molecular basis of their disease and treat them accordingly. Third, the network of Partners in Personal Oncology is distributed, both with respect to aggregating technology and expertise irrespective of institutional boundaries, and by allowing patients with cancer to access world-class resources irrespective of their geographic location. Fourth, patients with cancer who choose to participate in this program will be made aware of the highly experimental nature of this approach and, as such, will be true partners in discovery-based cancer research. Finally, by combining intensively monitored individual patient experiences with mechanisms for learning, it should be possible, through iteration, to increase the accuracy of our predictions over time.

CONCLUDING REMARKS

It is transformative to consider the possibility of linking the efforts of physicians, researchers, and patients in advancing cancer research. On the one hand, researchers cannot progress unless the hypotheses that they develop can be tested. On the other hand, patients with cancer who are not curable by available therapies are poorly served by clinical trials that ignore tumor-specific characteristics and by physicians who have only their intuitions to guide therapy. Cancer research is also poorly served because of the many existing clinical trials from which we currently learn almost nothing. The careful and comprehensive documentation of the molecular and clinical changes that occur within a patient with cancer and their tumor at baseline and following interventions designed to interfere with tumor growth would contribute immensely to understanding of cancer. Even if the interventions initially fail more than they succeed, the approaches described aim toward a new definition of 'doing our best'. This new type of clinical trial is likely to be a necessary step on the road to deconstructing cancer.

ACKNOWLEDGMENTS

We thank Lee Huntsman, George Stamatoyannopoulos, Stan Fields, Junfeng Wang, and Zhijun Duan for helpful comments and discussions. Supported through R01 CA 135357 and gifts from Norman Metcalfe and the Tietze Family Foundation.

REFERENCES

1 Curtis, C. *et al.* (2012) The genomic and transcriptomic architecture of 2,000 breast tumours reveals novel subgroups. *Nature* 486, 346–352

2 Gerlinger, M. *et al.* (2012) Intratumor heterogeneity and branched evolution revealed by multiregion sequencing. *N. Engl. J. Med.* 366, 883–892

3 Shibata, D. (2012) Heterogeneity and tumor history. *Science* 336, 304–305

4 Marusyk, A. *et al.* (2012) Intra-tumour heterogeneity: a looking glass for cancer? *Nat. Rev. Cancer* 12, 323–334

5 Gerlinger, M. and Swanton, C. (2010) How Darwinian models inform therapeutic failure initiated by clonal heterogeneity in cancer medicine. *Br. J. Cancer* 103, 1139–1143

6 McCarthy, P.L. *et al.* (2012) Lenalidomide after stem-cell transplantation for multiple myeloma. *N. Engl. J. Med.* 366, 1770–1781

7 Sonneveld, P. *et al.* (2012) Bortezomib induction and maintenance treatment in patients with newly diagnosed multiple myeloma: results of the randomized phase III HOVON-65/GMMG-HD4 Trial. *J. Clin. Oncol.* 30, 2946–2955

8 Harmon, A. (2010) *New York Times* 19 September, p. A1

9 Chapman, P.B. *et al.* (2011) Improved survival with vemurafenib in melanoma with BRAF V600E mutation. *N. Engl. J. Med.* 364, 2507–2516

10 Comis, R.L. *et al.* (2009) Physician-related factors involved in patient decisions to enroll onto cancer clinical trials. *J. Oncol. Pract.* 5, 50–56

11 Smith, G.C. and Pell, J.P. (2003) Parachute use to prevent death and major trauma related to gravitational challenge: systematic review of randomised controlled trials. *Br. Med. J.* 327, 1459–1461

12 Fox, E.J. *et al.* (2009) Cancer genome sequencing: an interim analysis. *Cancer Res.* 69, 4948–4950

13 Welch, J.S. *et al.* (2012) The origin and evolution of mutations in acute myeloid leukemia. *Cell* 150, 264–278

14 Hodis, E. *et al.* (2012) A landscape of driver mutations in melanoma. *Cell* 150, 251–263

15 Edelman, L.B. *et al.* (2010) *In silico* models of cancer. *Wiley Interdiscip. Rev. Syst. Biol. Med.* 2, 438–459

16 Wu, M. *et al.* (2010) Interaction between Ras(V12) and scribbled clones induces tumor growth and invasion. *Nature* 463, 545–548

17 Raaijmakers, M. *et al.* (2010) Bone progenitor dysfunction induces myelodysplasia and secondary leukaemia. *Nature* 464, 852–857

18 Ding, L. *et al.* (2012) Clonal evolution in relapsed acute myeloid leukaemia revealed by whole-genome sequencing. *Nature* 481, 506–510

19 Anderson, K. *et al.* (2011) Genetic variegation of clonal architecture and propagating cells in leukaemia. *Nature* 469, 356–361

20 Leek, J.T. *et al.* (2010) Tackling the widespread and critical impact of batch effects in high-throughput data. *Nat. Rev. Genet.* 11, 733–739

21 Huang, J. *et al.* (2001) Effects of ischemia on gene expression. *J. Surg. Res.* 99, 222–227

22 Jones, S.J. *et al.* (2010) Evolution of an adenocarcinoma in response to selection by targeted kinase inhibitors. *Genome Biol.* 11, R82

23 Kolata, G. (2012) *New York Times* 8 July, p. A1

24 Kolata, G. (2012) *New York Times* 9 July, p. A1

25 Cancer Genome Atlas Network. (2012) Comprehensive molecular characterization of human colon and rectal cancer. *Nature* 487, 330–337

26 Chen, R. *et al.* (2012) Personal omics profiling reveals dynamic molecular and medical phenotypes. *Cell* 148, 1293–1307

27 del Sol, A. *et al.* (2010) Diseases as network perturbations. *Curr. Opin. Biotechnol.* 21, 566–571

28 Goozner, M. (2011) Duke scandal highlights need for genomics research criteria. *J. Natl. Cancer Inst.* 103, 916–917

29 Micheel, C.M. *et al.* (2012) *Evolution of Translational Omics: Lessons Learned and the Path Forward*. National Academies Press

CHAPTER 3

Cell

The Genetic Basis for Cancer Treatment Decisions

Janet E. Dancey[1,2], Philippe L. Bedard[3,4], Nicole Onetto[1], Thomas J. Hudson[1,5,6,*]

[1]Ontario Institute for Cancer Research, Toronto, ON M5G 0A3, Canada, [2]NCIC-Clinical Trials Group, Queen's University, Kingston, ON K7L 3N6, Canada, [3]Princess Margaret Hospital, Division of Medical Oncology and Hematology, University Health Network, University of Toronto, Toronto, ON M5S 1A1, Canada, [4]Department of Medicine, University of Toronto, Toronto, ON M5S 1A1, Canada, [5]Department of Medical Biophysics, University of Toronto, Toronto, ON M5S 1A1, Canada, [6]Department of Molecular Genetics, University of Toronto, Toronto, ON M5S 1A1, Canada
*Correspondence: tom.hudson@oicr.on.ca

Cell, Vol. 148, No. 3, February 3, 2012 © 2012 Elsevier Inc.
http://dx.doi.org/10.1016/j.cell.2012.01.014

SUMMARY

Personalized cancer medicine is based on increased knowledge of the cancer mutation repertoire and availability of agents that target altered genes or pathways. Given advances in cancer genetics, technology, and therapeutics development, the timing is right to develop a clinical trial and research framework to move future clinical decisions from heuristic to evidence-based decisions. Although the challenges of integrating genomic testing into cancer treatment decision making are wide-ranging and complex, there is a scientific and ethical imperative to realize the benefits of personalized cancer medicine, given the overwhelming burden of cancer and the unprecedented opportunities for advancements in outcomes for patients.

INTRODUCTION

Numerous models have been proposed to explain the complex nature of cancer at molecular, cellular, and pathological levels (Hanahan and Weinberg, 2000, 2011). One model that explains cancer initiation, progression, dissemination, treatment response, and emergence of drug resistance is based on the progressive accumulation of mutations throughout the history of a tumor and its downstream colonies (Figure 1). Though incomplete, the somatic mutation

25

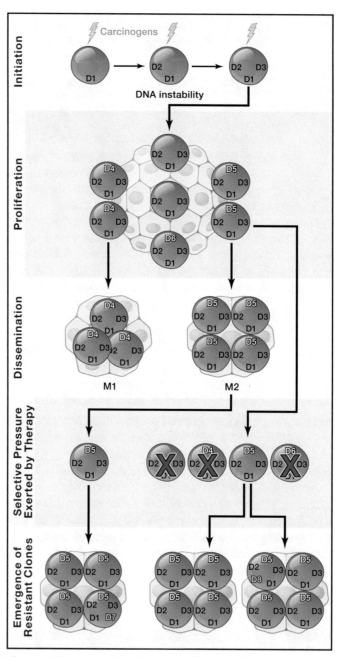

FIGURE 1 Accumulation of Driver Mutations in the History of a Tumor

Exposure to carcinogens, failure of DNA repair, and progressive genetic instability lead to accumulation of mutations that drive cancer development, growth, and metastases. Subclones with new mutations may become dominant within metastases or within persistent or recurrent cancer deposits through selective pressures exerted by cytotoxic or targeted chemotherapies.

model does incorporate one of the most consistent hallmarks of cancer: DNA mutations are found in all cancers. In addition, specific mutations have been linked to one or more forms of cancers, and mutant gene products have been associated with biological characteristics of cancer.

The spectrum of cancer mutations is diverse in terms of type, number, and functional consequences. Examples include single base changes that alter protein activity, amplifications, and deletions that modify the abundance of a gene and its products and alternative splicing or translocations that can create novel proteins. Mutations are abundant in cancer cells, numbering between thousands and hundreds of thousands per tumor (Stratton, 2011; Wong et al., 2011). However, most mutations in cancer cells do not appear to play a role in cancer progression; rather, they are indicative of the high mutation rate resulting from carcinogens and DNA instability (Pleasance et al., 2010a, 2010b). Such mutations have been called "passengers." A minority of cancer mutations are thought to be "drivers," defined as mutations involved in the development or progression of a tumor. A subset of these drivers and their component cellular pathways may be "actionable," i.e., have significant diagnostic, prognostic, or therapeutic implications in subsets of cancer patients and for specific therapies. A subset of mutations may also be druggable, i.e., targets for therapeutic development. Given current knowledge on gene function, classifying mutations into drivers and passengers—actionable and/or druggable—is difficult. It is still too early to deduce how many mutations are active at any given stage of a tumor. Moreover, the constant accumulation of mutations, with or without exogenous selective pressures of therapy, implies that tumors evolve to encompass many subpopulations that have distinct differences in mutation load within and between patients (Figure 1). Although there is great diversity in the types and numbers of mutations in human cancer, our ability to annotate and to assess functional and clinical consequence has expanded remarkably.

DNA sequencing technologies now allow whole-genome, exome, and transcriptome sequencing at rates that are dramatically faster and cheaper than traditional Sanger-based methods. Quantifying and cataloguing mutations, transcriptomes, and methylomes for many forms of cancer are underway in dozens of countries through coordinated projects of the International Cancer Genome Consortium (ICGC) (Hudson et al., 2010). Already, partial cancer genome data sets are available for several thousands of tumors with protein-altering mutations affecting more than 7,500 genes inventoried to date (ICGC Dataset Version 6; http://www.icgc.org). The availability of these large cancer data sets in the public domain will foster significant follow-up research by academia and industry and will lead to the validation of many new driver mutations, drug targets, and clinically useful biomarkers.

Table 1 Selected Genetic Markers and Their Application in Cancer Treatment

Genetic Marker	Application	Drug
BCR-ABL	Ph+ CML; Ph+ ALL	Imatinib, dasatinib, nilotinib
BCR-ABL/T315I	Resistance to anti-BCR-ABL agents	Imatinib, dasatinib, nilotinib
BRAF V600E	Metastatic melanoma	Vemurafenib
BRCA1/2	Metastatic ovarian cancer and breast cancer with BRCA 1/2 mutations	Olaparib, veliparib, iniparib
c-Kit	Kit (CD117)-positive malignant GIST	Imatinib
EGFR	Locally advanced, unresectable, or metastatic NSCLC	Erlotinib, gefitinib
EGFR T790M	Resistance to EGFR tyrosine kinase inhibitors in advanced NSCLC	Erlotinib, gefitinib
EML4-ALK	ALK kinase inhibitor for metastatic NSCLC with this fusion gene	Crizotinib
HER2 amplification	HER2-positive breast cancer or metastatic gastric or gastroesophageal junction adenocarcinoma	Trastuzumab
KRAS	Resistance to EGFR antibodies in metastatic colorectal cancer	Cetuximab, panitumumab
PML/RAR	Acute promyelocytic leukemia	ATRA, arsenic trioxide
TPMT	Deficiency is associated with increased risk of myelotoxicity	Mercaptopurine, azathioprine
UGT1A1	Homozygosity for UGT1A1*28 is associated with risk of toxicity	Irinotecan
DPD	Deficiency is associated with risk of severe toxicity	5-Fluorouracil

ATRA, all trans retinoic acid; Ph+, Philadelphia-positive chromosome; DPD, dihydropyrimine dehydrogenase; EGFR, epidermal growth factor receptor; EML4-ALK, echinoderm microtubule-associated protein-like 4 anaplastic lymphoma kinase; HER2, human epidermal growth receptor 2; GIST, gastrointestinal stromal tumors; ALL, acute lymphocytic leukemia; NSCLC, non-small cell lung cancer; TPMT, thiopurine S-methyltransferase.

A subset of mutations are being branded as "actionable" by clinicians, based on evidence from clinical studies that the presence or absence of gene mutations in tumor (and occasionally germline) DNA can be used to inform clinical management (Table 1). Some examples include *KRAS* mutations that correlate with resistance to epidermal growth factor receptor (EGFR) antibodies in colorectal cancer and *BCR-ABL* fusion gene products that are pathognomonic of chronic myelogenous leukemia and can be inhibited by agents such as imatinib. The list of potential actionable mutations that may impact on treatment recommendations for predictive or prognostic reasons, or those with known prognostic or diagnostic implications, is growing. Notwithstanding the historical links between certain actionable mutations and specific cancer histologies, further exploration has revealed that specific mutations are often observed across a range of tumor histologies, albeit at different frequencies. Figure 2 highlights many genes, including some with known actionable mutations, which are altered in several common cancers. One testable

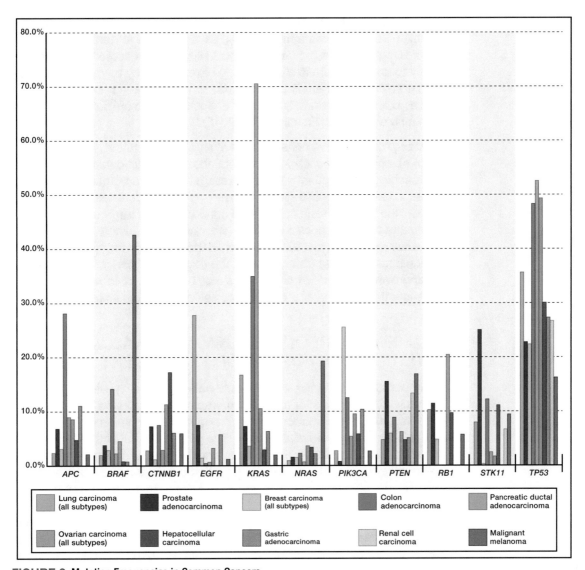

FIGURE 2 Mutation Frequencies in Common Cancers
Selected mutations are those found on Snapshop and OncoCarta panels. The mutation data were obtained from the Sanger Institute
Catalogue of Somatic Mutations in Cancer website at http://www.sanger.ac.uk/cosmic, COSMIC v54 Release (Forbes et al., 2011).

hypothesis is that mutations act as "drivers" in most if not all tumors where
they are observed. Moreover, if a mutation is predictive of a drug response in
one form of tumor (for example, BRAF V600E and vemurafenib for melanoma;
Chapman et al., 2011), then there may be some likelihood that the same drug
could affect tumors from other origins with the same mutation (for example,

BRAF V600E and ovarian cancer; Sieben et al., 2004). It is clear, however, that this hypothesis requires formal testing, as experience to date suggests that the presence of a specific genetic abnormality may not confer the same sensitivity to an agent across all cancers, as exemplified by trastuzumab, which has been shown to benefit patients with HER2-amplified breast and gastric cancer, but not those with ovarian or endometrial cancer (Bang et al., 2010; Bookman et al., 2003; Fleming et al., 2010).

If, in fact, the functional consequence of a specific mutation is similar across different cancers, the clinical implications are unavoidable. Rather than approaching each patient's tumor investigation with an organ-based list of mutation tests, one could systematically perform a global search for all such "actionable" mutations in any type of cancer and test targeted therapeutics in patients with the specific mutation(s) regardless of cancer histology. One of the key reasons to test this approach in the clinical setting now is the need to elucidate further whether many of the targeted anticancer therapies that are approved for specific cancer types may benefit patients with other forms of cancer that share similar genetic profiles and biologic features.

FRAMEWORK REQUIREMENTS FOR EVALUATING THE GENETIC BASIS FOR CANCER TREATMENT

It is important to recognize early that obtaining convincing evidence to guide future clinical decisions for matching therapies to mutations affecting unique patients with different tumor types will require large numbers of patients in meticulously conducted clinical trials. The goal of these initial trials will be to determine which mutation profiles correlate with sensitivity or resistance to specific therapies and whether the mutation profile and treatment outcome is consistent among different cancer histologies. As opposed to a classical randomized controlled trial in which a novel therapeutic strategy is assessed against current standard practice, a genomics-based clinical trial offers the potential for many different therapeutic options to be selected on the basis of genomic profiling. Each subgroup of patients harboring a specific mutation and receiving a targeted therapy (or assigned to a control group) will represent a minority of patients recruited. To achieve power to determine whether outcome is improved in subgroups, genomics-based clinical trials require both large sample sizes and large treatment effects within the mutation-defined subgroups. For example, a 1,000 patient genomic profiling trial could recruit 100 subjects harboring mutations in a target gene at 10% frequency, allowing a two-armed nested phase II study comparing a new therapy in selected cases and controls. The same genomics trials could support several nested phase II studies testing different agents in patients with different mutations. However, the frequency of mutated cancer genes is often less than 10%, and there will be a need to know the influence of the tissue of origin on outcome. Thus, the sample size requirement for genomics trials may be larger by at least one order of magnitude if there is interest in assessing treatment effect across and within patients with different cancer histologies that share the same mutations. Although the number of patients profiled may be greater, the numbers of patients per treatment arm may be smaller, as the magnitude of treatment effect should be greater to justify this complex approach. To achieve large patient numbers, genomics trials need to recruit patients at multiple centers and, ultimately, leverage several large multi-institutional trial networks. It is also likely that, in the future, the scientific community will want to synthesize data from multiple studies performed across the globe. This will be enabled not only by instituting appropriate data sharing policies (see below), but also by using similar standard operating procedures (SOPs) for sample collection, processing, analyses, mutation calling, and data collection and management as much as possible. The net outcome of the approach will be a new system of cancer classification that will include genomic factors that make a difference in patient prognosis and treatment.

Efficient workflows are required that incorporate all steps: initial invitations to participate, consent, sample collection, genomic analyses, validation of actionable mutations, expert deliberations, reporting to clinicians, intervention(s) including access to appropriate therapies, and follow-up (Figure 3). The addition of complex genetic or genomic testing and

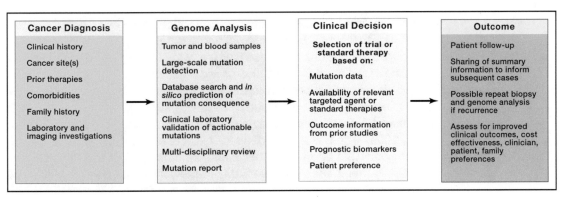

FIGURE 3 The Genetic Basis for Cancer Treatment
The key steps for the application and evaluation of clinical genomics for cancer treatment include the following. The recruitment of patients and acquisition of relevant clinical information. Sample collection and analyses for cancer genes. Interpretation of results of genomic analysis based on known functional and clinical significance of mutation. Provision of information to clinicians and patients for management. Clinical trials of novel treatments offered to cancer patients who are unlikely to benefit from standard of care and thus have a relatively poor prognosis and/or are more likely to benefit from a novel therapy due to the presence of tumor genetic abnormalities that predict sensitivity, lack of resistance, or toxicity to a treatment. Assessment of outcomes and sharing of results across cancer networks to accelerate clinical cancer genomic knowledge.

interpretation to clinical trials imposes some time-consuming activities that could delay the start of therapies. This is particularly true at this time, as most high-throughput genomic technologies require weeks and/or months for data generation and analysis. One way to minimize the potential impact on genomic testing to create a delay in treatment initiation is to sequence patient tumors early in the management of their disease—for example, at the time when metastatic disease is diagnosed and patients begin their first-line standard of care regimen—as the genomic information will inform subsequent choices of therapies. The caveat is, of course, that, over time and with each treatment, new mutations can be acquired. An alternative to early genomic testing is to sequence at the time of progression when a change in therapy is considered. To achieve a turnaround time of less than 3 weeks (a threshold suggested by clinical trials leaders), the choice of sequencing technologies and streamlining of data analysis steps are important.

PATIENT RECRUITMENT AND INFORMED CONSENT

Participation in genomics trials requires that prospective patients be informed of genomic testing and the potential of future therapies based on genomics results. Whereas the latter can be administered after genomics results are known and the informed consent document can be customized according to the specific intervention being considered, all participants need to be aware that extensive genomic information will be generated and that, in addition to generating data regarding "actionable" mutations that may modify treatment decisions, there may be "incidental findings," such as germline mutations associated with risk to other diseases (i.e., long QT syndrome). Furthermore, germline mutation data could also provide risk information relevant to family members (i.e., mutations in breast cancer type 1 susceptibility [BRCA1] or cystic fibrosis genes). The issue of returning such data to research partic-ipants and patients is currently a controversial topic in bioethics (McGuire et al., 2008). There is a clear need for experts and stakeholders to develop a framework that addresses ethical and legal obligations to inform subjects and family members. In addition, this framework should consider the pref-erences and concerns of research subjects and family members to receive information on germline variants of risk of cancer and other diseases.

In addition to the "risks" of identifying germline or somatic mutations that may affect patients or family members, patients should be aware of the extent that data will be shared and of the risks and potential consequences of breach of confidentiality. To support data sharing across participating net-works and ideally across the scientific community (described later), the con-sent process should notify participants that data will be made accessible to national and international researchers under robust mechanisms to protect confidentiality of participants (Toronto International Data Release Workshop Authors et al., 2009).

TISSUE REQUIREMENTS: QUANTITY, QUALITY, PROCESSING, AND TIMING

Any framework for clinical decision making on the basis of somatic genomic alterations requires timely access to tumor tissue for high-throughput molecular profiling that is readily available, of sufficient quality and quantity for successful analysis, and obtained at a time that the generated mutation profile remains relevant for the potential available treatments (Dias-Santagata et al., 2010; MacConaill et al., 2009). Most cancer patients have archival tissue available for molecular profiling from either their primary tumor and/or metastasis obtained from diagnostic biopsy or surgical excision. Local regulations dictate the minimum time period that hospital pathology departments must retain archival tumor tissue for the benefit of the medical care of the patient. In North America, most hospitals collect and archive tumor specimens as formalin-fixed, paraffin-embedded (FFPE) blocks to optimize histological assessment. Although collected on all patients, the diagnostic samples may or may not be representative of the mutations that subsequently arise in metastases or as a consequence of treatment. In addition, DNA and RNA preservation in FFPE tissue is challenging, as formalin fixation causes crosslinking and degradation into smaller fragments (Wang et al., 2009). Snap freezing tumor tissues in liquid nitrogen is the optimal method of nucleic acid preservation; however, this is not routinely performed outside of select European cancer centers. FFPE does provide preservation of histological features that allows for pathological review of hematoxylin and eosin stained slides to assess tumor cellularity and to mark the regions of tumor for macrodissection to isolate regions of nonnecrotic tumor from surrounding stroma.

Quantity and quality of tumor DNA are key sample considerations. Unfortunately, key parameters of tumor cellularity and optimal quantity of DNA remain unknown. Some authors suggest > 70% tumor cellularity with < 10% necrotic tumor tissue as guidelines (MacConaill et al., 2009), although less stringent thresholds may be employed if there is more tumor tissue available for macrodissection and DNA extraction or a more limited panel of gene mutation will be assayed. The minimum tumor tissue requirement and optimal method of DNA extraction remain unknown. As few as four 5 micron sections to isolate 15 ng or less of genomic DNA has been described as the requirement for successful sequencing using a customized multiplex colorectal cancer mutation (Colocarta) panel derived from the Oncocarta v1.0 platform (Sequenom, San Diego, CA) (Fumagalli et al., 2010). Increased genomic coverage requires greater quantities of tumor DNA. The Oncocarta v1.0 and OncoMap panels, which include, respectively, 238 mutations in 19 oncogenes and ~400 mutations in 33 oncogenes and tumor suppressors, recommend at least 500 ng of DNA (MacConaill et al., 2009; Thomas et al., 2007). There will always be a direct relationship between the extent of genetic testing and the quantity of DNA required; however, methods to allow expanded and deeper genomic sequencing, likely to be the mainstay technologies in the future for clinical laboratory testing, using small quantities of DNA from tumor, circulating tumor cells, or DNA would greatly enhance the successful evaluation of genomic testing in cancer management.

An unresolved issue is whether archival tissue from the primary tumor or a fresh biopsy of a metastatic lesion should be profiled for treatment selection for patients with advanced refractory disease. It is well recognized that cancers are genomically unstable and new mutations arise during the process of metastasis to distant sites, and/or treatment-resistant clones emerge over time (Campbell et al., 2010; Jones et al., 2010; Lee et al., 2010; Shah et al., 2009). Although metastatic tumor biopsies are increasingly acceptable to patients and their physicians if they may inform treatment decision making (Agulnik et al., 2007), it is not feasible in the current clinical practice environment to perform a metastatic tumor biopsy at the time of treatment resistance in all patients with advanced cancer, and at each point, a new treatment may be considered. Clonal evolution may differ across metastatic sites within an individual patient (Yachida et al., 2010), suggesting that genomic profiling of biopsy material from a single metastatic lesion may not be sufficient to completely capture the genomic diversity of advanced solid cancers.

Nevertheless, available data suggest that individual mutations may be highly concordant between primary and metastatic sites and that mutations identified in primary tumors predict benefit to certain drugs in patients with metastatic disease. For instance, concordance of *KRAS* mutations in colorectal primary cancers and metastases was 96% in two published series (Knijn et al., 2011; Santini et al., 2008). In non-small cell lung cancer, one report of a small cohort of 25 cases demonstrated concordance rates for *EGFR* and *KRAS* mutations of 76% (Kalikaki et al., 2008). Furthermore, the effectiveness of currently available targeted treatments for advanced cancer patients such as gefitinib or erlotinib for *EGFR*-mutated lung cancer or trastuzumab for *HER2*-amplified breast cancer has largely been demonstrated from trials that have identified genetic mutations in archived diagnostic samples rather than new biopsies from metastatic lesions (Mok et al., 2009; Slamon et al., 2001). In contrast to high concordance seen for *KRAS* and *EGFR* mutations, a recent study in breast cancer reported discordant *PIK3CA* mutations in 32% of 103 cases (Dupont Jensen et al., 2011), indicating that concordance may be mutation and/or tissue-type specific and may be influenced by prior therapy. At this time, whether to use archived diagnostic samples versus samples obtained at the time a new treatment is indicated is driven by convenience, costs, and standard practices rather than by data. Additional studies are needed to address the feasibility of biopsying of metastatic lesions for genomic profiling and whether treatment decisions based on this approach lead to improved outcomes compared with genomic profiling of archival samples of the corresponding primary tumor. In the current environment, serial biopsying of patients is not scalable to large clinical trials or current clinical practice environments. However, this important question can and should be addressed through a coordinated effort of committed investigators and academic cancer research centers.

GENOMIC TECHNOLOGIES AND DATA MANAGEMENT

At all stages of development and adoption, companion diagnostics used to identify somatic mutations to inform real-time clinical decisions need to meet clinical workflow speed requirements and high levels of test accuracy to not only detect mutant alleles, but also provide quantitative measure of their abundance. So-called "second-generation" deep-sequencing instruments (Natrajan and Reis-Filho, 2011; Wong et al., 2011) currently used by cancer genome centers to sequence entire genomes, exomes, transcriptomes, and methylomes often require weeks for sample template preparation, sequence generation, and data analyses. Generating and assembling the massive number of relatively short sequence reads into usable data that specify genes and mutations remains complex, which partly explains why this generation of instruments is mostly used in research facilities with sophisticated databases and highly qualified and diversified scientific staff. Because these technologies have had minimal use in diagnostic settings, additional validation of potential candidate mutations is required using clinical-grade sequencing assays in certified diagnostic laboratories. The addition of extensive genomic sequencing and follow-up mutation validation introduces significant stress to the clinical workflow (Figure 3). The advent of "third-generation" sequencers such as Pacific Biosciences PacBio RS and Life Technologies' Ion Torrent PGM provides increased speed of sequencing due to their use of sensors that detect nucleotides as they are added to DNA molecules in synthesis, although parallelization and machine throughput currently is much lower than with second-generation technologies (Eid et al., 2009; Korlach et al., 2010; Rothberg et al., 2011).

In contrast to germline DNA mutations, which represent 50% or 100% of alleles in heterozygous or homozygous individuals, respectively, clinically relevant somatic mutations may only be present in a small percentage of cells and thus represent less than 5% of sequence reads, either as a result of high percentages of nontumor cells in biospecimens or because some mutations are only present in a subpopulation of tumor cells. To achieve this needed sensitivity, protocols can be adapted to obtain high depth (i.e., to generate many overlapping sequence reads such that every nucleotide is detected multiple times). Although the typical depth requirement for normal diploid genomes is usually 20–30×, tumor coverage requirements may need to exceed 100× to detect clinically relevant somatic mutations. Important factors for determining depth include the relative proportions of tumor versus nontumor cells in the sample extracted for DNA analyses (which can be quite low, for example, in pancreatic cancer due to high stromal cell content) and tumor heterogeneity. The latter reflects the mosaic nature of tumors, whereby multiple subclones diverge in their mutation load (Figure 1), leading to different proportions of mutant alleles in the same tumor. The clinical implications of low-abundance mutations remain unclear.

Capturing sequence information on all nucleotides in genomes, exomes, or large sets of target genes is challenging using all current technologies, and the extent of cancer genome sequence that is needed to inform clinical decisions is debatable. There is a small subset of genes that is currently deemed to be actionable because mutations in these genes already have diagnostic, prognostic, or predictive implications. Potentially actionable cancer genes should be sequenced in their entirety. Published coverage estimates for whole-genome and exome datasets are below 90% (Cancer Genome Atlas Research Network, 2011), which is inadequate for genes associated with actionable mutations. Near complete coverage of all protein-coding bases in important genes can usually be achieved using polymerase chain reaction (PCR)-based strategies and optimized through trial and error. However, PCR-based approaches consume relatively more tumor DNA and do not scale well. In addition, there are thousands of genes that are known to harbor somatic mutations (see ICGC database). Though the consequence of most of these mutations is unknown, it would be useful to prospectively archive all mutation data in databases that can be shared among cancer organizations to accelerate the expansion of knowledge regarding clinical and functional significance of these new mutations. The cost of sequencing a few hundred genes, exomes, and whole genomes has and will continue to decrease, with high-throughput laboratories currently achieving costs in the $1,000–$2,000 range for large gene panels as well as exomes when using pooling strategies (Kozarewa and Turner, 2011). Whole-genome sequencing is approximately five to ten times more expensive; given the complexity in their analyses and that most of the clinically interpretable mutations are confined to protein-coding genes, whole-genome data sets will likely not become routine studies to be conducted in clinical trials and patient management in the next decade. The trade-offs between rapidity of analyses, depth, coverage, cost, and acquisition of new information on somatic mutations will continue to change in lockstep with continued improvements in technologies.

Sequencing and other genomic technologies used to detect somatic mutations are data intensive. The management and delivery of clinically useful and easily interpretable information to healthcare providers will need to address several issues, including data standards, integration and linkages with clinical and laboratory data and other external data warehouses, and data security. Some of these issues are generic; for example, the rapid increase in genomic data generation rate exceeds the corresponding growth rates in data storage technologies, network bandwidth, or processing speeds. Robust data pipelines are needed to track data associated with the sequence information, including instruments, protocols, mutation calling software, quality metrics, etc. Resolving privacy concerns around integrating clinical sequence data sets with electronic medical records requires further efforts.

Informatics challenges related to cancer sequencing arise as a result of tumor heterogeneity and interpretation of mutation data. All bases (with or without variants) need to be tracked in regard to depth, coverage, and base-calling method and confidence score. Because sequence data is ideally generated in tumor and matched normal samples, parallel data capture and analysis is needed to classify variants as germline and somatic. Germline variants can also be screened electronically via polymorphism databases such as dbSNP (http://www.ncbi.nlm.nih.gov/projects/SNP/). Interpreting the clinical significance of somatic mutations is challenging unless the mutation has previously been shown to be recurrent and actionable. Rapid access to curated information on cancer mutations and genes is the logical first step in this process, as a match with a previously characterized mutation that is known to predict response to a targeted therapy is the simplest scenario. Informatics systems are thus needed to query mutation and cancer gene databases (such as COSMIC and the NCI Gene Index) and large-scale data sets generated by The Cancer Genome Atlas (Cancer Genome Atlas Research Network, 2011) and the ICGC, as well as the literature to determine what is known about identified variants. Novel somatic mutations require careful interpretation that relies on informatics systems that provide information on the identity of the gene, the functional domain, and the extent of evolutionary conservation of the affected amino acid. Software such as SIFT (sorting intolerant from tolerant) (Ng and Henikoff, 2003) can be used to predict whether an amino acid substitution affects protein function. Though each clinical study or cancer center may benefit from storing novel mutations in its local databases to inform subsequent cases with similar mutations, many mutations will be too rare to recur in the same organization, which argues for the establishment of international databases with mutation, function, patient demographic, treatment, and outcome data to ensure robust statistical analyses.

REPORTING DATA TO CLINICIANS AND PATIENTS

Current cancer treatment decisions are informed by knowledge of a limited number of disease-modifying genes. As the cost of genome sequencing technologies rapidly declines, it is conceivable that oncologists will have knowledge of an individual patient's complete cancer genome in the near future. The ultimate goal of such comprehensive profiling is to benefit the individual patient being profiled and, for mutations in germline, family members. However, for the foreseeable future, the ability to generate genomic data will supersede the capacity to decipher patterns across complex data sets, draw inferences from prior experiences, and make informed treatment recommendations that will benefit the profiled individual patient. Novel tools to integrate genomic information with traditional clinical and pathological data in an iterative manner are needed, as are tools that present complex results to clinicians and patients in understandable formats.

Given the complexity of the data, the high number of somatic mutations that can be detected using large-scale sequencing, and the many unique situations that will be encountered, there is a place for establishing expert panels to review the mutation data, deliberate clinical significance, and offer a multidisciplinary perspective regarding the consequences of mutation profiles observed in patients. Multidisciplinary representation allows for input from experts having different training and background, including genome scientists, clinicians, ethicists, clinical geneticists, and genetic counselors to provide balanced interpretations of the potential functional and clinical significance of mutations in the foreseeable future when information from diverse sources will rapidly evolve. Expert panel reports to clinicians should include the rationale for decisions, the degree of consensus, and the level of evidence supporting the decision. Clinical significance should be based on publications reporting on prognostic or predictive role and whether there are clinical trials of targeted agents for the protein product of the mutation, the gene, or pathway. This approach is scalable if it leverages existing and emerging databases and informatics tools that generate draft physician reports that can be reviewed and modified by the expert panel as new information arises. Cursory review will be needed for frequent actionable mutations, and more time will be devoted to deal with novel mutations of possible significance and incidental findings.

The content and format of reports to clinicians are important considerations. Data reports to clinicians must be understandable. Critical pieces of clinical and diagnostic information need to be prioritized according to their clinical utility and level of validation. There is a need for easily accessible smart user interfaces that provide the support for clinical decisions. These need to be structured around best practices and tailored to the level of expertise of the decision maker.

MONITORING AND EVALUATING THE UTILITY OF CLINICAL GENOMICS

It will be critically important to evaluate the utility of genomic results. For the foreseeable future, genomic sequencing will be largely a research approach, and its value must be demonstrated prior to its broad adoption. The genomic information should be not only understood, but also used by clinicians to inform their discussions with patients and to modify treatment recommendation. Such treatment recommendations should result in improved clinical outcomes at affordable costs. Thus, genomic clinical trials should ascertain whether the genomic analyses and mutation-based treatment decisions result in greater survival, improved quality of life, and avoidance of toxicity.

It is important to highlight that genomic results will include somatic mutations, identifying (or failing to identify) a druggable tumor marker, and (where relevant) the receipt of germline genetic results identifying inherited risk of cancer, of other diseases, and of drug toxicity. Information on inherited risks of disease may have minimal impact on a treatment and outcome of a patient with advanced cancer; however, such information may be relevant to patients with potentially curable cancer and to their family members. What information should be conveyed—how, when, and to whom—are areas that require additional research to assess preferences of patients, clinicians, family members, bioethicists, and policy makers.

DATA SHARING

Data from these diverse inputs, linked to information on treatment selection and response, should be broadly leveraged across research centers to generalize knowledge and increase the likelihood that genomic profiling will benefit individual patients in the future (Figure 4). This will require some degree of altruism among patients to make their personal genomic and medical information publicly available and a spirit of collaboration between researchers to share their data prior to publication (Mousses et al., 2008; Toronto International Data Release Workshop Authors et al., 2009). Balancing timely access to data with protection of sensitive personal health information of patients and their families is challenging. Four core bioethical principles have been established by the International Cancer Genome Consortium (ICGC) to guide data sharing: (1) participation of individual patients

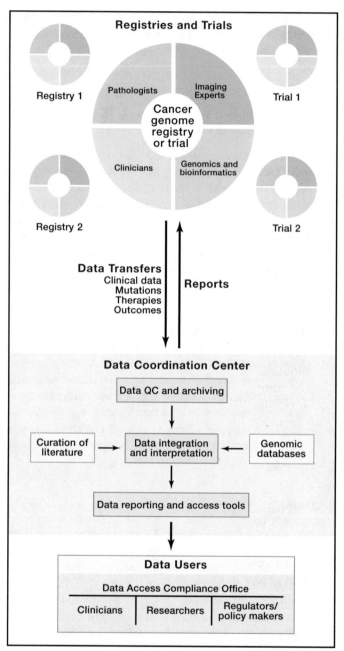

FIGURE 4 Model for Sharing Cancer Genome Data Sets from Registries and Clinical Trials with Clinicians, Researchers, Regulators, and Policymakers

Novel tools and data repositories are needed to integrate genomic information with traditional clinical and pathological data and to present complex results to clinicians and patients in understandable formats. Genomic clinical trials and registries provide patient demographics, germline and somatic variants, is voluntary; (2) a patient's care will not be affected by his/her decision regarding participation; (3) samples and data collected will be used for cancer research, which may include whole-genome sequencing; and (4) data generated will be made accessible to researchers through either an open or controlled access database under terms and conditions that will maximize participant confidentiality (Hudson et al., 2010).

IMPLICATIONS FOR DRUG DEVELOPMENT

Advances in the understanding of specific somatic mutations or amplifications and incorporation of single gene tests have had demonstrable impact on drug development and cancer treatment. The characterization of actionable mutations has already allowed: (1) selection of subsets of patients for clinical trial participation and ultimately for marketing authorization based on greater treatment benefit (i.e., trastuzumab for HER2 breast cancer); (2) restriction of labeled indications for targeted agents to avoid treating patients that do not benefit (i.e., lack of efficacy of EGFR antibodies for colorectal cancer patients with KRAS mutations); and (3) prediction of toxicity risk (i.e., neutropenia and diarrhea associated with irinotecan in patients for UGT 1A1*28 homozygosity). Molecular testing will continue to impact patient eligibility for clinical trials, study design, drug approval, market utilization, and reimbursement; however, the challenges to rational and practical utilization of complex genomic data are still not fully understood. The amount of genomic information becoming available is adding a high level of complexity to the process of drug development. Information technologies to manage extensive and diverse biological, chemical, and clinical data sets and computational methods to identify the most pertinent information are essential to guide the development of new drugs and establish priorities to generate scientific hypotheses that warrant clinical testing. Clinical development plans for new agents should aim at documenting major and unequivocal treatment benefit validating the hypothesis.

To date, pilot trials of molecular profiling have focused on patients with advanced disease to provide a molecular-based rationale for enrollment of cancer patients in phase I/II clinical trials and have used exploratory methodologies (Tsimberidou et al., ASCO, abstract; Von Hoff et al., 2010). The

---

treatment, and outcomes. Informatics systems query mutation and cancer gene databases and curated literature to determine what is known about identified variants. Interpretation of novel somatic mutations may be based on information on the identity of the gene and the functional domain and extent of evolutionary conservation of the affected amino acid. Data reports to clinicians include clinical and diagnostic information on the gene(s) and mutation(s) according to their clinical utility and level of validation.

next wave of trials incorporating tumor DNA sequencing data should establish that genomic testing is associated with improvements in drug development processes that ultimately improve patient management and provide clinical benefit. Efficient clinical trial designs are needed to discriminate new agents and/or combination strategies that warrant further development in patients selected by genomic testing rather than solely by histological characteristics. Efficiencies in clinical trial conduct are gained by identifying patients who are unlikely to benefit from currently available treatment and thus have a relatively poor outcome and/or are more likely to benefit from the investigational agent due to the presence of tumor genetic abnormalities that confer drug sensitivity. In either of these situations, fewer numbers of patients are required to demonstrate an improved treatment effect. This approach requires the discipline to quickly discontinue development when a sufficient signal of activity is not detected in the trial population thought most likely to benefit from a new treatment or if such a subgroup cannot be identified. Definitive clinical trials leading to drug approval should be based on strong scientific hypotheses and robust signals of activity and should aim to show efficacy and safety in selected populations based on complex genomic testing. This will have a major impact on the drug approval process and will modify marketing expectations for new agents at least at the time of initial approval. This is, however, probably the only viable and sustainable approach to allow the rapid translation of the new genomic advances to the cancer patients.

Obviously, there are costs to incorporating genomic sequencing analyses in cancer therapeutic trials. The information gained from broad and deep assessments of the cancer genome may be greater than needed for a clinical trial evaluating the safety and activity of an agent of interest to achieve trial specific objectives. For a given therapeutic agent, the numbers of genes and associated mutations known to be relevant to the disease, drug target, and pathway may be relatively limited. Financial constraints and ethical concerns of administering agents to patients that may be inactive may also limit the ability to screen large numbers of patients and enroll them on an investigational drug trial to determine activity in rare genetic subgroups. Given the current costs of exome sequencing, an initial strategy is to ensure coverage of specific genes relevant to the drug/target, to expand to the top few hundreds of genes of possible clinical and biological interest, and then, as costs fall further, to include other emerging genes of interest. As the information gained from broad molecular screening is of value beyond the industry sponsors, there should be a willingness among public agencies, health insurers, philanthropy, and industry to fund these activities, with the proviso that the information gained will be made publicly available.

IMPLICATIONS FOR REGULATORY BODIES

In many jurisdictions, regulatory authorities have adapted their drug development guidelines, approval decisions, and pharmacovigilance processes to incorporate the knowledge of specific gene alterations in individual patients. A recent review of the approval of 33 new oncology agents approved by the CHMP in Europe between 2000 and 2008 noted that pharmacogenetic biomarkers were mentioned in nine cases (EMEA Committee for Medicinal Products, 2008). Of interest, genetic testing was associated with prescription restriction of the new agent to a subset of patients expressing the biomarker. In addition, many of the regulatory decisions were based on the utilization of nonapproved tests and, in many cases, on retrospective analyses of subsets of patients enrolled in large phase III trials conducted in nonselected patient populations. This illustrates that, in the early days of the incorporation of genomic testing in oncology drug development, regulatory authorities had to be reactive to a rapidly moving field.

Several new guidance documents have been issued or are in development to inform the design of clinical studies incorporating genetic biomarkers and the development of companion diagnostic tests (FDA, 2005). An example of drug and companion diagnostic codevelopment paradigm is the recent approval of vemurafenib for patients with advanced melanoma harboring the BRAF V600 mutation. This represents the first FDA approval of a drug and a companion diagnostic mutation test that stipulates within the package insert the use of the approved test to determine patient eligibility for treatment. However, institutions may prefer to perform mutation analyses using more extensive profiling technologies in their own Clinical Laboratory Improvement Amendments (CLIA) facilities. Individual tests for single or limited numbers of mutations are unlikely to be efficient or cost effective for cancer patients or for drug development. Genomic profiling will generate more extensive data that could impact patient management to a greater extent than very selective tests approved by regulatory agencies. In the case of *BRAF*, the approved test only documents the presence of a single point mutation (V600E); however, assessing other *BRAF* mutations (i.e., V600K and V600D) and mutations in other genes may be relevant to understanding treatment resistance and could ultimately help patients get access to second-generation inhibitors of different *BRAF* variants or other relevant targets. There is a need for collaboration among regulatory agencies, industry, and academics at the forefront of genomic technology to develop new approaches for comprehensive genomic testing in the drug and test approval processes.

In addition to developing new approaches for genomic testing that will be not only acceptable to regulatory agencies, but also reflective of the constant scientific progress, other key regulatory questions related to the evidence of safety and efficacy for drug and test approval need to be addressed. For example, in highly selected patient populations identified by genomics testing, are randomized trials needed for initial approval of a new target therapy, or should the concept of "targeted approval" be considered based on striking results from open label phase II studies (Chabner, 2011; FDA, 2011)? What is the size of the safety database that will be required for initial approval, knowing that genomic-based therapy will lead to relatively small patient subsets? Generating the large safety data set in accordance with regulatory guidelines will become more challenging prior to approval, and the solution may be the adoption of a risk management program in highly selected populations after a product has been approved for marketing. Finally, because tumor growth is usually driven by complex genomic alterations that cannot be controlled by inhibiting a single pathway, how should combinations of investigational agents be developed to document activity and safety as required by regulatory agencies? The FDA draft guidance on the development of investigational drug combinations is a first step to accelerating the early clinical testing of novel targeted agents (FDA, 2010). It is clear that regulatory authorities are adapting and providing guidance on issues associated with genomic testing and drug and diagnostic approvals. However, further dialog is needed to ensure an appropriate and dynamic regulatory framework as technological and scientific advances in cancer genomics and its role in drug development continue to evolve.

IMPLICATIONS FOR PATIENTS, HEALTHCARE PROVIDERS, AND PAYERS

The implications of personalized cancer medicine are complex and can be viewed from the perspectives of the patient, the healthcare provider, and the society. From the cancer patient's perspective, the prospect of receiving specific targeted agents that match "driver" mutations offers an attractive therapeutic strategy even though toxicity due to off-target effects may remain relevant. Until molecular profiling becomes a standard of clinical care, there are many challenges for the healthcare provider, such as the access to CLIA-certified laboratories that can perform validated genomic evaluations, the assurance of timely turnaround of results to minimize treatment delay, and the responsibility of finding appropriate treatment for patients based on the returned results. There will be an expanded need for clinicians knowledgeable in cancer genetics and the interpretation of genomic results and for clinical geneticists to work alongside oncologists in multidisciplinary clinics to advise patients and family members on inheritable risks. This will require new curricula, training, and facile knowledge transfer and addressing critical shortages of geneticists and counselors (Cooksey et al., 2005; Cooksey et al., 2006). From the societal view point, the economic balance of personalized cancer medicine must take into consideration the benefits derived from the cost savings of avoiding empiric prescription of expensive medicines versus the expenditures of training personnel with appropriate expertise, setting up certified laboratories with close monitoring of quality control, and high-throughput screening.

CONCLUSIONS

The recent advances in DNA sequencing allow for the characterization of a large number of genes and, ultimately, of the entire cancer genome in a timeframe that is compatible with treatment decisions for the patient. This creates opportunities for the development of new agents but also results in challenges that will only be solved if scientists, clinical investigators, pharmaceutical companies, regulatory agencies, and third-party payers collaborate closely. A rigorous approach to developing a complete clinical workflow in which every component of the process is optimized prior to scale-up is essential.

Genomic analyses and results need to be accessible to guide management and clinical trial decisions throughout a patient's disease course. This means that the information should be available irrespective of the party who covered the cost of genomic testing and the initial research study that led to the analyses. It is essential that access to individual agents and to rational drug combinations be easier for both investigational agents as well as marketed drugs. This will require collaboration between pharmaceutical companies that control access to most of the new agents available for clinical testing and third-party payers. Data generated through repetitive genomic studies from individual patients at different stages of disease should be made publicly available to better understand the genomic evolution according to disease stage and therapeutic intervention. This information is required to define the clinical setting in which a therapy will be most effective and to elucidate mechanisms of therapeutic resistance. Many may argue that such complex and far reaching collaborations are not attainable; we would ask "how can we not?" given the overwhelming burden of cancer and the unprecedented opportunities for advancements in outcomes for cancer patients.

ACKNOWLEDGMENTS

We thank our numerous clinical and research colleagues in Ontario who have been involved in the design and execution of a pilot study to assess genomic testing of actionable mutations in a multicenter clinical trial setting and, in particular, Dr. Lillian Siu at the University Health Network's Princess Margaret Hospital for critical advice during the preparation of this manuscript. J.E.D. and T.J.H. are supported by Investigator Awards from the Ontario Institute for Cancer Research and through generous support from the Ontario Ministry of Economic Development and Innovation. The authors also wish to thank Cancer Care Ontario, the Birmingham Cancer Research Fund, and the Princess Margaret Hospital Foundation for their support.

REFERENCES

Agulnik, M., da Cunha Santos, G., Hedley, D., Nicklee, T., Dos Reis, P.P., Ho, J., Pond, G.R., Chen, H., Chen, S., Shyr, Y., et al. (2007). Predictive and pharmacodynamic biomarker studies in tumor and skin tissue samples of patients with recurrent or metastatic squamous cell carcinoma of the head and neck treated with erlotinib. J. Clin. Oncol. 25, 2184–2190.

Bang, Y.J., Van Cutsem, E., Feyereislova, A., Chung, H.C., Shen, L., Sawaki, A., Lordick, F., Ohtsu, A., Omuro, Y., Satoh, T., et al; ToGA Trial Investigators. (2010). Trastuzumab in combination with chemotherapy versus chemotherapy alone for treatment of HER2-positive advanced gastric or gastro-oesophageal junction cancer (ToGA): a phase 3, open-label, randomised controlled trial. Lancet 376, 687–697.

Bookman, M.A., Darcy, K.M., Clarke-Pearson, D., Boothby, R.A., and Horowitz, I.R. (2003). Evaluation of monoclonal humanized anti-HER2 antibody, trastuzumab, in patients with recurrent or refractory ovarian or primary peritoneal carcinoma with overexpression of HER2: a phase II trial of the Gynecologic Oncology Group. J. Clin. Oncol. 21, 283–290.

Campbell, P.J., Yachida, S., Mudie, L.J., Stephens, P.J., Pleasance, E.D., Stebbings, L.A., Morsberger, L.A., Latimer, C., McLaren, S., Lin, M.-L., et al. (2010). The patterns and dynamics of genomic instability in metastatic pancreatic cancer. Nature 467, 1109–1113.

Cancer Genome Atlas Research Network. (2011). Integrated genomic analyses of ovarian carcinoma. Nature 474, 609–615.

Chabner, B.A. (2011). Early accelerated approval for highly targeted cancer drugs. N. Engl. J. Med. 364, 1087–1089.

Chapman, P.B., Hauschild, A., Robert, C., Haanen, J.B., Ascierto, P., Larkin, J., Dummer, R., Garbe, C., Testori, A., Maio, M., et al; BRIM-3 Study Group. (2011). Improved survival with vemurafenib in melanoma with BRAF V600E mutation. N. Engl. J. Med. 364, 2507–2516.

Cooksey, J.A., Forte, G., Benkendorf, J., and Blitzer, M.G. (2005). The state of the medical geneticist workforce: findings of the 2003 survey of American Board of Medical Genetics certified geneticists. Genet. Med. 7, 439–443.

Cooksey, J.A., Forte, G., Flanagan, P.A., Benkendorf, J., and Blitzer, M.G. (2006). The medical genetics workforce: an analysis of clinical geneticist subgroups. Genet. Med. 8, 603–614.

Dias-Santagata, D., Akhavanfard, S., David, S.S., Vernovsky, K., Kuhlmann, G., Boisvert, S.L., Stubbs, H., McDermott, U., Settleman, J., Kwak, E.L., et al. (2010). Rapid targeted mutational analysis of human tumours: a clinical platform to guide personalized cancer medicine. EMBO Mol. Med. 2, 146–158.

Dupont Jensen, J., Laenkholm, A.V., Knoop, A., Ewertz, M., Bandaru, R., Liu, W., Hackl, W., Barrett, J.C., and Gardner, H. (2011). PIK3CA mutations may be discordant between primary and corresponding metastatic disease in breast cancer. Clin. Cancer Res. 17, 667–677.

Eid, J., Fehr, A., Gray, J., Luong, K., Lyle, J., Otto, G., Peluso, P., Rank, D., Baybayan, P., Bettman, B., et al. (2009). Real-time DNA sequencing from single polymerase molecules. Science 323, 133–138.

EMEA Committee for Medicinal Products. (2008). Reflection paper on pharmacogenomics in oncology. EMEA/CHMP/PGxWP/128435/2006.

Fleming, G.F., Sill, M.W., Darcy, K.M., McMeekin, D.S., Thigpen, J.T., Adler, L.M., Berek, J.S., Chapman, J.A., DiSilvestro, P.A., Horowitz, I.R., and Fiorica, J.V. (2010). Phase II trial of trastuzumab in women with advanced or recurrent, HER2-positive endometrial carcinoma: a Gynecologic Oncology Group study. Gynecol. Oncol. 116, 15–20.

FDA (Food and Drug Administration). (2005). Guidance for industry pharmacogenomic data submissions. March 2005. http://www.fda.gov/downloads/RegulatoryInformation/Guidances/ucm126957.pdf.

FDA (Food and Drug Administration). (2010). Draft guidance for industry codevelopment of two or more unmarketed investigational drugs for use in combination. December 2010. http://www.fda.gov/downloads/Drugs/GuidanceComplianceRegulatoryInformation/Guidances/UCM236669.pdf.

FDA (Food and Drug Administration). (2011). FDA approves Xalkori with companion diagnostic for a type of late-stage lung cancer. August 25, 2011. http://www.fda.gov/NewsEvents/Newsroom/PressAnnouncements/ucm269856.htm.

Forbes, S.A., Bindal, N., Bamford, S., Cole, C., Kok, C.Y., Beare, D., Jia, M., Shepherd, R., Leung, K., Menzies, A., et al. (2011). COSMIC: mining complete cancer genomes in the Catalogue of Somatic Mutations in Cancer. Nucleic Acids Res. *39*(Database issue), D945–D950.

Fumagalli, D., Gavin, P.G., Taniyama, Y., Kim, S.-I., Choi, H.-J., Paik, S., and Pogue-Geile, K.L. (2010). A rapid, sensitive, reproducible and cost-effective method for mutation profiling of colon cancer and metastatic lymph nodes. BMC Cancer *10*, 101.

Hanahan, D., and Weinberg, R.A. (2000). The hallmarks of cancer. Cell *100*, 57–70.

Hanahan, D., and Weinberg, R.A. (2011). Hallmarks of cancer: the next generation. Cell *144*, 646–674.

Hudson, T.J., Anderson, W., Artez, A., Barker, A.D., Bell, C., Bernabé, R.R., Bhan, M.K., Calvo, F., Eerola, I., Gerhard, D.S., et al; International Cancer Genome Consortium. (2010). International network of cancer genome projects. Nature *464*, 993–998.

Jones, S., Laskin, J., Li, Y.Y., Griffith, O.L., An, J., Bilenky, M., Butterfield, Y.S., Cezard, T., Chuah, E., Corbett, R., et al. (2010). Evolution of an adenocarcinoma in response to selection by targeted kinase inhibitors. Genome Biol. *11*, R82.

Kalikaki, A., Koutsopoulos, A., Trypaki, M., Souglakos, J., Stathopoulos, E., Georgoulias, V., Mavroudis, D., and Voutsina, A. (2008). Comparison of EGFR and K-RAS gene status between primary tumours and corresponding metastases in NSCLC. Br. J. Cancer *99*, 923–929.

Knijn, N., Mekenkamp, L.J., Klomp, M., Vink-Börger, M.E., Tol, J., Teerenstra, S., Meijer, J.W., Tebar, M., Riemersma, S., van Krieken, J.H., et al. (2011). KRAS mutation analysis: a comparison between primary tumours and matched liver metastases in 305 colorectal cancer patients. Br. J. Cancer *104*, 1020–1026.

Korlach, J., Bjornson, K.P., Chaudhuri, B.P., Cicero, R.L., Flusberg, B.A., Gray, J.J., Holden, D., Saxena, R., Wegener, J., and Turner, S.W. (2010). Real-time DNA sequencing from single polymerase molecules. Methods Enzymol. *472*, 431–455.

Kozarewa, I., and Turner, D.J. (2011). 96-plex molecular barcoding for the Illumina Genome Analyzer. Methods Mol. Biol. *733*, 279–298.

Lee, W., Jiang, Z., Liu, J., Haverty, P.M., Guan, Y., Stinson, J., Yue, P., Zhang, Y., Pant, K.P., Bhatt, D., et al. (2010). The mutation spectrum revealed by paired genome sequences from a lung cancer patient. Nature *465*, 473–477.

MacConaill, L.E., Campbell, C.D., Kehoe, S.M., Bass, A.J., Hatton, C., Niu, L., Davis, M., Yao, K., Hanna, M., Mondal, C., et al. (2009). Profiling critical cancer gene mutations in clinical tumor samples. PLoS ONE *4*, e7887.

McGuire, A.L., Caulfield, T., and Cho, M.K. (2008). Research ethics and the challenge of whole-genome sequencing. Nat. Rev. Genet. *9*, 152–156.

Mok, T.S., Wu, Y.L., Thongprasert, S., Yang, C.H., Chu, D.T., Saijo, N., Sunpaweravong, P., Han, B., Margono, B., Ichinose, Y., et al. (2009). Gefitinib or carboplatin-paclitaxel in pulmonary adenocarcinoma. N. Engl. J. Med. *361*, 947–957.

Mousses, S., Kiefer, J., Von Hoff, D., and Trent, J. (2008). Using biointelligence to search the cancer genome: an epistemological perspective on knowledge recovery strategies to enable precision medical genomics. Oncogene *27*(Suppl 2), S58–S66.

Natrajan, R., and Reis-Filho, J.S. (2011). Next-generation sequencing applied to molecular diagnostics. Expert Rev. Mol. Diagn. *11*, 425–444.

Ng, P.C., and Henikoff, S. (2003). SIFT: Predicting amino acid changes that affect protein function. Nucleic Acids Res. *31*, 3812–3814.

Pleasance, E.D., Cheetham, R.K., Stephens, P.J., McBride, D.J., Humphray, S.J., Greenman, C.D., Varela, I., Lin, M.L., Ordóñez, G.R., Bignell, G.R., et al. (2010). A comprehensive catalogue of somatic mutations from a human cancer genome. Nature *463*, 191–196.

Pleasance, E.D., Stephens, P.J., O'Meara, S., McBride, D.J., Meynert, A., Jones, D., Lin, M.L., Beare, D., Lau, K.W., Greenman, C., et al. (2010). A small-cell lung cancer genome with complex signatures of tobacco exposure. Nature *463*, 184–190.

Rothberg, J.M., Hinz, W., Rearick, T.M., Schultz, J., Mileski, W., Davey, M., Leamon, J.H., Johnson, K., Milgrew, M.J., Edwards, M., et al. (2011). An integrated semiconductor device enabling non-optical genome sequencing. Nature *475*, 348–352.

Santini, D., Loupakis, F., Vincenzi, B., Floriani, I., Stasi, I., Canestrari, E., Rulli, E., Maltese, P.E., Andreoni, F., Masi, G., et al. (2008). High concordance of KRAS status between primary colorectal tumors and related metastatic sites: implications for clinical practice. Oncologist *13*, 1270–1275.

Shah, S.P., Morin, R.D., Khattra, J., Prentice, L., Pugh, T., Burleigh, A., Delaney, A., Gelmon, K., Guliany, R., Senz, J., et al. (2009). Mutational evolution in a lobular breast tumour profiled at single nucleotide resolution. Nature *461*, 809–813.

Sieben, N.L., Macropoulos, P., Roemen, G.M., Kolkman-Uljee, S.M., Jan Fleuren, G., Houmadi, R., Diss, T., Warren, B., Al Adnani, M., De Goeij, A.P., et al. (2004). In ovarian neoplasms, BRAF, but not KRAS, mutations are restricted to low-grade serous tumours. J. Pathol. *202*, 336–340.

Slamon, D.J., Leyland-Jones, B., Shak, S., Fuchs, H., Paton, V., Bajamonde, A., Fleming, T., Eiermann, W., Wolter, J., Pegram, M., et al. (2001). Use of chemotherapy plus a monoclonal antibody against HER2 for metastatic breast cancer that overexpresses HER2. N. Engl. J. Med. *344*, 783–792.

Stratton, M.R. (2011). Exploring the genomes of cancer cells: progress and promise. Science *331*, 1553–1558.

Thomas, R.K., Baker, A.C., Debiasi, R.M., Winckler, W., Laframboise, T., Lin, W.M., Wang, M., Feng, W., Zander, T., MacConaill, L.E., et al. (2007). High-throughput oncogene mutation profiling in human cancer. Nat. Genet. *39*, 347–351.

Toronto International Data Release Workshop Authors, Birney, E., Hudson, T.J., Green, E.D., Gunter, C., Eddy, S., Rogers, J., Harris, J.R., Ehrlich, S.D., and Apweiler, R. (2009). Prepublication data sharing. Nature *461*, 168–170.

Von Hoff, D.D., Stephenson, J.J., Jr., Rosen, P., Loesch, D.M., Borad, M.J., Anthony, S., Jameson, G., Brown, S., Cantafio, N., Richards, D.A., et al. (2010). Pilot study using molecular profiling of patients' tumors to find potential targets and select treatments for their refractory cancers. J. Clin. Oncol. *28*, 4877–4883.

Wang, Y., Carlton, V.E., Karlin-Neumann, G., Sapolsky, R., Zhang, L., Moorhead, M., Wang, Z.C., Richardson, A.L., Warren, R., Walther, A., et al. (2009). High quality copy number and genotype data from FFPE samples using Molecular Inversion Probe (MIP) microarrays. BMC Med. Genomics *2*, 8.

Wong, K.M., Hudson, T.J., and McPherson, J.D. (2011). Unraveling the genetics of cancer: Genome sequencing and beyond. Annu. Rev. Genomics Hum. Genet. *12*, 407–430.

Yachida, S., Jones, S., Bozic, I., Antal, T., Leary, R., Fu, B., Kamiyama, M., Hruban, R.H., Eshleman, J.R., Nowak, M.A., et al. (2010). Distant metastasis occurs late during the genetic evolution of pancreatic cancer. Nature *467*, 1114–1117.

ell

Cancer Epigenetics: From Mechanism to Therapy

Mark A. Dawson[1,2], Tony Kouzarides[1,*]

[1]Gurdon Institute and Department of Pathology, University of Cambridge, Tennis Court Road, Cambridge CB2 1QN, UK, [2]Department of Haematology, Cambridge Institute for Medical Research and Addenbrooke's Hospital, University of Cambridge, Hills Road, Cambridge CB2 0XY, UK
*Correspondence: t.kouzarides@gurdon.cam.ac.uk

Cell, Vol. 150, No. 1, July 6, 2012 © 2012 Elsevier Inc.
http://dx.doi.org/10.1016/j.cell.2012.06.013

SUMMARY

The epigenetic regulation of DNA-templated processes has been intensely studied over the last 15 years. DNA methylation, histone modification, nucleosome remodeling, and RNA-mediated targeting regulate many biological processes that are fundamental to the genesis of cancer. Here, we present the basic principles behind these epigenetic pathways and highlight the evidence suggesting that their misregulation can culminate in cancer. This information, along with the promising clinical and preclinical results seen with epigenetic drugs against chromatin regulators, signifies that it is time to embrace the central role of epigenetics in cancer.

INTRODUCTION

Chromatin is the macromolecular complex of DNA and histone proteins, which provides the scaffold for the packaging of our entire genome. It contains the heritable material of eukaryotic cells. The basic functional unit of chromatin is the nucleosome. It contains 147 base pairs of DNA, which is wrapped around a histone octamer, with two each of histones H2A, H2B, H3, and H4. In general and simple terms, chromatin can be subdivided into two major regions: (1) heterochromatin, which is highly condensed, late to replicate, and primarily contains inactive genes; and (2) euchromatin, which is relatively open and contains most of the active genes. Efforts to study the coordinated regulation of the nucleosome have demonstrated that all of its components are

49

CellPress

subject to covalent modification, which fundamentally alters the organization and function of these basic tenants of chromatin (Allis et al., 2007).

The term "epigenetics" was originally coined by Conrad Waddington to describe heritable changes in a cellular phenotype that were independent of alterations in the DNA sequence. Despite decades of debate and research, a consensus definition of epigenetics remains both contentious and ambiguous (Berger et al., 2009). Epigenetics is most commonly used to describe chromatin-based events that regulate DNA-templated processes, and this will be the definition we use in this review.

Modifications to DNA and histones are dynamically laid down and removed by chromatin-modifying enzymes in a highly regulated manner. There are now at least four different DNA modifications (Baylin and Jones, 2011; Wu and Zhang, 2011) and 16 classes of histone modifications (Kouzarides, 2007; Tan et al., 2011). These are described in Table 1. These modifications can alter chromatin structure by altering noncovalent interactions within and between nucleosomes. They also serve as docking sites for specialized proteins with unique domains that specifically recognize these modifications. These chromatin readers recruit additional chromatin modifiers and remodeling enzymes, which serve as the effectors of the modification.

The information conveyed by epigenetic modifications plays a critical role in the regulation of all DNA-based processes, such as transcription, DNA repair, and replication. Consequently, abnormal expression patterns or genomic alterations in chromatin regulators can have profound results and can lead to the induction and maintenance of various cancers. In this Review, we highlight recent advances in our understanding of these epigenetic pathways and discuss their role in oncogenesis. We provide a comprehensive list of all the recurrent cancer mutations described thus far in epigenetic pathways regulating modifications of DNA (Figure 2), histones (Figures 3, 4, and 5), and chromatin remodeling (Figure 6). Where relevant, we will also emphasize existing and emerging drug therapies aimed at targeting epigenetic regulators (Figure 1).

CHARACTERIZING THE EPIGENOME

Our appreciation of epigenetic complexity and plasticity has dramatically increased over the last few years following the development of several global proteomic and genomic technologies. The coupling of next-generation sequencing (NGS) platforms with established chromatin techniques such as chromatin immunoprecipitation (ChIP-Seq) has presented us with a previously unparalleled view of the epigenome (Park, 2009). These technologies have provided comprehensive maps of nucleosome positioning (Segal and Widom, 2009), chromatin conformation (de Wit and de Laat, 2012), transcription factor binding sites (Farnham, 2009), and the localization of histone

(Rando and Chang, 2009) and DNA (Laird, 2010) modifications. In addition, NGS has revealed surprising facts about the mammalian transcriptome. We now have a greater appreciation of the fact that most of our genome is transcribed and that noncoding RNA may play a fundamental role in epigenetic regulation (Amaral et al., 2008).

Table 1 Chromatin Modifications, Readers, and Their Function

Chromatin Modification	Nomenclature	Chromatin-Reader Motif	Attributed Function
DNA Modifications			
5-methylcytosine	5mC	MBD domain	transcription
5-hydroxymethylcytosine	5hmC	unknown	transcription
5-formylcytosine	5fC	unknown	unknown
5-carboxylcytosine	5caC	unknown	unknown
Histone Modifications			
Acetylation	K-ac	BromodomainTandem, PHD fingers	transcription, repair, replication, and condensation
Methylation (lysine)	K-me1, K-me2, K-me3	Chromodomain, Tudor domain, MBT domain, PWWP domain, PHD fingers, WD40/β propeller	transcription and repair
Methylation (arginine)	R-me1, R-me2s, R-me2a	Tudor domain	transcription
Phosphorylation (serine and threonine)	S-ph, T-ph	14-3-3, BRCT	transcription, repair, and condensation
Phosphorylation (tyrosine)	Y-ph	SH2[a]	transcription and repair
Ubiquitylation	K-ub	UIM, IUIM	transcription and repair
Sumoylation	K-su	SIM[a]	transcription and repair
ADP ribosylation	E-ar	Macro domain, PBZ domain	transcription and repair
Deimination	R→Cit	unknown	transcription and decondensation
Proline isomerisation	P-cis⇌P-trans	unknown	transcription
Crotonylation	K-cr	unknown	transcription
Propionylation	K-pr	unknown	unknown
Butyrylation	K-bu	unknown	unknown
Formylation	K-fo	unknown	unknown
Hyroxylation	Y-oh	unknown	unknown
O-GlcNAcylation (serine and threonine)	S-GlcNAc; T-GlcNAc	unknown	transcription

Modifications: me1, monomethylation; me2, dimethylation; me3, trimethylation; me2s, symmetrical dimethylation; me2a, asymmetrical dimethylation; and Cit, citrulline. Reader domains: MBD, methyl-CpG-binding domain; PHD, plant homeodomain; MBT, malignant brain tumor domain; PWWP, proline-tryptophan-tryptophan-proline domain; BRCT, BRCA1 C terminus domain; UIM, ubiquitin interaction motif; IUIM, inverted ubiquitin interaction motif; SIM, sumo interaction motif; and PBZ, poly ADP-ribose binding zinc finger.
[a]These are established binding modules for the posttranslational modification; however, binding to modified histones has not been firmly established.

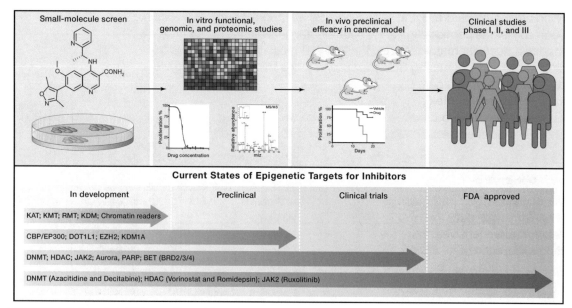

FIGURE 1 Epigenetic Inhibitors as Cancer Therapies

This schematic depicts the process for epigenetic drug development and the current status of various epigenetic therapies. Candidate small molecules are first tested in vitro in malignant cell lines for specificity and phenotypic response. These may, in the first instance, assess the inhibition of proliferation, induction of apoptosis, or cell-cycle arrest. These phenotypic assays are often coupled to genomic and proteomic methods to identify potential molecular mechanisms for the observed response. Inhibitors that demonstrate potential in vitro are then tested in vivo in animal models of cancer to ascertain whether they may provide therapeutic benefit in terms of survival. Animal studies also provide valuable information regarding the toxicity and pharmacokinetic properties of the drug. Based on these preclinical studies, candidate molecules may be taken forward into the clinical setting. When new drugs prove beneficial in well-conducted clinical trials, they are approved for routine clinical use by regulatory authorities such as the FDA. KAT, histone lysine acetyltransferase; KMT, histone lysine methyltransferase; RMT, histone arginine methyltransferase; and PARP, poly ADP ribose polymerase.

Most of the complexity surrounding the epigenome comes from the modification pathways that have been identified. Recent improvements in the sensitivity and accuracy of mass spectrometry (MS) instruments have driven many of these discoveries (Stunnenberg and Vermeulen, 2011). Moreover, although MS is inherently not quantitative, recent advances in labeling methodologies, such as stable isotope labeling by amino acids in cell culture (SILAC), isobaric tags for relative and absolute quantification (iTRAQ), and isotope-coded affinity tag (ICAT), have allowed a greater ability to provide quantitative measurements (Stunnenberg and Vermeulen, 2011).

These quantitative methods have generated "protein recruitment maps" for histone and DNA modifications, which contain proteins that recognize chromatin modifications (Bartke et al., 2010; Vermeulen et al., 2010). Many of these chromatin readers have more than one reading motif, so it is important to understand how they recognize several modifications either simultaneously or sequentially. The concept of multivalent engagement by chromatin-binding modules has recently been explored by using either modified histone peptides (Vermeulen et al., 2010) or in-vitro-assembled and -modified nucleosomes (Bartke et al., 2010; Ruthenburg et al., 2011). The latter approach in particular has uncovered some of the rules governing the recruitment of protein complexes to methylated DNA and modified histones in a nucleosomal context. The next step in our understanding will require a high-resolution in vivo genomic approach to detail the dynamic events on any given nucleosome during the course of gene expression.

EPIGENETICS AND THE CANCER CONNECTION

The earliest indications of an epigenetic link to cancer were derived from gene expression and DNA methylation studies. These studies are too numerous to comprehensively detail in this review; however, the reader is referred to an excellent review detailing the history of cancer epigenetics (Feinberg and Tycko, 2004). Although many of these initial studies were purely correlative, they did highlight a potential connection between epigenetic pathways and cancer. These early observations have been significantly strengthened by recent results from the International Cancer Genome Consortium (ICGC). Whole-genome sequencing in a vast array of cancers has provided a catalog of recurrent somatic mutations in numerous epigenetic regulators (Forbes et al., 2011; Stratton et al., 2009). A central tenet in analyzing these cancer genomes is the identification of "driver" mutations (causally implicated in the process of oncogenesis). A key feature of driver mutations is that they are recurrently found in a variety of cancers, and/or they are often present at a high prevalence in a specific tumor type. We will mostly concentrate our discussions on suspected or proven driver mutations in epigenetic regulators.

For instance, malignancies such as follicular lymphoma contain recurrent mutations of the histone methyltransferase *MLL2* in close to 90% of cases (Morin et al., 2011). Similarly, *UTX*, a histone demethylase, is mutated in up to 12 histologically distinct cancers (van Haaften et al., 2009). Compilation of the epigenetic regulators mutated in cancer highlights histone acetylation and methylation as the most widely affected epigenetic pathways (Figures 3 and 4). These and other pathways that are affected to a lesser extent will be described in the following sections.

Deep sequencing technologies aimed at mapping chromatin modifications have also begun to shed some light on the origins of epigenetic abnormalities in cancer. Cross-referencing of DNA methylation profiles in human cancers with ChIP-Seq data for histone modifications and the binding of chromatin regulators have raised intriguing correlations between cancer-associated DNA hypermethylation and genes marked with "bivalent" histone modifications in multipotent cells (Easwaran et al., 2012; Ohm et al., 2007). These bivalent genes are marked by active (H3K4me3) and repressive (H3K27me3) histone modifications (Bernstein et al., 2006) and appear to identify transcriptionally poised genes that are integral to development and lineage commitment. Interestingly, many of these genes are targeted for DNA methylation in cancer. Equally intriguing are recent comparisons between malignant and normal tissues from the same individuals. These data demonstrate broad domains within the malignant cells that contain significant alterations in DNA methylation. These regions appear to correlate with late-replicating regions of the genome associated with the nuclear lamina (Berman et al., 2012). Although there remains little mechanistic insight into how and why these regions of the genome are vulnerable to epigenetic alterations in cancer, these studies highlight the means by which global sequencing platforms have started to uncover avenues for further investigation.

Genetic lesions in chromatin modifiers and global alterations in the epigenetic landscape not only imply a causative role for these proteins in cancer but also provide potential targets for therapeutic intervention. A number of small-molecule inhibitors have already been developed against chromatin regulators (Figure 1). These are at various stages of development, and three of these (targeting DNMTs, HDACs, and JAK2) have already been granted approval by the US Food and Drug Administration (FDA). This success may suggest that the interest in epigenetic pathways as targets for drug discovery had been high over the past decade. However, the reality is that the field of drug discovery had been somewhat held back due to concerns over the pleiotropic effects of both the drugs and their targets. Indeed, some of the approved drugs (against HDACs) have little enzyme specificity, and their mechanism of action remains contentious (Minucci and Pelicci, 2006).

The belief and investment in epigenetic cancer therapies may now gain momentum and reach a new level of support following the recent preclinical success of inhibitors against BRD4, an acetyl-lysine chromatin-binding protein (Dawson et al., 2011; Delmore et al., 2011; Filippakopoulos et al., 2010; Mertz et al., 2011; Zuber et al., 2011). The molecular mechanisms governing these impressive preclinical results have also been largely uncovered and are discussed below. This process is pivotal for the successful progression of these inhibitors into the clinic. These results, along with the growing list of genetic lesions in epigenetic regulators, highlight the fact that we have now entered an era of epigenetic cancer therapies.

EPIGENETIC PATHWAYS CONNECTED TO CANCER

DNA Methylation

The methylation of the 5-carbon on cytosine residues (5mC) in CpG dinucleotides was the first described covalent modification of DNA and is perhaps the most extensively characterized modification of chromatin. DNA methylation is primarily noted within centromeres, telomeres, inactive X-chromosomes, and repeat sequences (Baylin and Jones, 2011; Robertson, 2005). Although global hypomethylation is commonly observed in malignant cells, the best-studied epigenetic alterations in cancer are the methylation changes that occur within CpG islands, which are present in ~70% of all mammalian promoters. CpG island methylation plays an important role in transcriptional regulation, and it is commonly altered during malignant transformation (Baylin and Jones, 2011; Robertson, 2005). NGS platforms have now provided genome-wide maps of CpG methylation. These have confirmed that between 5%–10% of normally unmethylated CpG promoter islands become abnormally methylated in various cancer genomes. They also demonstrate that CpG hypermethylation of promoters not only affects the expression of protein coding genes but also the expression of various noncoding RNAs, some of which have a role in malignant transformation (Baylin and Jones, 2011). Importantly, these genome-wide DNA methylome studies have also uncovered intriguing alterations in DNA methylation within gene bodies and at CpG "shores," which are conserved sequences upstream and downstream of CpG islands. The functional relevance of these regional alterations in methylation are yet to be fully deciphered, but it is interesting to note that they have challenged the general dogma that DNA methylation invariably equates with transcriptional silencing. In fact, these studies have established that many actively transcribed genes have high levels of DNA methylation within the gene body, suggesting that the context and spatial distribution of DNA methylation is vital in transcriptional regulation (Baylin and Jones, 2011).

Three active DNA methyltransferases (DNMTs) have been identified in higher eukaryotes. DNMT1 is a maintenance methyltransferase that recognizes hemimethylated DNA generated during DNA replication and then methylates newly synthesized CpG dinucleotides, whose partners on the parental strand are already methylated (Li et al., 1992). Conversely, DNMT3a and DNMT3b, although also capable of methylating hemimethylated DNA, function primarily as de novo methyltransferases to establish DNA methylation during embryogenesis (Okano et al., 1999). DNA methylation provides a platform for several methyl-binding proteins. These include MBD1, MBD2, MBD3, and MeCP2. These in turn function to recruit histone-modifying enzymes to coordinate the chromatin-templated processes (Klose and Bird, 2006).

Although mutations in DNA methyltransferases and MBD proteins have long been known to contribute to developmental abnormalities (Robertson, 2005), we have only recently become aware of somatic mutations of these key genes in human malignancies (Figure 2). Recent sequencing of cancer genomes has identified recurrent mutations in *DNMT3A* in up to 25% of patients with acute

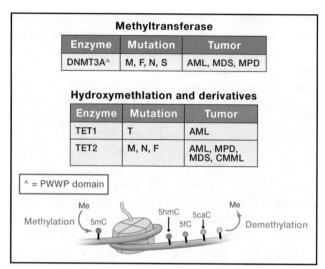

FIGURE 2 Cancer Mutations Affecting Epigenetic Regulators of DNA Methylation
The 5-carbon of cytosine nucleotides are methylated (5mC) by a family of DNMTs. One of these, DNMT3A, is mutated in AML, myeloproliferative diseases (MPD), and myelodysplastic syndromes (MDS). In addition to its catalytic activity, DNMT3A has a chromatin-reader motif, the PWWP domain, which may aid in localizing this enzyme to chromatin. Somatically acquired mutations in cancer may also affect this domain. The TET family of DNA hydroxylases metabolizes 5mC into several oxidative intermediates, including 5-hydroxymethylcytosine (5hmC), 5-formylcytosine (5fC), and 5-carboxylcytosine (5caC). These intermediates are likely involved in the process of active DNA demethylation. Two of the three TET family members are mutated in cancers, including AML, MPD, MDS, and CMML. Mutation types are as follows: M, missense; F, frameshift; N, nonsense; S, splice site mutation; and T, translocation.

myeloid leukemia (AML) (Ley et al., 2010). Importantly, these mutations are invariably heterozygous and are predicted to disrupt the catalytic activity of the enzyme. Moreover, their presence appears to impact prognosis (Patel et al., 2012). However, at present, the mechanisms by which these mutations contribute to the development and/or maintenance of AML remains elusive.

Understanding the cellular consequences of normal and aberrant DNA methylation remains a key area of interest, especially because hypomethylating agents are one of the few epigenetic therapies that have gained FDA approval for routine clinical use (Figure 1). Although hypomethylating agents such as azacitidine and decitabine have shown mixed results in various solid malignancies, they have found a therapeutic niche in the myelodysplastic syndromes (MDS). Until recently, this group of disorders was largely refractory to therapeutic intervention, and MDS was primarily managed with supportive care. However, several large studies have now shown that treatment with azacitidine, even in poor prognosis patients, improves their quality of life and extends survival time. Indeed, azacitidine is the first therapy to have demonstrated a survival benefit for patients with MDS (Fenaux et al., 2009).

The molecular mechanisms governing the impressive responses seen in MDS are largely unknown. However, recent evidence would suggest that low doses of these agents hold the key to therapeutic benefit (Tsai et al., 2012). It is also emerging that the combinatorial use of DNMT and HDAC inhibitors may offer superior therapeutic outcomes (Gore, 2011).

DNA Hydroxy-Methylation and Its Oxidation Derivatives

Historically, DNA methylation was generally considered to be a relatively stable chromatin modification. However, early studies assessing the global distribution of this modification during embryogenesis had clearly identified an active global loss of DNA methylation in the early zygote, especially in the male pronucleus. More recently, high-resolution genome-wide mapping of this modification in pluripotent and differentiated cells has also confirmed the dynamic nature of DNA methylation, evidently signifying the existence of an enzymatic activity within mammalian cells that either erases or alters this chromatin modification (Baylin and Jones, 2011). In 2009, two seminal manuscripts describing the presence of 5-hydroxymethylcytosine (5hmC) offered the first insights into the metabolism of 5mC (Kriaucionis and Heintz, 2009; Tahiliani et al., 2009).

The ten-eleven translocation (TET 1–3) family of proteins have now been demonstrated to be the mammalian DNA hydroxylases responsible for catalytically converting 5mC to 5hmC. Indeed, iterative oxidation of 5hmC by the TET family results in further oxidation derivatives, including 5-formylcytosine (5fC) and 5-carboxylcytosine (5caC). Although the biological significance of the 5mC oxidation derivatives is yet to be established, several lines of evidence highlight their importance in transcriptional regulation: (1) they are likely to be an essential intermediate in the process of both active and passive DNA demethylation, (2) they preclude or enhance the binding of several MBD proteins and, as such, will have local and global effects by altering the recruitment of chromatin regulators, and (3) genome-wide mapping of 5hmC has identified a distinctive distribution of this modification at both active and repressed genes, including its presence within gene bodies and at the promoters of bivalently marked, transcriptionally poised genes (Wu and Zhang, 2011). Notably, 5hmC was also mapped to several intergenic cis-regulatory elements that are either functional enhancers or insulator elements. Consistent with the notion that 5hmC is likely to have a role in both transcriptional activation and silencing, the TET proteins have also been shown to have activating and repressive functions (Wu and Zhang, 2011). Genome-wide mapping of TET1 has demonstrated it to have a strong preference for CpG-rich DNA and, consistent with its catalytic function, it also been localized to regions enriched for 5mC and 5hmC.

The TET family of proteins derive their name from the initial description of a recurrent chromosomal translocation, t(10;11)(q22;q23), which juxtaposes the

MLL gene with *TET1* in a subset of patients with AML (Lorsbach et al., 2003). Notably, concurrent to the initial description of the catalytic activity for the TET family of DNA hydroxylases, several reports emerged describing recurrent mutations in *TET2* in numerous hematological malignancies (Cimmino et al., 2011; Delhommeau et al., 2009; Langemeijer et al., 2009) (Figure 2). Interestingly, TET2-deficient mice develop a chronic myelomonocytic leukemia (CMML) phenotype, which is in keeping with the high prevalence of *TET2* mutations in patients with this disease (Moran-Crusio et al., 2011; Quivoron et al., 2011). The clinical implications of *TET2* mutations have largely been inconclusive; however, in some subsets of AML patients, *TET2* mutations appear to confer a poor prognosis (Patel et al., 2012). Early insights into the process of TET2-mediated oncogenesis have revealed that the patient-associated mutations are largely loss-of-function mutations that consequently result in decreased 5hmC levels and a reciprocal increase in 5mC levels within the malignant cells that harbor them. Moreover, mutations in *TET2* also appear to confer enhanced self-renewal properties to the malignant clones (Cimmino et al., 2011).

Histone Modifications

In 1964, Vincent Allfrey prophetically surmised that histone modifications might have a functional influence on the regulation of transcription (Allfrey et al., 1964). Nearly half a century later, the field is still grappling with the task of unraveling the mechanisms underlying his enlightened statement. In this time, we have learned that these modifications have a major influence, not just on transcription, but in all DNA-templated processes (Kouzarides, 2007). The major cellular processes attributed to each of these modifications are summarized in Table 1.

The great diversity in histone modifications introduces a remarkable complexity that is slowly beginning to be elucidated. Using transcription as an example, we have learned that multiple coexisting histone modifications are associated with activation, and some are associated with repression. However, these modification patterns are not static entities but a dynamically changing and complex landscape that evolves in a cell context-dependent fashion. Moreover, active and repressive modifications are not always mutually exclusive, as evidenced by "bivalent domains." The combinatorial influence that one or more histone modifications have on the deposition, interpretation, or erasure of other histone modifications has been broadly termed "histone crosstalk," and recent evidence would suggest that crosstalk is widespread and is of great biological significance (Lee et al., 2010).

It should be noted that the cellular enzymes that modify histones may also have nonhistone targets and, as such, it has been difficult to divorce the cellular consequences of individual histone modifications from the broader targets of many of these enzymes. In addition to their catalytic function, many chromatin modifiers also possess "reader" domains allowing them to bind to specific regions of the genome and respond to information conveyed by upstream signaling cascades. This is important, as it provides two avenues for therapeutically targeting these epigenetic regulators. The residues that line the binding pocket of reader domains can dictate a particular preference for specific modification states, whereas residues outside the binding pocket contribute to determining the histone sequence specificity. This combination allows similar reader domains to dock at different modified residues or at the same amino acid displaying different modification states. For example, some methyl-lysine readers engage most efficiently with di/tri-methylated lysine (Kme2/3), whereas others prefer mono- or unmethylated lysines. Alternatively, when the same lysines are now acetylated, they bind to proteins containing bromodomains (Taverna et al., 2007). The main modification binding pockets contained within chromatin-associated proteins is summarized in Table 1.

Many of the proteins that modify or bind these histone modifications are misregulated in cancer, and in the ensuing sections, we will discuss the most extensively studied histone modifications in relation to oncogenesis and novel therapeutics.

Histone Acetylation

The N^ε-acetylation of lysine residues is a major histone modification involved in transcription, chromatin structure, and DNA repair. Acetylation neutralizes lysine's positive charge and may consequently weaken the electrostatic interaction between histones and negatively charged DNA. For this reason, histone acetylation is often associated with a more "open" chromatin conformation. Consistent with this, ChIP-Seq analyses have confirmed the distribution of histone acetylation at promoters and enhancers and, in some cases, throughout the transcribed region of active genes (Heintzman et al., 2007; Wang et al., 2008). Importantly, lysine acetylation also serves as the nidus for the binding of various proteins with bromodomains and tandem plant homeodomain (PHD) fingers, which recognize this modification (Taverna et al., 2007).

Acetylation is highly dynamic and is regulated by the competing activities of two enzymatic families, the histone lysine acetyltransferases (KATs) and the histone deacetylases (HDACs). There are two major classes of KATs: (1) type-B, which are predominantly cytoplasmic and modify free histones, and (2) type-A, which are primarily nuclear and can be broadly classified into the GNAT, MYST, and CBP/p300 families.

KATs were the first enzymes shown to modify histones. The importance of these findings to cancer was immediately apparent, as one of these enzymes, CBP, was identified by its ability to bind the transforming portion of the viral oncoprotein E1A (Bannister and Kouzarides, 1996). It is now clear that many, if not most, of the KATs have been implicated in neoplastic transformation, and a number of viral oncoproteins are known to associate with them. There are numerous examples of recurrent chromosomal translocations (e.g., *MLL-CBP* [Wang et al., 2005] and *MOZ-TIF2* [Huntly et al., 2004]) or coding mutations (e.g., *p300/CBP* [Iyer et al., 2004; Pasqualucci et al., 2011]) involving various KATs in a broad range of solid and hematological malignancies (Figure 3). Furthermore, altered expression levels of several of the KATs have also been noted in a range of cancers (Avvakumov and Côté, 2007; Iyer et al., 2004). In some cases, such as the leukemia-associated fusion gene *MOZ-TIF2*, we know a great deal about the cellular consequences of this translocation involving a MYST family member. MOZ-TIF2 is sufficient to recapitulate an aggressive leukemia in murine models; it can confer stem cell properties and reactivate a self-renewal program when introduced into committed hematopoietic progenitors, and much of this oncogenic potential is dependent on its inherent and recruited KAT activity as well as its ability to bind to nucleosomes (Deguchi et al., 2003; Huntly et al., 2004).

Despite these insights, the great conundrum with regards to unraveling the molecular mechanisms by which histone acetyltransferases contribute to malignant transformation has been dissecting the contribution of altered patterns in acetylation on histone and nonhistone proteins. Although it is clear that global histone acetylation patterns are perturbed in cancers (Fraga et al., 2005; Seligson et al., 2005), it is also well established that several nonhistone proteins, including many important oncogenes and tumor suppressors such as MYC, p53, and PTEN, are also dynamically acetylated (Choudhary et al., 2009). A pragmatic view on this issue is that both histone and nonhistone acetylation are likely to be important and, in most part, the abundance of substrates has not deterred the enthusiasm for the development of histone acetyltransferase inhibitors (KAT-I). Although there is only modest structural homology between the different families of KATs, developing specific inhibitors has proven to be fraught with frustration (Cole, 2008). However, recent progress with derivatives of the naturally occurring KAT-I, such as curcumin, anacardic acid, and garcinol, as well as the synthesis of novel chemical probes, suggest that therapeutically targeting the various KATs with some specificity is likely to be achieved in the near future (Cole, 2008).

Histone Deacetylation

HDACs are enzymes that reverse lysine acetylation and restore the positive charge on the side chain. There are 18 such enzymes identified, and these are subdivided into four major classes, depending on sequence homology. Class I (HDAC 1-3 and HDAC8) and class II (HDAC 4-7 and HDAC 9-10) represent the HDACs most closely related to yeast scRpd3 and scHda1, respectively, whereas class IV comprises only one enzyme, HDAC11. Class I, II, and IV HDACs share a related catalytic mechanism that requires a zinc metal ion but does not involve the use of a cofactor. In contrast, class III HDACs (sirtuin 1–7) are homologous to yeast *sc*Sir2 and employ a distinct catalytic mechanism that is NAD$^+$-dependent. Analogous to the KATs, HDACs target both histone and nonhistone proteins. Substrate specificity for these enzymes is largely mediated by components of multisubunit complexes in which HDACs are found, such as Mi2/NuRD, Sin3A, and Co-REST (Bantscheff et al., 2011; Xhemalce et al., 2011).

In the context of malignancy, chimeric fusion proteins that are seen in leukemia, such as PML-RARα, PLZF-RARα, and AML1-ETO, have been shown to recruit HDACs to mediate aberrant gene silencing, which contributes to leukemogenesis (Johnstone and Licht, 2003). HDACs can also interact with nonchimeric oncogenes such as BCL6, whose repressive activity is controlled by dynamic acetylation (Bereshchenko et al., 2002). Importantly, inhibitors of histone deactylases (HDAC-I) are able to reverse some of the aberrant gene repression seen in these malignancies and induce growth arrest, differentiation, and apoptosis in the malignant cells (Federico and Bagella, 2011; Johnstone and Licht, 2003). Based on impressive preclinical and clinical data, two pan-HDAC inhibitors, Vorinostat and Romidepsin, have recently been granted FDA approval (Olsen et al., 2007; Piekarz et al., 2009) for clinical use in patients with cutaneous T cell lymphoma (Figure 1). Although somatic mutations in HDACs do not appear to be prominent in cancer (Figure 3), the expression levels of various HDACs appear to be altered in numerous malignancies. Consequently, several novel HDAC inhibitors are currently under investigation for clinical use in a broad range of cancers (Federico and Bagella, 2011; Johnstone and Licht, 2003). However, the pleiotropic effects of HDACs continue to pose significant challenges in dissecting the specific effects on histone and nonhistone proteins (Bantscheff et al., 2011).

Histone Acetylation Readers

The primary readers of N$^\varepsilon$-acetylation of lysine residues are families of proteins that contain an evolutionarily conserved binding motif termed a bromodomain. There are over 40 described human proteins with bromodomains (Chung and Witherington, 2011). These comprise a diverse group of proteins that function as chromatin remodelers, histone acetyltransferases, histone methyltransferases, and transcriptional coactivators. Many of these proteins also contain several separate evolutionarily conserved "chromatin-reading" motifs such as PHD fingers, which recognize distinct histone posttranslational modifications (Table 1).

Until recently, it had not been feasible to therapeutically target protein-protein interactions with small molecules. However, several recent studies have shown that it is possible to develop highly specific and chemically distinct small molecules against the BET family (BRD2, BRD3, BRD4, and BRDt) of

Acetyltransferases

Enzyme	Mutation	Tumor
KAT3A (CBP)*	T, N, F, M	AML, ALL, DLBCL, B-NHL, TCC
KAT3B (p300)*	T, N, F, M	AML, ALL, DLBCL, TCC, Colorectal, Breast, Pancreatic
KAT6A (MOZ)+	T	AML, MDS
KAT6B (MORF)+	T	AML, Uterine leiomyoma

Readers

Reader	Mutation	Tumor
BRD1**	T	ALL
BRD3*	T	Midline carcinoma
BRD4*	T	Midline carcinoma
TRIM33*+	T	Papillary thyroid
PBRM1*	N, F, M, S, D	Renal, Breast

* = Bromodomain
+ = PHD Finger

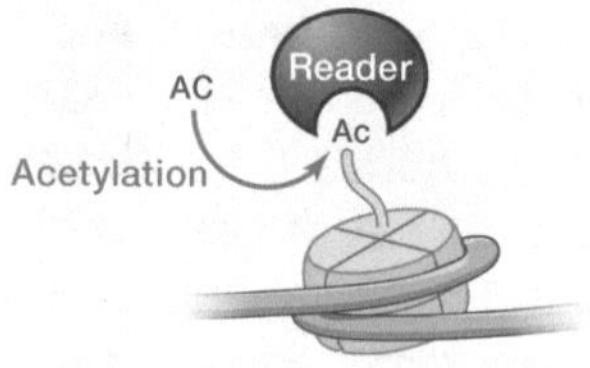

FIGURE 3 **Cancer Mutations Affecting Epigenetic Regulators Involved in Histone Acetylation**
These tables provide somatic cancer-associated mutations identified in histone acetyltransferases and proteins that contain bromodomains (which recognize and bind acetylated histones). Several histone acetyltransferases possess chromatin-reader motifs and, thus, mutations in the proteins may alter both their catalytic activities as well as the ability of these proteins to scaffold multiprotein complexes to chromatin. Interestingly, sequencing of cancer genomes to date has not identified any recurrent somatic mutations in histone deacetylase enzymes. Abbreviations for the cancers are as follows: AML, acute myeloid leukemia; ALL, acute lymphoid leukemia; B-NHL, B-cell non-Hodgkin's lymphoma; DLBCL, diffuse large B-cell lymphoma; and TCC, transitional cell carcinoma of the urinary bladder. Mutation types are as follows: M, missense; F, frameshift; N, nonsense; S, splice site mutation; T, translocation; and D, deletion.

bromodomain proteins (Dawson et al., 2011; Filippakopoulos et al., 2010; Nicodeme et al., 2010). The BET family shares a common structural composition featuring tandem amino-terminal bromodomains that exhibit high levels of sequence conservation. BET proteins play a fundamental role in transcriptional elongation and cell-cycle progression. Moreover, recurrent translocations involving *BRD3/4* are associated with the aggressive and invariably fatal NUT-midline carcinoma (Filippakopoulos et al., 2010) (Figure 3).

Targeting the BET bromodomains is a promising therapeutic avenue in cancer. The BET inhibitors have recently been shown to have excellent efficacy in NUT-midline carcinoma (Filippakopoulos et al., 2010) and in a range of hematological malignancies (Dawson et al., 2011; Delmore et al., 2011; Mertz et al., 2011; Zuber et al., 2011). A central theme reported in all of the studies thus far is the downregulation of *MYC* transcription following BET inhibition. MYC is a master regulator of cell proliferation and survival; it is also one of the most common genes dysregulated in cancer (Meyer and Penn, 2008). Following BET inhibition with either RNAi or specific BET inhibitors, the expression of *MYC* was noted to be substantially decreased in a variety of malignant hematopoietic cell lines, including MLL-translocated acute myeloid leukemia (Dawson et al., 2011; Zuber et al., 2011), multiple myeloma (Delmore et al., 2011), and Burkitt's lymphoma (Mertz et al., 2011). Furthermore, murine models of these diseases confirmed the excellent therapeutic efficacy of BET inhibition in vivo.

Although MYC has a prominent role in these diseases, it is unlikely that the profound effects observed by BET inhibition are solely mediated by MYC inhibition. There are many malignant cell lines that overexpress MYC yet fail to respond to BET inhibition (Mertz et al., 2011); MYC expression is not always affected by BET inhibition (Mertz et al., 2011); MYC is often equally downregulated in responsive and nonresponsive malignant cell lines (Dawson et al., 2011; Zuber et al., 2011); and, importantly, MYC overexpression fails to rescue the apoptosis induced by BET inhibition (Zuber et al., 2011). The molecular mechanisms governing the efficacy of BET inhibition are slowly being deciphered. What seems to be clear from the current analyses is that BET inhibitors specifically regulate a small number of genes, and inhibition of transcriptional elongation may be a primary mode of action.

Histone Methylation

Histones are methylated on the side chains of arginine, lysine, and histidine residues. Methylation, unlike acetylation and phosphorylation, does not alter the overall charge of the molecule. Lysines may be mono-, di-, or tri-methylated, and arginine residues may be symmetrically or asymmetrically methylated. The best-characterized sites of histone methylation are those that occur on lysine residues and, therefore, these will be the focus of this section. Although many lysine residues on the various histones are methylated, the best studied are H3K4, H3K9, H3K27, H3K36, H3K79, and H4K20. Some of these (H3K4, H3K36, and H3K79) are often associated with active genes in euchromatin, whereas others (H3K9, H3K27, and H4K20) are associated with heterochromatic regions of the genome (Barski et al., 2007). Different methylation states on the same residue can also localize differently. For instance, H3K4me2/3 usually spans the transcriptional start site (TSS) of active genes (Barski et al., 2007), whereas H3K4me1 is a modification associated with active enhancers (Heintzman et al., 2009). Similarly, whereas monomethylation of H3K9 may be seen at active genes, trimethylation of H3K9 is associated with gene repression (Barski et al., 2007).

The enzymatic protagonists for lysine methylation contain a conserved SET domain, which possesses methyltransferase activity. The only exception to this is hDOT1L, the enzyme that methylates H3K79. In contrast to the KATs, the histone lysine methyltransferases (KMT) tend to be highly specific enzymes that specifically target certain lysine residues. Cytogenetic studies, as well as NGS of various cancer genomes, have demonstrated recurrent translocations and/or coding mutations in a large number of KMT, including *MMSET*, *EZH2*, and *MLL* family members (Figure 4).

Whereas the oncogenic effects exerted by the MLL fusions have been extensively studied and reviewed (Krivtsov and Armstrong, 2007), an emerging area of interest is the dichotomous role of EZH2 in human malignancies. EZH2 is the catalytic component of the PRC2 complex, which is primarily responsible for the methylation of H3K27. Early gene-expression studies implicated the overexpression of EZH2 as a progressive event that conferred a poor prognosis in prostate and breast cancer (Margueron and Reinberg, 2011). These initial studies suggested that *EZH2* was an oncogene. However, NGS and targeted resequencing of cancer genomes have recently identified coding mutations within EZH2 in various lymphoid and myeloid neoplasms that have somewhat muddied the waters by suggesting both oncogenic and tumor-suppressive roles for EZH2. Heterozygous missense mutations resulting in the substitution of tyrosine 641 (Y641) within the SET domain of EZH2 were noted in 22% of patients with diffuse large B-cell lymphoma (Morin et al., 2010). Functional characterization of this mutation demonstrated that it conferred increased catalytic activity and a preference for converting H3K27me1 to H3K27me2/3, again supporting the contention that *EZH2* is an oncogene (Sneeringer et al., 2010). In contrast, loss-of-function mutations in *EZH2* gene, conferring a poor prognosis, have been described in the myeloid malignancies (Ernst et al., 2010; Nikoloski et al., 2010) and T-ALL (Ntziachristos et al., 2012; Zhang et al., 2012), suggesting a tumor-suppressive role for EZH2 in these cell lineages.

The precise mechanisms by which gain and loss of EZH2 activity culminate in cancers are an area of active investigation. In light of the varied roles

HNSCC, head and neck squamous cell carcinoma; FL, follicular lymphoma; MDS, myelodysplastic syndromes; MPD, myeloproliferative diseases; and TCC, transitional cell carcinoma of the urinary bladder. Mutation types are as follows: M, missense; F, frameshift; N, nonsense; S, splice site mutation; T, translocation; D, deletion; and PTD, partial tandem duplication.

Methyltransferases

Enzyme	Mutation	Tumor
KMT2A (MLL1[*+])	T, PTD	AML, ALL, TCC
KMT2B (MLL2[*])	N, F, M	Medulloblastoma, Renal, DLBCL, FL
KMT2C (MLL3[*])	N	Medulloblastoma, TCC, Breast
KMT3A (SETD2)	N, F, S, M	Renal, Breast
KMT3B (NSD1[+^])	T	AML
NSD2[+^]	T	Multiple myeloma
NSD3[^]	T	AML
KMT6 (EZH2)	M	DLBCL, MPD, MDS

Readers

Reader	Mutation	Tumor
TRIM33[**]	T	Papillary thyroid
ING1[+]	M, D	Melanoma, Breast
ING4[+]	D	HNSCC
MSH6[^]	M, N, F, S	Colorectal

Demethylases

Enzyme	Mutation	Tumor
KDM5A (JARID1A)[+]	T	AML
KDM5C (JARID1C)[+]	N, F, S	Renal
KDM6A (UTX)	D, N, F, S	AML, TCC, Renal, Oesophageal, Multiple myeloma

Me — Reader — Me
Me
Methylation — Demethylation

* = Bromodomain
+ = PHD Finger
^ = PWWP domain

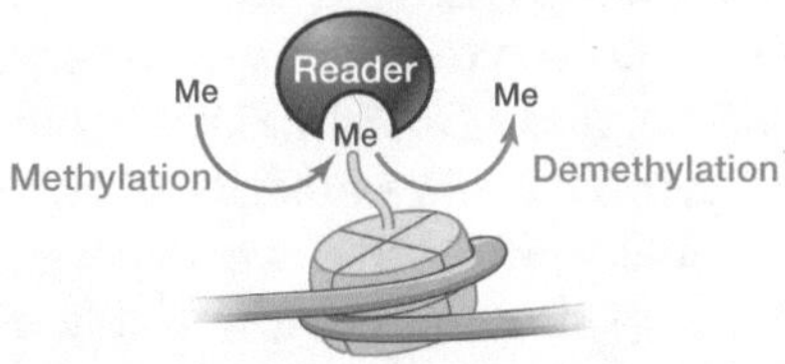

FIGURE 4 Cancer Mutations Affecting Epigenetic Regulators Involved in Histone Methylation
Recurrent mutations in histone methyltransferases, demethylases, and methyllysine binders have been identified in a large number of cancers. These mutations may significantly alter the catalytic activity of the methyltransferases or demethylases. In addition, as many of these enzymes also contain chromatin-reader motifs, they may also affect the ability of these proteins to survey and bind epigenetic modifications. Abbreviations for the cancers are as follows: AML, acute myeloid leukemia; ALL, acute lymphoid leukemia; B-NHL, B-cell non-Hodgkin's lymphoma; DLBCL, diffuse large B-cell lymphoma;

that polycomb proteins play in self-renewal and differentiation (Margueron and Reinberg, 2011), solution of this problem will necessitate vigilance and appreciation of the cellular context within which the mutations arise. The increased awareness of the involvement of KMTs in cancer has heightened efforts to identify specific inhibitors. These efforts will only be encouraged by the recent demonstration that small-molecule inhibition of DOT1L shows preclinical promise as a targeted therapy in MLL leukemia (Daigle et al., 2011), a disease in which aberrant DOT1L activity is ill defined but clearly involved (Krivtsov and Armstrong, 2007).

Histone Demethylation

The initial notion that histone lysine methylation was a highly stable, nondynamic modification has now been irrefutably overturned by the identification of two classes of lysine demethylases (Mosammaparast and Shi, 2010). The prima facie example, LSD1 (KDM1A), belongs to the first class of demethylases that demethylates lysines via an amine oxidation reaction with flavin adenine dinucleotide (FAD) as a cofactor. As this family of enzymes requires a protonated nitrogen to initiate demethylation, they are limited to demethylating mono- and dimethyllysine. The second and more expansive class of enzymes is broadly referred to as the Jumonji demethylases. They have a conserved JmjC domain, which functions via an oxidative mechanism and radical attack (involving Fe(II) and α-ketoglutarate). The Jumonji family does not require a free electron pair on the nitrogen atom to initiate catalysis and, therefore, unlike LSD1, they can demethylate all three methyl lysine states. Unsurprisingly, the multisubunit complexes within which these enzymes reside confer much of their target specificity. As an example, LSD1 can function as a transcriptional repressor by demethylating H3K4me1/2 as part of the corepressor for RE1-silencing transcription factor (Co-REST) complex, but its activity is linked to gene activation when it associates with the androgen receptor to demethylate H3K9me2 (Mosammaparast and Shi, 2010). Thus far, recurrent coding mutations have been noted in *KDM5A* (*JARID1A*), *KDM5C* (*JARID1C*), and *KDM6A* (*UTX*) (Figure 4). Mutations in *UTX*, in particular, are prevalent in a large number of solid and hematological cancers. Small-molecule inhibitors of the two families of histone demethylases are at various stages of development, and this interest will be spurred on by emerging preclinical data showing the therapeutic potential of compounds that inhibit LSD1/KDM1A in AML (Barretina et al., 2012; Schenk et al., 2012).

Interestingly, recent findings related to recurrent mutations in the genes encoding the metabolic enzymes isocitrate dehydrogenase-1 (*IDH1*) and *IDH2* have broad implications for the Jumonji class of demethylases, which use α-ketoglutarate (α-KG). IDH1/2 are nicotinamide adenine dinucleotide phosphate (NADP)-dependent enzymes that normally catalyze the oxidative decarboxylation of isocitrate to α-KG, which is associated with the production of NADPH. Mutations in *IDH1* and *IDH2* are seen in up to 70% of patients with secondary glioblastoma mutiforme and are also noted as recurrent mutations in a range of myeloid malignancies, most notably AML (Cimmino et al., 2011). These mutations manifest in a neomorphic enzymatic activity that results in the NADPH-dependent reduction of α-KG to 2-hydroxyglutarate (2-HG). Consequently, malignant cells with *IDH1/2* mutations may harbor 2-HG levels that are up to 100-fold higher than normal (Cimmino et al., 2011). 2-HG is a competitive inhibitor of the α-KG-dependent dioxygenases; in fact, 2-HG has been shown to adopt a near-identical orientation within the catalytic core of the JmjC domain (Xu et al., 2011). As 2-HG levels accumulate within the malignant cells, there is a purported blanket inhibition of the Jumonji class of histone demethylases. Accordingly, there is a discernable increase in histone methylation levels (Xu et al., 2011). These remarkable findings are yet to be fully investigated, and it will be important to determine whether all the Jumonji family members are equally susceptible to 2-HG inhibition. A similar question can be posed for the TET family of enzymes (see above), which also use α-ketoglutarate.

Histone Methylation Readers

The various states of lysine methylation result in considerable physicochemical diversity of lysine; these modification states are read and interpreted by proteins containing different specialized recognition motifs. Broadly speaking, the aromatic cages that engage methyllysine can be divided into two major families, the Royal Family (Tudor domains, Chromo domains, and malignant brain tumor [MBT] domains) and PHD fingers. The structural composition of these domains that allows for this diversity has recently been expertly reviewed (Taverna et al., 2007).

Analogous to the situation with bromodomain proteins, several methyllysine readers have also been implicated in cancer (Figure 4). For instance, all three isoforms of the chromodomain protein HP1 have altered expression in numerous cancers (Dialynas et al., 2008). However, thus far, no cancer-specific somatic mutations have been identified in HP1. In contrast, ING family members have had coding mutations identified in malignancies such as melanoma and breast cancer, including those that specifically target the PHD finger, which recognizes H3K4me3 (Coles and Jones, 2009). Despite these findings, neither of the aforementioned examples establishes a causal relationship between cancer and the abrogation of methyllysine binding at chromatin. The best example of this, and indeed a proof of principle for therapeutically targeting methyllysine binders, has recently been shown in a specific form of AML (Wang et al., 2009). Leukemia, induced by the fusion of NUP98 with the PHD finger containing part of JARID1A or PHF23, can be abrogated by mutations that negate the ability of the PHD finger to bind H3K4me3. Functional compensation of this effect can be provided by other PHD fingers that recognize this modification, but not those that do not bind H3K4me3. Moreover, mechanistic insights were provided, demonstrating that chromatin binding of the fusion protein inhibits the deposition of H3K27me3, which leads to the continued expression of critical hematopoietic oncogenes such as *HoxA9*, *Meis1*, and *Pbx1* (Wang et al., 2009). In light of these findings, and as result of the structural diversity present in methyllysine-binding modules, it is likely that small molecules that disrupt this important protein-protein interaction may be effective anticancer agents.

Histone Phosphorylation

The phosphorylation of serine, threonine, and tyrosine residues has been documented on all core and most variant histones. Phosphorylation alters the charge of the protein, affecting its ionic properties and influencing the overall structure and function of the local chromatin environment. The phosphorylation of histones is integral to essential cellular processes such as mitosis, apoptosis, DNA repair, replication, and transcription. Generally speaking, the specific histone phosphorylation sites on core histones can be divided into two broad categories: (1) those involved in transcription regulation, and (2) those involved in chromatin condensation. Notably, several of these histone modifications, such as H3S10, are associated with both categories (Baek, 2011).

Kinases are the main orchestrators of signal transduction pathways conveying extracellular cues within the cell. Altered expression, coding mutations, and recurrent translocations involving signaling kinases are some of the most frequent oncogenic phenomena described in cancer (Hanahan and Weinberg, 2011). Many of these kinases have established roles as signal transducers in the cytoplasm; however, it has recently been recognized that some kinases may also have nuclear functions, which include the phosphorylation of histones (Baek, 2011; Bungard et al., 2010; Dawson et al., 2009) (Figure 5). One such enzyme is the nonreceptor tyrosine kinase, JAK2, which is frequently amplified or mutated in the hematological malignancies. Within the nucleus, JAK2 specifically phosphorylates H3Y41, disrupts the binding of the chromatin repressor HP1α, and activates the expression of hematopoietic oncogenes such as *Lmo2* (Dawson et al., 2009). These findings have now been given a broader application in other malignancies, such as Hodgkin's disease and primary mediastinal B-cell lymphoma, in which this mechanism has been shown to contribute to oncogenesis (Rui et al., 2010). Given that many small-molecule inhibitors against kinases are clinically used as anticancer therapies, it is interesting to note that several of these (e.g., JAK2 and Aurora inhibitors) result in a global reduction in the histone modifications laid down by these enzymes. These agents can therefore be considered as potential epigenetic therapies.

Kinases

Enzyme	Mutation	Tumor
ATM	D, M, N, F, S	T-PLL, AML, ALL, Medulloblastoma, Glioma
JAK2	T, M	AML, ALL, MPD, CML
PIM1	T	NHL

Readers

Reader	Mutation	Tumor
BRCA1	D, M, N, F, S	Ovarian, Breast, Prostate

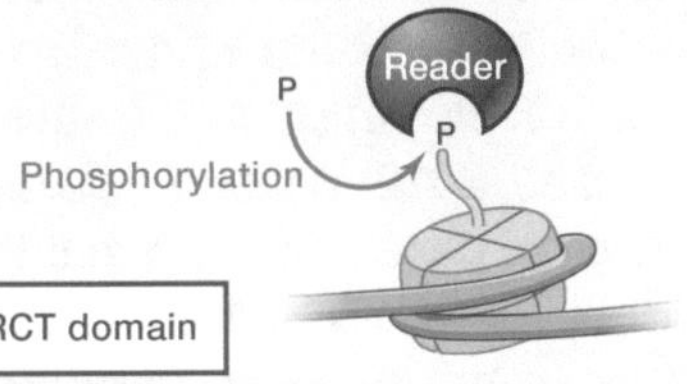

§ = BRCT domain

FIGURE 5 Cancer Mutations Affecting Epigenetic Regulators Involved in Histone Phosphorylation

Recurrent mutations in signaling kinases are one of the most frequent oncogenic events found in cancer. Some of these kinases signal directly to chromatin. Activating and inactivating mutations of these have been noted in a range of malignancies. Thus far, BRCA1, which contains a BRCT domain, is the only potential phosphochromatin reader recurrently mutated in cancer. It should be noted, however, that BRCA1 binding to modified histones via its BRCT domain has not yet been firmly established. As our knowledge about histone phosphatases and phosphohistone binders increases, we are likely to find mutations in many of these proteins that contribute to oncogenesis. Abbreviations for the cancers are as follows: AML, acute myeloid leukemia; ALL, acute lymphoid leukemia; CML, chronic myeloid leukemia; NHL, non-Hodgkin's lymphoma; MPD, myeloproliferative diseases; and T-PLL, T cell prolymphocytic leukemia. Mutation types are as follows: M, missense; F, frameshift; N, nonsense; S, splice site mutation; T, translocation; and D, deletion.

Histone phosphorylation is a highly dynamic posttranslational modification, which is reciprocally controlled by the competing activities of protein kinases and protein phosphatases. Phosphatases, like protein kinases, demonstrate specificity for either serine/threonine residues or tyrosine residues, or they may have dual specificity; they are further subdivided based on their requirement for a metallic ion for their catalytic activity. Although there is little doubt that histone phosphatases are integral to chromatin biology, outside of the realm of DNA repair and regulation of mitosis, little is currently known about the function of these enzymes at chromatin and their potential misadventures in cancer (Xhemalce et al., 2011).

The phosphorylation sites on serine, threonine, and tyrosine residues may serve as the binding site for a range of cellular proteins. Proteins such as

MDC1 bind at sites of double-strand breaks by tethering to γH2AX via its tandem BRCT domain (Stucki et al., 2005). Furthermore, the 14-3-3 family of proteins, of which there are seven mammalian isoforms, contain highly conserved phosphoserine-binding modules which some, such as 14-3-3ζ, use to bind H3S10ph and H3S28ph. Many of these proteins, including 14-3-3ζ, are abnormally expressed in various human malignancies and, consequently, therapeutically targeting them may prove beneficial (Yang et al., 2012).

Cancer Mutations in Histone Genes

Two recent studies have demonstrated recurrent somatic mutations in genes encoding the replication-independent histone H3 variant H3.3 (*H3F3A*) and the canonical histone H3.1 (*HIST1H3B*) in up to one-third of pediatric glioblastomas (Schwartzentruber et al., 2012; Wu et al., 2012). These mutations are invariably heterozygous and are clustered such that they primarily result in amino acid substitutions at two critical residues in the tail of histone H3 (K27M, G34R/G34V). By virtue of the residues they disrupt, these mutations are likely to have an important influence on chromatin structure and transcription. The K27M mutation alters the ability of this critical residue to be both methylated and acetylated. These posttranslational modifications of H3K27 have different genomic distributions within euchromatin and heterochromatin; they are recognized by different epigenetic readers and are ultimately associated with different transcriptional outcomes. Similarly, it is also likely that the G34 mutations, due to their proximity to H3K36, will also influence transcription. In support of this contention is the fact that tumors carrying the K27M and G34R/G34V mutations had distinct gene-expression profiles, and tumors with the G34V mutation demonstrated a global increase in H3K36me3 (Schwartzentruber et al., 2012).

These studies also raise several interesting mechanistic questions. For instance, given that there are several copies of genes encoding for histone H3.1/3.3 within our genome, why do these mutated histone proteins get incorporated into nucleosomes? How do these mutated proteins influence the function of histone chaperones, nucleosome assembly, stability, and mobility? One possibility uncovered from these studies suggests that telomere maintenance and heterochromatin stability may be compromised as a consequence of the H3.3 mutations. Several of these pediatric glioblastoma multiforme (GBMs) also harbored mutations in the ATRX/DAXX chromatin-remodeling complex, which is responsible for the deposition of H3.3. These tumors with mutations in *H3F3A/ATRX/DAXX* were associated with increased alternative lengthening of telomeres and genomic instability (Schwartzentruber et al., 2012). The *ATRX/DAXX* mutations described here are also a seminal feature of pancreatic neuroendocrine tumors (Jiao et al.,

2011) and highlight emerging evidence suggesting that mutations in members of chromatin-remodeling complexes are a common feature in human malignancy.

Chromatin Remodelers

The myriad of covalent modifications on the nucleosome often provides the scaffold and context for dynamic ATP-dependent chromatin remodeling. Based on their biochemical activity and subunit composition, the mammalian chromatin-remodeling complexes can be broadly split into four major families: the switching defective/sucrose nonfermenting (SWI/SNF) family, the imitation SWI (ISWI) family, the nucleosome remodeling and deacetylation (NuRD)/Mi-2/chromodomain helicase DNA-binding (CHD) family, and the inositol requiring 80 (INO80) family. These enzymes are evolutionarily conserved and use ATP as an energy source to mobilize, evict, and exchange histones. Each of these families has distinct domain structures and is populated by members that contain various chromatin reader motifs (SANT domains, bromodomains, and chromodomains) that confer some regional and context specificity to their chromatin-remodeling activities (Wang et al., 2007).

Several members from the various chromatin-remodeling families, such as SNF5 (Versteege et al., 1998), BRG1 (Wilson and Roberts, 2011), and MTA1 (Li et al., 2012), were known to be mutated in malignancies, raising the possibility that they may be bone fide tumor suppressors (Figure 6). Strong evidence in support of this contention has now emerged from the sequencing of cancer genomes. These efforts have highlighted high-frequency mutations in several SWI/SNF complex members in a range of hematological (Chapman et al., 2011; Morin et al., 2011) and solid malignancies (Gui et al., 2011; Jones et al., 2010; Tan et al., 2011; Varela et al., 2011; Wang et al., 2011). The prevalence of these mutations would suggest that many of the members of these complexes are involved in the development and maintenance of cancer; however, functional insights into the mechanisms of oncogenesis are only just beginning to emerge. It is clear that the SWI/SNF complexes have several lineage-specific subunits and interact with tissue-specific transcription factors to regulate differentiation. They also have a reciprocal and antagonistic relationship with the polycomb complexes. One possibility, which remains to be formally established, is that mutations in SWI/SNF members potentiate malignancy by skewing the balance between self-renewal and differentiation. Recent data would also suggest a role for the SWI/SNF complexes in regulating cell-cycle progression, cell motility, and nuclear hormone signaling (Wilson and Roberts, 2011).

Genetic evidence from mouse models has confirmed that altered expression of these purported tumor suppressors can increase the propensity to develop cancer. In the case of *BRG1*, even haploinsufficiency results in

SWI/SNF		
Gene	Mutation	Tumor
BRG1*	N, M, F, D	Lung, Rhabdoid, Medulloblastoma, Breast, Prostate, Pancreas, HNSCC
BRM*	N, M,F	HNSCC
ARID1A	N, F, M, T	OCC, Endometroid, Renal, Gastric, Breast, Medulloblastoma, TCC
ARID1B	F, M, D	Breast
ARID2	N, F, S	Hepatocellular carcinoma
SNF5	D, N, F, S, T	Rhabdoid, Familial Schwannomatosis, Chondrosarcoma, Epethioloid sarcoma, Meningioma, Chordoma, Undifferentiated sarcoma
PBRM1*	N, F, M, S, D	Renal, Breast
BCL7A	T, M	B-NHL, Multiple myeloma
BAF60A	M	Breast

* = Bromodomain

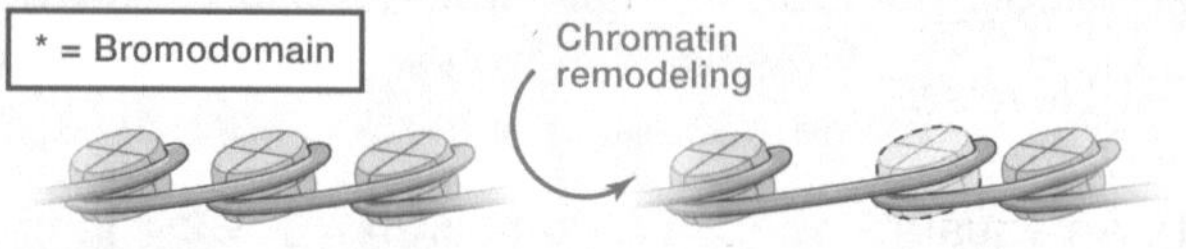

Chromatin remodeling

FIGURE 6 Cancer Mutations Affecting Members of the SWI/SNF Chromatin-Remodeling Complex

SWI/SNF is a multisubunit complex that binds chromatin and disrupts histone-DNA contacts. The SWI/SNF complex alters nucleosome positioning and structure by sliding and evicting nucleosomes to make the DNA more accessible to transcription factors and other chromatin regulators. Recurrent mutations in several members of the SWI/SNF complex have been identified in a large number of cancers. Abbreviations for the cancers are as follows: B-NHL, B-cell non-Hodgkin's lymphoma; HNSCC, head and neck squamous cell carcinoma; OCC, ovarian clear cell carcinoma; and TCC, transitional cell carcinoma of the urinary bladder. Mutation types are as follows: M, missense; F, frameshift; N, nonsense; S, splice site mutation; T, translocation; and D, deletion.

increased tumors (Wilson and Roberts, 2011). However, despite the wealth of information implicating the SWI/SNF complexes in cancer (Figure 6), there is no mechanistic evidence to demonstrate that altered chromatin remodeling due to aberrant chromatin binding or loss of ATPase activity is involved.

Noncoding RNAs

The high-throughput genomic platforms have established that virtually the entire genome is transcribed; however, only ~2% of this is subsequently translated (Amaral et al., 2008). The remaining "noncoding" RNAs (ncRNAs) can be roughly categorized into small (under 200 nucleotides) and large ncRNAs. These RNAs are increasingly recognized to be vital for normal development and may be compromised in diseases such as cancer. The small ncRNAs include small nucleolar RNAs (snoRNAs), PIWI-interacting RNAs (piRNAs), small interfering RNAs (siRNAs), and microRNAs (miRNAs). Many of these families show a high degree of sequence conservation across species and are involved in transcriptional and posttranscriptional gene silencing through specific base pairing with their targets. In contrast, the long ncRNAs (lncRNAs) demonstrate poor cross-species sequence conservation, and their mechanism of action in transcriptional regulation is more varied. Notably, these lncRNAs appear to have a critical function at chromatin, where they may act as molecular chaperones or scaffolds for various chromatin regulators, and their function may be subverted in cancer (Wang and Chang, 2011).

One of the best-studied lncRNAs that emerges from the mammalian HOXC cluster but invariably acts in *trans* is HOTAIR. HOTAIR provides a concurrent molecular scaffold for the targeting and coordinated action of both the PRC2 complex and the LSD1-containing CoREST/REST complex (Wang and Chang, 2011). HOTAIR is aberrantly overexpressed in advanced breast and colorectal cancer (Kogo et al., 2011; Wang and Chang, 2011), and manipulation of HOTAIR levels within malignant cells can functionally alter the invasive potential of these cancers by changing PRC2 occupancy (Wang and Chang, 2011). An equally intriguing example that has broad implications for both normal development and aberrant targeting of chromatin complexes in cancer is the lncRNA HOTTIP. In contrast to HOTAIR, HOTTIP is expressed from the mammalian *HOXA* cluster and acts in *cis* to aid in the transcriptional activation of the 5′ *HOXA* genes (Wang and Chang, 2011). HOTTIP, by means of chromatin looping, is brought into close proximity of the 5′ *HOXA* genes and recruits MLL1 complexes to lay down H3K4me3 and potentiate transcription. Given that the 5′ *HOXA* cluster plays a seminal role in development and maintenance of a large number of leukemias, these findings raise the possibility that abnormal expression and/or function of HOTTIP may be a feature of these diseases.

Discerning the molecular mechanisms and nuances of RNA-protein interactions is a pivotal area of chromatin research, as the stereochemical nature of these interactions may in the future lend itself to specific targeting by innovative small molecules as cancer therapies.

PERSPECTIVE AND CONCLUSIONS

Information from global proteomic and genomic techniques has confirmed many of the hypotheses regarding the molecular causes of cancer, but it has challenged others. The principal tenet in oncology—that cancer is a disease initiated and driven by genetic anomalies—remains uncontested, but it is now clear that epigenetic pathways also play a significant role in oncogenesis. One concern had been that the endpoint of these pathways may not necessarily be epigenetic. However, these concerns are ameliorated by the multiplicity of mutations in epigenetic regulators, including chromatin-remodeling complexes, and the observation that histones themselves are mutated at sites of key modifications in cancer. In fact, it is now irrefutable that many of the hallmarks of cancer, such as malignant self-renewal, differentiation blockade, evasion of cell death, and tissue invasiveness are profoundly influenced by changes in the epigenome.

Despite these assertions, there are still many questions to be answered before we can use our current basic knowledge in the clinical arena. The first important issue is that of selectivity. How can ubiquitously expressed epigenetic regulators serve as selective targets? The answer may lie in the fact that epigenetic components control a small number of genes instead of having global effects on gene expression. For example, the BET protein inhibitors alter only a few hundred genes, and these genes differ depending on cell type (Dawson et al., 2011; Nicodeme et al., 2010). Thus, these drugs can disrupt a selective set of genes. What remains uncertain and imperative to now learn is how these epigenetic regulators are targeted to these "essential" genes and what makes these genes solely reliant on certain epigenetic regulators.

Related to this issue is the observation that epigenetic inhibitors lead to dramatic effects in malignant cells, though their normal counterparts remain largely unaltered. This suggests that, during normal homeostasis, epigenetic regulators function in a multitiered and semiredundant manner, but in cancer, they may be required to maintain the expression of a few key target genes. A slight tip in the balance of this regulation is sufficient to result in a cell catastrophe. This "epigenetic vulnerability" of certain cancer cells in many ways mirrors the age old axiom of "oncogene addiction" (Weinstein, 2002). Some cancer cells are reliant on specific epigenetic pathways, whereas normal cells have alternative compensating pathways to rely on.

Finally, it is now also evident from both clinical and preclinical studies that hematopoietic malignancies are clearly more vulnerable to epigenetic interventions than solid malignancies. Thus, not all cancers are equally susceptible to epigenetic therapies. The biology underpinning this observation urgently warrants our attention if epigenetic therapies are to be more widely applicable. Broadly speaking, even aggressive hematopoietic malignancies, such as AML, appear to harbor as few as ten coding mutations; in contrast, the cancer genomes of solid malignancies appear to be vastly more complex. Furthermore, the in vivo niche occupied by hematopoietic cells offers a very different environment for drug exposure, and hematopoietic cells may metabolize these drugs differently than other tissues. Could these intrinsic cellular differences account for the varied efficacy of these agents? Are these therapies being used appropriately in the solid malignancies?

This latter question raises the more fundamental issue of rationally designed combination epigenetic therapies. It is likely that many of these new epigenetic drugs offer synergistic benefits, and these new therapies may also synergize with conventional chemotherapies. This strategy of combination therapy may not only increase therapeutic efficacy but also reduce the likelihood of drug resistance.

The plethora of genetic lesions in epigenetic regulators offers many possible targets for drug discovery and will no doubt attract the attention of the pharmaceutical industry. However, given the expense of the drug discovery process, what should guide the choice of target? The "drugability" of enzymes has traditionally biased this choice, but the current success of targeting acetyl-readers may propel other modification readers (e.g., methyl-readers) as the candidates of choice. In addition, one should not rely solely on the existence of genetic lesions to guide the target for drug discovery. There are no genetic lesions reported in histone deacetylases, yet clinically safe and effective drugs have been developed against these enzymes. A potential way forward is to use high-throughput genotype/phenotype drug discovery programs in cancer cells, as has been recently reported (Barretina et al., 2012; Garnett et al., 2012).

Although the biography of cancer will continue to evolve and surprise us, the prevailing mood within the field of cancer epigenetics is one of optimism. Clearly, the roads leading to effective cancer therapies are long and treacherous, and we do not have a map to lead us to success. However, what we may now have is a promising path to follow.

ACKNOWLEDGMENTS

The scope of this review and its space limitations have unfortunately meant that we have not been able to separately cite many of the original publications that have contributed substantially to the field. We sincerely apologize to the authors of these publications. We would like to thank Drs. Andy Bannister and Brian Huntly for valued input and critical appraisal of the manuscript. We would also like to thank Dr. Peter Campbell for sharing data from the International Cancer Genome Consortium and Prof. Gerald Crabtree for helpful discussions. Mark Dawson is supported by a Wellcome-Beit Intermediate Clinical Fellowship, and the Kouzarides lab is funded by a program grant from Cancer Research UK (CRUK).

REFERENCES

Allfrey, V.G., Faulkner, R., and Mirsky, A.E. (1964). Acetylation and methylation of histones and their possible role in the regulation of RNA synthesis. Proc. Natl. Acad. Sci. USA *51*, 786–794.

Allis, C.D., Jenuwein, T., and Reinberg, D. (2007). Epigenetics (Cold Spring Harbor, NY: Cold Spring Harbor Laboratory Press).

Amaral, P.P., Dinger, M.E., Mercer, T.R., and Mattick, J.S. (2008). The eukaryotic genome as an RNA machine. Science *319*, 1787–1789.

Avvakumov, N., and Côté, J. (2007). The MYST family of histone acetyltransferases and their intimate links to cancer. Oncogene 26, 5395–5407.

Baek, S.H. (2011). When signaling kinases meet histones and histone modifiers in the nucleus. Mol. Cell 42, 274–284.

Bannister, A.J., and Kouzarides, T. (1996). The CBP co-activator is a histone acetyltransferase. Nature 384, 641–643.

Bantscheff, M., Hopf, C., Savitski, M.M., Dittmann, A., Grandi, P., Michon, A.-M., Schlegl, J., Abraham, Y., Becher, I., Bergamini, G., et al. (2011). Chemoproteomics profiling of HDAC inhibitors reveals selective targeting of HDAC complexes. Nat. Biotechnol. 29, 255–265.

Barretina, J., Caponigro, G., Stransky, N., Venkatesan, K., Margolin, A.A., Kim, S., Wilson, C.J., Lehár, J., Kryukov, G.V., Sonkin, D., et al. (2012). The Cancer Cell Line Encyclopedia enables predictive modelling of anticancer drug sensitivity. Nature 483, 603–607.

Barski, A., Cuddapah, S., Cui, K., Roh, T.-Y., Schones, D.E., Wang, Z., Wei, G., Chepelev, I., and Zhao, K. (2007). High-resolution profiling of histone methylations in the human genome. Cell 129, 823–837.

Bartke, T., Vermeulen, M., Xhemalce, B., Robson, S.C., Mann, M., and Kouzarides, T. (2010). Nucleosome-interacting proteins regulated by DNA and histone methylation. Cell 143, 470–484.

Baylin, S.B., and Jones, P.A. (2011). A decade of exploring the cancer epigenome - biological and translational implications. Nat. Rev. Cancer 11, 726–734.

Bereshchenko, O.R., Gu, W., and Dalla-Favera, R. (2002). Acetylation inactivates the transcriptional repressor BCL6. Nat. Genet. 32, 606–613.

Berger, S.L., Kouzarides, T., Shiekhattar, R., and Shilatifard, A. (2009). An operational definition of epigenetics. Genes Dev. 23, 781–783.

Berman, B.P., Weisenberger, D.J., Aman, J.F., Hinoue, T., Ramjan, Z., Liu, Y., Noushmehr, H., Lange, C.P., van Dijk, C.M., Tollenaar, R.A., et al. (2012). Regions of focal DNA hypermethylation and long-range hypomethylation in colorectal cancer coincide with nuclear lamina-associated domains. Nat. Genet. 44, 40–46.

Bernstein, B.E., Mikkelsen, T.S., Xie, X., Kamal, M., Huebert, D.J., Cuff, J., Fry, B., Meissner, A., Wernig, M., Plath, K., et al. (2006). A bivalent chromatin structure marks key developmental genes in embryonic stem cells. Cell 125, 315–326.

Bungard, D., Fuerth, B.J., Zeng, P.Y., Faubert, B., Maas, N.L., Viollet, B., Carling, D., Thompson, C.B., Jones, R.G., and Berger, S.L. (2010). Signaling kinase AMPK activates stress-promoted transcription via histone H2B phosphorylation. Science 329, 1201–1205.

Chapman, M.A., Lawrence, M.S., Keats, J.J., Cibulskis, K., Sougnez, C., Schinzel, A.C., Harview, C.L., Brunet, J.P., Ahmann, G.J., Adli, M., et al. (2011). Initial genome sequencing and analysis of multiple myeloma. Nature 471, 467–472.

Choudhary, C., Kumar, C., Gnad, F., Nielsen, M.L., Rehman, M., Walther, T.C., Olsen, J.V., and Mann, M. (2009). Lysine acetylation targets protein complexes and co-regulates major cellular functions. Science 325, 834–840.

Chung, C.W., and Witherington, J. (2011). Progress in the discovery of small-molecule inhibitors of bromodomain—histone interactions. J. Biomol. Screen. 16, 1170–1185.

Cimmino, L., Abdel-Wahab, O., Levine, R.L., and Aifantis, I. (2011). TET family proteins and their role in stem cell differentiation and transformation. Cell Stem Cell 9, 193–204.

Cole, P.A. (2008). Chemical probes for histone-modifying enzymes. Nat. Chem. Biol. 4, 590–597.

Coles, A.H., and Jones, S.N. (2009). The ING gene family in the regulation of cell growth and tumorigenesis. J. Cell. Physiol. 218, 45–57.

Daigle, S.R., Olhava, E.J., Therkelsen, C.A., Majer, C.R., Sneeringer, C.J., Song, J., Johnston, L.D., Scott, M.P., Smith, J.J., Xiao, Y., et al. (2011). Selective killing of mixed lineage leukemia cells by a potent small-molecule DOT1L inhibitor. Cancer Cell *20*, 53–65.

Dawson, M.A., Bannister, A.J., Göttgens, B., Foster, S.D., Bartke, T., Green, A.R., and Kouzarides, T. (2009). JAK2 phosphorylates histone H3Y41 and excludes HP1alpha from chromatin. Nature *461*, 819–822.

Dawson, M.A., Prinjha, R.K., Dittmann, A., Giotopoulos, G., Bantscheff, M., Chan, W.I., Robson, S.C., Chung, C.W., Hopf, C., Savitski, M.M., et al. (2011). Inhibition of BET recruitment to chromatin as an effective treatment for MLL-fusion leukaemia. Nature *478*, 529–533.

de Wit, E., and de Laat, W. (2012). A decade of 3C technologies: insights into nuclear organization. Genes Dev. *26*, 11–24.

Deguchi, K., Ayton, P.M., Carapeti, M., Kutok, J.L., Snyder, C.S., Williams, I.R., Cross, N.C.P., Glass, C.K., Cleary, M.L., and Gilliland, D.G. (2003). MOZ-TIF2-induced acute myeloid leukemia requires the MOZ nucleosome binding motif and TIF2-mediated recruitment of CBP. Cancer Cell *3*, 259–271.

Delhommeau, F., Dupont, S., Della Valle, V., James, C., Trannoy, S., Massé, A., Kosmider, O., Le Couedic, J.P., Robert, F., Alberdi, A., et al. (2009). Mutation in TET2 in myeloid cancers. N. Engl. J. Med. *360*, 2289–2301.

Delmore, J.E., Issa, G.C., Lemieux, M.E., Rahl, P.B., Shi, J., Jacobs, H.M., Kastritis, E., Gilpatrick, T., Paranal, R.M., Qi, J., et al. (2011). BET bromodomain inhibition as a therapeutic strategy to target c-Myc. Cell *146*, 904–917.

Dialynas, G.K., Vitalini, M.W., and Wallrath, L.L. (2008). Linking Heterochromatin Protein 1 (HP1) to cancer progression. Mutat. Res. *647*, 13–20.

Easwaran, H., Johnstone, S.E., Van Neste, L., Ohm, J., Mosbruger, T., Wang, Q., Aryee, M.J., Joyce, P., Ahuja, N., Weisenberger, D., et al. (2012). A DNA hypermethylation module for the stem/progenitor cell signature of cancer. Genome Res. *22*, 837–849.

Ernst, T., Chase, A.J., Score, J., Hidalgo-Curtis, C.E., Bryant, C., Jones, A.V., Waghorn, K., Zoi, K., Ross, F.M., Reiter, A., et al. (2010). Inactivating mutations of the histone methyltransferase gene EZH2 in myeloid disorders. Nat. Genet. *42*, 722–726.

Farnham, P.J. (2009). Insights from genomic profiling of transcription factors. Nat. Rev. Genet. *10*, 605–616.

Federico, M., and Bagella, L. (2011). Histone deacetylase inhibitors in the treatment of hematological malignancies and solid tumors. J. Biomed. Biotechnol. *2011*, 475641.

Feinberg, A.P., and Tycko, B. (2004). The history of cancer epigenetics. Nat. Rev. Cancer *4*, 143–153.

Fenaux, P., Mufti, G.J., Hellstrom-Lindberg, E., Santini, V., Finelli, C., Giagounidis, A., Schoch, R., Gattermann, N., Sanz, G., List, A., et al; International Vidaza High-Risk MDS Survival Study Group. (2009). Efficacy of azacitidine compared with that of conventional care regimens in the treatment of higher-risk myelodysplastic syndromes: a randomised, open-label, phase III study. Lancet Oncol. *10*, 223–232.

Filippakopoulos, P., Qi, J., Picaud, S., Shen, Y., Smith, W.B., Fedorov, O., Morse, E.M., Keates, T., Hickman, T.T., Felletar, I., et al. (2010). Selective inhibition of BET bromodomains. Nature *468*, 1067–1073.

Forbes, S.A., Bindal, N., Bamford, S., Cole, C., Kok, C.Y., Beare, D., Jia, M., Shepherd, R., Leung, K., Menzies, A., et al. (2011). COSMIC: mining complete cancer genomes in the Catalogue of Somatic Mutations in Cancer. Nucleic Acids Res. *39* (Database issue), D945–D950.

Fraga, M.F., Ballestar, E., Villar-Garea, A., Boix-Chornet, M., Espada, J., Schotta, G., Bonaldi, T., Haydon, C., Ropero, S., Petrie, K., et al. (2005). Loss of acetylation at Lys16 and trimethylation at Lys20 of histone H4 is a common hallmark of human cancer. Nat. Genet. *37*, 391–400.

Garnett, M.J., Edelman, E.J., Heidorn, S.J., Greenman, C.D., Dastur, A., Lau, K.W., Greninger, P., Thompson, I.R., Luo, X., Soares, J., et al. (2012). Systematic identification of genomic markers of drug sensitivity in cancer cells. Nature *483*, 570–575.

Gore, S.D. (2011). New ways to use DNA methyltransferase inhibitors for the treatment of myelodysplastic syndrome. Hematology (Am. Soc. Hematol. Educ. Program) *2011*, 550–555.

Gui, Y., Guo, G., Huang, Y., Hu, X., Tang, A., Gao, S., Wu, R., Chen, C., Li, X., Zhou, L., et al. (2011). Frequent mutations of chromatin remodeling genes in transitional cell carcinoma of the bladder. Nat. Genet. *43*, 875–878.

Hanahan, D., and Weinberg, R.A. (2011). Hallmarks of cancer: the next generation. Cell *144*, 646–674.

Heintzman, N.D., Stuart, R.K., Hon, G., Fu, Y., Ching, C.W., Hawkins, R.D., Barrera, L.O., Van Calcar, S., Qu, C., Ching, K.A., et al. (2007). Distinct and predictive chromatin signatures of transcriptional promoters and enhancers in the human genome. Nat. Genet. *39*, 311–318.

Heintzman, N.D., Hon, G.C., Hawkins, R.D., Kheradpour, P., Stark, A., Harp, L.F., Ye, Z., Lee, L.K., Stuart, R.K., Ching, C.W., et al. (2009). Histone modifications at human enhancers reflect global cell-type-specific gene expression. Nature *459*, 108–112.

Huntly, B.J., Shigematsu, H., Deguchi, K., Lee, B.H., Mizuno, S., Duclos, N., Rowan, R., Amaral, S., Curley, D., Williams, I.R., et al. (2004). MOZ-TIF2, but not BCR-ABL, confers properties of leukemic stem cells to committed murine hematopoietic progenitors. Cancer Cell *6*, 587–596.

Iyer, N.G., Ozdag, H., and Caldas, C. (2004). p300/CBP and cancer. Oncogene *23*, 4225–4231.

Jiao, Y., Shi, C., Edil, B.H., de Wilde, R.F., Klimstra, D.S., Maitra, A., Schulick, R.D., Tang, L.H., Wolfgang, C.L., Choti, M.A., et al. (2011). DAXX/ATRX, MEN1, and mTOR pathway genes are frequently altered in pancreatic neuroendocrine tumors. Science *331*, 1199–1203.

Johnstone, R.W., and Licht, J.D. (2003). Histone deacetylase inhibitors in cancer therapy: is transcription the primary target? Cancer Cell *4*, 13–18.

Jones, S., Wang, T.L., Shih, IeM., Mao, T.L., Nakayama, K., Roden, R., Glas, R., Slamon, D., Diaz, L.A., Jr., Vogelstein, B., et al. (2010). Frequent mutations of chromatin remodeling gene ARID1A in ovarian clear cell carcinoma. Science *330*, 228–231.

Klose, R.J., and Bird, A.P. (2006). Genomic DNA methylation: the mark and its mediators. Trends Biochem. Sci. *31*, 89–97.

Kogo, R., Shimamura, T., Mimori, K., Kawahara, K., Imoto, S., Sudo, T., Tanaka, F., Shibata, K., Suzuki, A., Komune, S., et al. (2011). Long noncoding RNA HOTAIR regulates polycomb-dependent chromatin modification and is associated with poor prognosis in colorectal cancers. Cancer Res. *71*, 6320–6326.

Kouzarides, T. (2007). Chromatin modifications and their function. Cell *128*, 693–705.

Kriaucionis, S., and Heintz, N. (2009). The nuclear DNA base 5-hydroxymethylcytosine is present in Purkinje neurons and the brain. Science *324*, 929–930.

Krivtsov, A.V., and Armstrong, S.A. (2007). MLL translocations, histone modifications and leukaemia stem-cell development. Nat. Rev. Cancer *7*, 823–833.

Laird, P.W. (2010). Principles and challenges of genomewide DNA methylation analysis. Nat. Rev. Genet. *11*, 191–203.

Langemeijer, S.M., Kuiper, R.P., Berends, M., Knops, R., Aslanyan, M.G., Massop, M., Stevens-Linders, E., van Hoogen, P., van Kessel, A.G., Raymakers, R.A., et al. (2009). Acquired mutations in TET2 are common in myelodysplastic syndromes. Nat. Genet. *41*, 838–842.

Lee, J.S., Smith, E., and Shilatifard, A. (2010). The language of histone crosstalk. Cell *142*, 682–685.

Ley, T.J., Ding, L., Walter, M.J., McLellan, M.D., Lamprecht, T., Larson, D.E., Kandoth, C., Payton, J.E., Baty, J., Welch, J., et al. (2010). DNMT3A mutations in acute myeloid leukemia. N. Engl. J. Med. *363*, 2424–2433.

Li, E., Bestor, T.H., and Jaenisch, R. (1992). Targeted mutation of the DNA methyltransferase gene results in embryonic lethality. Cell *69*, 915–926.

Li, D.Q., Pakala, S.B., Nair, S.S., Eswaran, J., and Kumar, R. (2012). Metastasis-associated protein 1/nucleosome remodeling and histone deacetylase complex in cancer. Cancer Res. *72*, 387–394.

Lorsbach, R.B., Moore, J., Mathew, S., Raimondi, S.C., Mukatira, S.T., and Downing, J.R. (2003). TET1, a member of a novel protein family, is fused to MLL in acute myeloid leukemia containing the t(10;11)(q22;q23). Leukemia *17*, 637–641.

Margueron, R., and Reinberg, D. (2011). The Polycomb complex PRC2 and its mark in life. Nature *469*, 343–349.

Mertz, J.A., Conery, A.R., Bryant, B.M., Sandy, P., Balasubramanian, S., Mele, D.A., Bergeron, L., and Sims, R.J., III (2011). Targeting MYC dependence in cancer by inhibiting BET bromodomains. Proc. Natl. Acad. Sci. USA *108*, 16669–16674.

Meyer, N., and Penn, L.Z. (2008). Reflecting on 25 years with MYC. Nat. Rev. Cancer *8*, 976–990.

Minucci, S., and Pelicci, P.G. (2006). Histone deacetylase inhibitors and the promise of epigenetic (and more) treatments for cancer. Nat. Rev. Cancer *6*, 38–51.

Moran-Crusio, K., Reavie, L., Shih, A., Abdel-Wahab, O., Ndiaye-Lobry, D., Lobry, C., Figueroa, M.E., Vasanthakumar, A., Patel, J., Zhao, X., et al. (2011). Tet2 loss leads to increased hematopoietic stem cell self-renewal and myeloid transformation. Cancer Cell *20*, 11–24.

Morin, R.D., Johnson, N.A., Severson, T.M., Mungall, A.J., An, J., Goya, R., Paul, J.E., Boyle, M., Woolcock, B.W., Kuchenbauer, F., et al. (2010). Somatic mutations altering EZH2 (Tyr641) in follicular and diffuse large B-cell lymphomas of germinal-center origin. Nat. Genet. *42*, 181–185.

Morin, R.D., Mendez-Lago, M., Mungall, A.J., Goya, R., Mungall, K.L., Corbett, R.D., Johnson, N.A., Severson, T.M., Chiu, R., Field, M., et al. (2011). Frequent mutation of histone-modifying genes in non-Hodgkin lymphoma. Nature *476*, 298–303.

Mosammaparast, N., and Shi, Y. (2010). Reversal of histone methylation: biochemical and molecular mechanisms of histone demethylases. Annu. Rev. Biochem. *79*, 155–179.

Nicodeme, E., Jeffrey, K.L., Schaefer, U., Beinke, S., Dewell, S., Chung, C.W., Chandwani, R., Marazzi, I., Wilson, P., Coste, H., et al. (2010). Suppression of inflammation by a synthetic histone mimic. Nature *468*, 1119–1123.

Nikoloski, G., Langemeijer, S.M., Kuiper, R.P., Knops, R., Massop, M., Tönnissen, E.R., van der Heijden, A., Scheele, T.N., Vandenberghe, P., de Witte, T., et al. (2010). Somatic mutations of the histone methyltransferase gene EZH2 in myelodysplastic syndromes. Nat. Genet. *42*, 665–667.

Ntziachristos, P., Tsirigos, A., Van Vlierberghe, P., Nedjic, J., Trimarchi, T., Flaherty, M.S., Ferres-Marco, D., da Ros, V., Tang, Z., Siegle, J., et al. (2012). Genetic inactivation of the polycomb repressive complex 2 in T cell acute lymphoblastic leukemia. Nat. Med. *18*, 298–301.

Ohm, J.E., McGarvey, K.M., Yu, X., Cheng, L., Schuebel, K.E., Cope, L., Mohammad, H.P., Chen, W., Daniel, V.C., Yu, W., et al. (2007). A stem cell-like chromatin pattern may predispose tumor suppressor genes to DNA hypermethylation and heritable silencing. Nat. Genet. *39*, 237–242.

Okano, M., Bell, D.W., Haber, D.A., and Li, E. (1999). DNA methyltransferases Dnmt3a and Dnmt3b are essential for de novo methylation and mammalian development. Cell *99*, 247–257.

Olsen, E.A., Kim, Y.H., Kuzel, T.M., Pacheco, T.R., Foss, F.M., Parker, S., Frankel, S.R., Chen, C., Ricker, J.L., Arduino, J.M., and Duvic, M. (2007). Phase IIb multicenter trial of vorinostat in patients with persistent, progressive, or treatment refractory cutaneous T-cell lymphoma. J. Clin. Oncol. *25*, 3109–3115.

Park, P.J. (2009). ChIP-seq: advantages and challenges of a maturing technology. Nat. Rev. Genet. *10*, 669–680.

Pasqualucci, L., Dominguez-Sola, D., Chiarenza, A., Fabbri, G., Grunn, A., Trifonov, V., Kasper, L.H., Lerach, S., Tang, H., Ma, J., et al. (2011). Inactivating mutations of acetyltransferase genes in B-cell lymphoma. Nature *471*, 189–195.

Patel, J.P., Gönen, M., Figueroa, M.E., Fernandez, H., Sun, Z., Racevskis, J., Van Vlierberghe, P., Dolgalev, I., Thomas, S., Aminova, O., et al. (2012). Prognostic relevance of integrated genetic profiling in acute myeloid leukemia. N. Engl. J. Med. *366*, 1079–1089.

Piekarz, R.L., Frye, R., Turner, M., Wright, J.J., Allen, S.L., Kirschbaum, M.H., Zain, J., Prince, H.M., Leonard, J.P., Geskin, L.J., et al. (2009). Phase II multi-institutional trial of the histone deacetylase inhibitor romidepsin as monotherapy for patients with cutaneous T-cell lymphoma. J. Clin. Oncol. *27*, 5410–5417.

Quivoron, C., Couronné, L., Della Valle, V., Lopez, C.K., Plo, I., Wagner-Ballon, O., Do Cruzeiro, M., Delhommeau, F., Arnulf, B., Stern, M.H., et al. (2011). TET2 inactivation results in pleiotropic hematopoietic abnormalities in mouse and is a recurrent event during human lymphomagenesis. Cancer Cell *20*, 25–38.

Rando, O.J., and Chang, H.Y. (2009). Genome-wide views of chromatin structure. Annu. Rev. Biochem. *78*, 245–271.

Robertson, K.D. (2005). DNA methylation and human disease. Nat. Rev. Genet. *6*, 597–610.

Rui, L., Emre, N.C., Kruhlak, M.J., Chung, H.J., Steidl, C., Slack, G., Wright, G.W., Lenz, G., Ngo, V.N., Shaffer, A.L., et al. (2010). Cooperative epigenetic modulation by cancer amplicon genes. Cancer Cell *18*, 590–605.

Ruthenburg, A.J., Li, H., Milne, T.A., Dewell, S., McGinty, R.K., Yuen, M., Ueberheide, B., Dou, Y., Muir, T.W., Patel, D.J., and Allis, C.D. (2011). Recognition of a mononucleosomal histone modification pattern by BPTF via multivalent interactions. Cell *145*, 692–706.

Schenk, T., Chen, W.C., Göllner, S., Howell, L., Jin, L., Hebestreit, K., Klein, H.U., Popescu, A.C., Burnett, A., Mills, K., et al. (2012). Inhibition of the LSD1 (KDM1A) demethylase reactivates the all-trans-retinoic acid differentiation pathway in acute myeloid leukemia. Nat. Med. *18*, 605–611.

Schwartzentruber, J., Korshunov, A., Liu, X.Y., Jones, D.T., Pfaff, E., Jacob, K., Sturm, D., Fontebasso, A.M., Quang, D.A., Tönjes, M., et al. (2012). Driver mutations in histone H3.3 and chromatin remodelling genes in paediatric glioblastoma. Nature *482*, 226–231.

Segal, E., and Widom, J. (2009). From DNA sequence to transcriptional behaviour: a quantitative approach. Nat. Rev. Genet. *10*, 443–456.

Seligson, D.B., Horvath, S., Shi, T., Yu, H., Tze, S., Grunstein, M., and Kurdistani, S.K. (2005). Global histone modification patterns predict risk of prostate cancer recurrence. Nature *435*, 1262–1266.

Sneeringer, C.J., Scott, M.P., Kuntz, K.W., Knutson, S.K., Pollock, R.M., Richon, V.M., and Copeland, R.A. (2010). Coordinated activities of wild-type plus mutant EZH2 drive tumor-associated hypertrimethylation of lysine 27 on histone H3 (H3K27) in human B-cell lymphomas. Proc. Natl. Acad. Sci. USA *107*, 20980–20985.

Stratton, M.R., Campbell, P.J., and Futreal, P.A. (2009). The cancer genome. Nature *458*, 719–724.

Stucki, M., Clapperton, J.A., Mohammad, D., Yaffe, M.B., Smerdon, S.J., and Jackson, S.P. (2005). MDC1 directly binds phosphorylated histone H2AX to regulate cellular responses to DNA double-strand breaks. Cell *123*, 1213–1226.

Stunnenberg, H.G., and Vermeulen, M. (2011). Towards cracking the epigenetic code using a combination of high-throughput epigenomics and quantitative mass spectrometry-based proteomics. Bioessays *33*, 547–551.

Tahiliani, M., Koh, K.P., Shen, Y., Pastor, W.A., Bandukwala, H., Brudno, Y., Agarwal, S., Iyer, L.M., Liu, D.R., Aravind, L., and Rao, A. (2009). Conversion of 5-methylcytosine to 5-hydroxymethylcytosine in mammalian DNA by MLL partner TET1. Science *324*, 930–935.

Tan, M., Luo, H., Lee, S., Jin, F., Yang, J.S., Montellier, E., Buchou, T., Cheng, Z., Rousseaux, S., Rajagopal, N., et al. (2011). Identification of 67 histone marks and histone lysine crotonylation as a new type of histone modification. Cell *146*, 1016–1028.

Taverna, S.D., Li, H., Ruthenburg, A.J., Allis, C.D., and Patel, D.J. (2007). How chromatin-binding modules interpret histone modifications: lessons from professional pocket pickers. Nat. Struct. Mol. Biol. *14*, 1025–1040.

Tsai, H.C., Li, H., Van Neste, L., Cai, Y., Robert, C., Rassool, F.V., Shin, J.J., Harbom, K.M., Beaty, R., Pappou, E., et al. (2012). Transient low doses of DNA-demethylating agents exert durable antitumor effects on hematological and epithelial tumor cells. Cancer Cell *21*, 430–446.

van Haaften, G., Dalgliesh, G.L., Davies, H., Chen, L., Bignell, G., Greenman, C., Edkins, S., Hardy, C., O'Meara, S., Teague, J., et al. (2009). Somatic mutations of the histone H3K27 demethylase gene UTX in human cancer. Nat. Genet. *41*, 521–523.

Varela, I., Tarpey, P., Raine, K., Huang, D., Ong, C.K., Stephens, P., Davies, H., Jones, D., Lin, M.L., Teague, J., et al. (2011). Exome sequencing identifies frequent mutation of the SWI/SNF complex gene PBRM1 in renal carcinoma. Nature *469*, 539–542.

Vermeulen, M., Eberl, H.C., Matarese, F., Marks, H., Denissov, S., Butter, F., Lee, K.K., Olsen, J.V., Hyman, A.A., Stunnenberg, H.G., and Mann, M. (2010). Quantitative interaction proteomics and genome-wide profiling of epigenetic histone marks and their readers. Cell *142*, 967–980.

Versteege, I., Sévenet, N., Lange, J., Rousseau-Merck, M.F., Ambros, P., Handgretinger, R., Aurias, A., and Delattre, O. (1998). Truncating mutations of hSNF5/INI1 in aggressive paediatric cancer. Nature *394*, 203–206.

Wang, K.C., and Chang, H.Y. (2011). Molecular mechanisms of long noncoding RNAs. Mol. Cell *43*, 904–914.

Wang, J., Iwasaki, H., Krivtsov, A., Febbo, P.G., Thorner, A.R., Ernst, P., Anastasiadou, E., Kutok, J.L., Kogan, S.C., Zinkel, S.S., et al. (2005). Conditional MLL-CBP targets GMP and models therapy-related myeloproliferative disease. EMBO J. *24*, 368–381.

Wang, G.G., Allis, C.D., and Chi, P. (2007). Chromatin remodeling and cancer, Part II: ATP-dependent chromatin remodeling. Trends Mol. Med. *13*, 373–380.

Wang, Z., Zang, C., Rosenfeld, J.A., Schones, D.E., Barski, A., Cuddapah, S., Cui, K., Roh, T.Y., Peng, W., Zhang, M.Q., and Zhao, K. (2008). Combinatorial patterns of histone acetylations and methylations in the human genome. Nat. Genet. *40*, 897–903.

Wang, G.G., Song, J., Wang, Z., Dormann, H.L., Casadio, F., Li, H., Luo, J.-L., Patel, D.J., and Allis, C.D. (2009). Haematopoietic malignancies caused by dysregulation of a chromatin-binding PHD finger. Nature *459*, 847–851.

Wang, K., Kan, J., Yuen, S.T., Shi, S.T., Chu, K.M., Law, S., Chan, T.L., Kan, Z., Chan, A.S., Tsui, W.Y., et al. (2011). Exome sequencing identifies frequent mutation of ARID1A in molecular subtypes of gastric cancer. Nat. Genet. *43*, 1219–1223.

Weinstein, I.B. (2002). Cancer. Addiction to oncogenes—the Achilles heal of cancer. Science *297*, 63–64.

Wilson, B.G., and Roberts, C.W. (2011). SWI/SNF nucleosome remodellers and cancer. Nat. Rev. Cancer *11*, 481–492.

Wu, H., and Zhang, Y. (2011). Mechanisms and functions of Tet protein-mediated 5-methylcytosine oxidation. Genes Dev. *25*, 2436–2452.

Wu, G., Broniscer, A., McEachron, T.A., Lu, C., Paugh, B.S., Becksfort, J., Qu, C., Ding, L., Huether, R., Parker, M., et al; St. Jude Children's Research Hospital–Washington University Pediatric Cancer Genome Project. (2012). Somatic histone H3 alterations in pediatric diffuse intrinsic pontine gliomas and non-brainstem glioblastomas. Nat. Genet. *44*, 251–253.

Xhemalce, B., Dawson, M.A., and Bannister, A.J. (2011). Histone Modifications. In Encyclopedia of Molecular Cell Biology and Molecular Medicine. Weinheim: Wiley-VCH Verlag.

Xu, W., Yang, H., Liu, Y., Yang, Y., Wang, P., Kim, S.H., Ito, S., Yang, C., Wang, P., Xiao, M.T., et al. (2011). Oncometabolite 2-hydroxyglutarate is a competitive inhibitor of α-ketoglutarate-dependent dioxygenases. Cancer Cell *19*, 17–30.

Yang, X., Cao, W., Zhang, L., Zhang, W., Zhang, X., and Lin, H. (2012). Targeting 14-3-3zeta in cancer therapy. Cancer Gene Ther. *19*, 153–159.

Zhang, J., Ding, L., Holmfeldt, L., Wu, G., Heatley, S.L., Payne-Turner, D., Easton, J., Chen, X., Wang, J., Rusch, M., et al. (2012). The genetic basis of early T-cell precursor acute lymphoblastic leukaemia. Nature *481*, 157–163.

Zuber, J., Shi, J., Wang, E., Rappaport, A.R., Herrmann, H., Sison, E.A., Magoon, D., Qi, J., Blatt, K., Wunderlich, M., et al. (2011). RNAi screen identifies Brd4 as a therapeutic target in acute myeloid leukaemia. Nature *478*, 524–528.

ell

Converting Cancer Therapies into Cures: Lessons from Infectious Diseases

Michael S. Glickman[1,*], Charles L. Sawyers[2,3,*]

[1]Infectious Diseases Service and Immunology Program, Memorial Sloan Kettering Cancer Center, 1275 York Avenue, New York, NY 10065, USA, [2]Human Oncology and Pathogenesis Program, Memorial Sloan Kettering Cancer Center, 1275 York Avenue, New York, NY 10065, USA, [3]Howard Hughes Medical Institute
*Correspondence: glickmam@mskcc.org (M.S.G.), sawyersc@mskcc.org (C.L.S.)

Cell, 148, Vol. 6, No. March 16, 2012 © 2012 Elsevier Inc.
http://dx.doi.org/10.1016/j.cell.2012.02.015

SUMMARY

During the past decade, cancer drug development has shifted from a focus on cytotoxic chemotherapies to drugs that target specific molecular alterations in tumors. Although these drugs dramatically shrink tumors, the responses are temporary. Research is now focused on overcoming drug resistance, a frequent cause of treatment failure. Here we reflect on analogous challenges faced by researchers in infectious diseases. We compare and contrast the resistance mechanisms arising in cancer and infectious diseases and discuss how approaches for overcoming viral and bacterial infections, such as HIV and tuberculosis, are instructive for developing a more rational approach for cancer therapy. In particular, maximizing the effect of the initial treatment response, which often requires synergistic combination therapy, is foremost among these approaches. A remaining challenge in both fields is identifying drugs that eliminate drug-tolerant "persister" cells (infectious disease) or tumor-initiating/stem cells (cancer) to prevent late relapse and shorten treatment duration.

INTRODUCTION

The modern era of antimicrobial therapy is ~60 years old. It has produced agents that target bacterial cell wall biosynthesis (e.g., penicillins, cephalosporins, vancomycin, Isoniazid), protein synthesis (e.g., aminoglycosides, tetracyclines, chloramphenicol, oxazolidinones, macrolides),

83

RNA synthesis (e.g., rifampin), and DNA metabolism (e.g., sulfonamides, quinolones). Despite this diversity of targets, resistance remains a universal accompaniment to antimicrobial therapy. Microbes use remarkably diverse strategies to overcome selective pressure, and much is known about the mechanisms of antimicrobial resistance. Although antimicrobial resistance remains a major problem on a population level, the emergence of drug resistance in an individual patient with a chronic infection can be prevented by the administration of a highly effective combination therapy regimen, which either cures the patient or prevents death from previously lethal infections.

The history of targeted cancer therapy is much shorter than that of infectious diseases, but already it is replete with a similarly diverse range of resistance mechanisms. However, effective combinations leading to cures have not yet emerged. Oncology has a track record of prior success in developing curative combination chemotherapy for pediatric leukemia, germ cell tumors, and lymphoma, but this progress required decades of empirically mixing and matching available agents. There is optimism that this timeline can be shortened with targeted cancer drugs because our understanding of cancer biology today is markedly more advanced.

Here we compare and contrast examples of drug resistance in infectious diseases and cancer, with the hope that lessons learned in one field may inform the other. We acknowledge that this is a forced comparison. There are fundamental differences in the principles underlying the search for drugs that target a foreign invader (i.e., in infections) versus mutant cells that emerge from the host (i.e., in cancer), particularly with regard to anticipated toxicities. Yet, current drug-targeting strategies in both fields share the goal of exploiting the unique dependencies of each disease, such as tumor-specific mutations in cancers or microbe/virus-specific targets in infectious agents. Another challenge in comparing these disciplines are the different definitions of treatment success. Resolution of the illness in the patient is central to both, but the infectious diseases field must also consider the impact of drug resistance on public health. Overtreatment with broad spectrum antibiotics cures most patients but hastens the emergence and spread of multidrug-resistant strains, which can impact the health of currently uninfected individuals.

Rather than divide the discussion into separate sections on infectious diseases and cancer, we consider both fields together, beginning with mutational and nonmutational mechanisms of resistance (Table 1). We follow with a review of successful combination drug strategies in infectious diseases. We provide insights into why they worked and highlight a few instances when monotherapy is surprisingly effective in both disciplines. We conclude with the argument that molecular diagnostics, which already play a critical role

Table 1 Mechanisms of Resistance	
Mutational	**Nonmutational**
mutation of drug target	drug-tolerant persister cells
amplification of drug target	tumor-initiating cells/cancer stem cells
bypass of drug targeted pathway	signaling pathway feedback
drug inactivation	lineage switching

in defining drug-sensitive subsets of cancer patients, could also transform current infectious diseases treatment. To learn more about the mechanistic details of and treatment options for drug resistance in HIV, tuberculosis (TB), and malaria, see Review by Goldberg et al. on page 1271 of this issue.

MUTATIONAL MECHANISMS OF RESISTANCE

Mutation of the Drug Target

A common resistance mechanism shared across antimicrobial and anticancer agents is mutation in the gene that encodes the drug target (Figure 1A). HIV serves as an illustrative example of this resistance mechanism. The goal of HIV therapy is long-term suppression of viral replication with combinations of antiretroviral agents targeting viral reverse transcriptase (RT), protease, or integrase enzymes. Loss of viral suppression is often associated with emergence of HIV-1 variants, which express drug-resistant alleles of the viral RT, protease, or integrase gene due to mutations (Blanco et al., 2011; Zolopa et al., 1999). Mutations in drug targets are also common in antibacterial resistance. For example, rifampicin binds and inhibits the β subunit of the bacterial RNA polymerase enzyme complex (Campbell et al., 2001). β subunit mutations that impair drug binding confer rifampicin resistance. Resistance to β-lactam antibiotics can also occur through drug target mutation. β-lactam antibiotics, such as penicillins and cephalosporins, inhibit bacterial peptidoglycan biosynthesis by binding and inhibiting transpeptidases (i.e., penicillin-binding proteins or PBPs), enzymes that crosslink the peptidoglycan peptide side chains. PBP mutations that diminish the affinity for β-lactam confer β-lactam resistance in a wide variety of gram-positive pathogens, including *Staphylococcus aureus*, *Streptococcus pneumoniae*, and *Enterococcus* (Zapun et al., 2008).

Drug target mutations are also common with many anticancer agents, particularly in the growing class of kinase inhibitors that target oncogenic driver mutations. This mechanism was first demonstrated in patients with chronic myeloid leukemia (CML) who developed resistance to the ABL kinase inhibitor imatinib. Now this mechanism has been observed with nearly every kinase

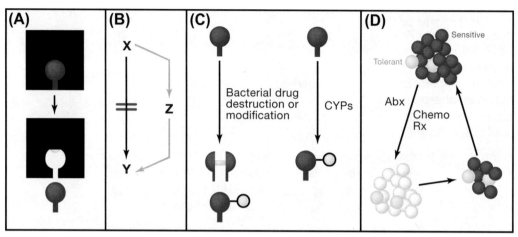

FIGURE 1   Mechanisms of Resistance to Antimicrobials and Targeted Anticancer Agents
(A) Resistance via target mutation. This mechanism has been well described in both antimicrobial resistance and tumor cell resistance. The red drug binds tightly to its target (black square). A mutational event leads to alteration in the binding site for the drug (yellow circle), leading to loss of drug binding. This mechanism governs resistance to β-lactam antimicrobials (and other antimicrobial classes) as well as resistance to kinase inhibitors.
(B) Resistance via bypass pathways. Treatment with antimicrobial or anticancer agents (red lines) leads to a block in the pathway converting X to Y. Conceptually, Y can be a metabolite or a phenotypic state (e.g., cell proliferation). Resistance to the effect of the drug can be mediated by upregulation of a parallel pathway that allows Y to be restored. This mechanism of resistance has been documented in anticancer therapy, for example, amplification of MET to bypass a drug-induced block in EGFR signaling.
(C) Resistance by drug destruction or modification. Bacterial enzymes, such as β-lactamases or aminoglycoside-modifying enzymes, mediate antimicrobial resistance by drug destruction or modification. This mechanism has not been described for resistance to anticancer agents, although modification of anticancer agents through CYPs can affect efficacy (although this effect is not mediated by the tumor cell).
(D) Intrinsic, nonmutational, resistance. Here the population of cells (either tumor or microbial) are genotypically identical. The red cells are drug sensitive and are rapidly killed by antimicrobial or anticancer therapy, but the yellow cells are poorly killed by antimicrobial or cancer therapy (tolerant). These tolerant cells are called microbial persisters or cancer stem cells. Therapy with antimicrobials or anticancer agents leads to substantial killing, but the persister cells are able to resist treatment and can repopulate the infection or tumor with drug-sensitive cells, causing disease relapse.

inhibitor tested to date. In the case of CML, mutations in the kinase domain of BCR-ABL impair imatinib binding, although preserving the catalytic activity of the enzyme (ATP hydrolysis) that is required for oncogenicity (Gorre et al., 2001). Some mutations confer resistance by blocking interactions between the drug and target through steric hindrance. Other mutations restrict the flexibility of the enzyme to conformations that are unsuitable for drug binding (Burgess et al., 2005; Gorre et al., 2001; Shah et al., 2002). Similar mechanisms account for resistance to epidermal growth factor receptor (EGFR) kinase inhibitors and ALK kinase inhibitors in lung cancer, KIT kinase inhibitors in gastrointestinal stromal tumors (GIST), platelet-derived growth factor receptor (PDGFR) kinase inhibitors in hypereosinohilic syndrome, and BRAF kinase inhibitors in melanoma (Antonescu et al., 2005; Choi et al., 2010; Cools et al., 2003; Pao et al., 2005; Poulikakos et al., 2011). In CML and lung cancer, drug-resistance mutations can be detected in some patients prior to treatment with the kinase inhibitor and can impact prognosis by shortening the time to disease progression. The drug-resistant allele is generally present in a small minority of cells, but in a few cases, drug-resistant clones have undergone substantial expansion in the absence of drug. This observation raises the question of whether some resistance mutations also confer a tumor fitness advantage (Shah et al., 2002; Bean et al., 2007; Godin-Heymann et al., 2007; Maheswaran et al., 2008; Skaggs et al., 2006). This contrasts with HIV infections, in which virions bearing drug-resistant RT mutations generally have reduced viral fitness (Martinez-Picado and Martínez, 2008).

Gene Amplification of the Drug Target or a Bypass Pathway

Drug resistance can also occur through amplification of the drug target gene in the absence of mutation (Figure 1B). A prime example of this mechanism in cancer is the androgen receptor (AR). The AR gene is amplified in ~30% of prostate cancers that have acquired resistance to standard androgen deprivation therapy with drugs (e.g., leuprolide) that lower testosterone production and AR antagonists (e.g., bicalutamide or flutamide) that block ligand binding (Scher and Sawyers, 2005). BCR-ABL gene amplification can also drive resistance to ABL kinase inhibitors in CML, although this mechanism is less common than mutations in the kinase domain (Gorre et al., 2001). Amplification of the dihydrofolate reductase (DHFR) gene confers resistance to the chemotherapeutic agent methotrexate in cancer cell lines (Schimke et al., 1978), but this mechanism has not emerged as a common resistance mechanism in patients. However, amplification of the bacterial DHFR gene can cause resistance to the antibiotic trimethoprim (Steen and Sköld, 1985).

Approximately 20% of patients with EGFR mutant lung cancer develop resistance to EGFR inhibitors by amplification of another receptor tyrosine kinase, MET (Engelman et al., 2007). This mechanism has been termed "oncogene bypass" because the primary drug target remains unaltered and continues to be inhibited by drug. Resistance occurs because MET activates downstream components of the EGFR signaling pathway, bypassing the need for EGFR (Figure 1B). A conceptually similar mechanism has been documented with thymidine auxotrophs of *E. coli* and *Enterococcus*. These strains can become resistant to sulfonamides because the inhibition of thymidine biosynthesis by the antibiotic can be bypassed by the acquisition of thymidine from the environment (Maskell et al., 1978) (Figure 1B).

Drug Inactivation

The most common mechanism of antibacterial resistance is drug destruction by bacterial enzymes (Figure 1C). β-lactamases, which cleave the amide bond of the β-lactam ring, confer resistance to the antibiotic through drug destruction. Progressive chemical modification of the β-lactam nucleus to prevent destruction by β-lactamases has yielded compounds that are active against β-lactamase-producing organisms. However, progressively broader spectrum β-lactamases, including some capable of hydrolyzing all β-lactams and carbapenems, have consequently become widespread. These broad spectrum β-lactamases are an escalating concern and threaten the utility of this class of antibiotics (Cornaglia et al., 2011).

Resistance through antibiotic modification is not limited to the β-lactam class of antibiotics. Aminoglycoside antibiotics inhibit protein synthesis by binding to the 30S subunit of the bacterial ribosome. The most common mechanism of resistance to these drugs is through bacterial acquisition of enzymes that covalently modify the aminoglycoside by phosphorylation, acetylation, or adenylation. The modified aminoglycoside no longer binds its target on the ribosome.

In contrast to antibacterials, drug inactivation has not emerged as a major cause of resistance to anticancer agents. Cytochrome P450 enzymes (CYPs) play a critical role in the metabolism of many drugs through oxidation reactions that can inactivate the compound or lead to its rapid elimination (Figure 1C). Although this metabolic degradation was problematic for many classes of drugs in the past, it is less relevant today because most drug candidates are routinely counterscreened early in development against panels of CYPs. Compounds that score as potent CYP substrates are typically eliminated or chemically modified to "dial out" the CYP activity while preserving the desired anticancer function.

Nonetheless, there are examples of drugs in which CYP modification may affect therapeutic response. The antiestrogen tamoxifen, which is widely used in the treatment of estrogen receptor-positive breast cancer, is metabolized by CYP2D6 to its primary active metabolite endoxifen. Some genetic variants of CYP2D6 confer reduced levels of enzyme activity, and there is evidence that women with these variants may not respond as well to tamoxifen because they generate lower levels of endoxifen. Furthermore, CYP2D6 is inhibited by selective serotonin reuptake inhibitors (SSRIs). SSRIs are commonly prescribed to ameliorate "hot flashes," but they may counteract the clinical benefit of tamoxifen by blocking production of its primary metabolite (Borges et al., 2006).

Resistance to the antimycobacterial drugs isoniazid and pyrazinamide, both important agents in the treatment of TB, also occurs because of the failure to generate the active metabolite of the drug. Both antibiotics are prodrugs that must be activated by the TB bacterial enzymes KatG (for isoniazid) and PncA (for pyrazinamide). Resistance to both drugs often occurs through mutations in their respective activator enzymes (Altamirano et al., 1994; Scorpio and Zhang, 1996; Zhang et al., 1992).

Although conceptually similar, these mechanisms differ from the example with tamoxifen/CYP2D6 because resistance to isoniazid and pyrazinamide is cell autonomous. We are not aware of examples of cancer drug inactivation mediated specifically by tumor cells. This mechanism has not been considered by most cancer scientists due to challenges in measuring concentrations of the drug and potential metabolites in tumor biopsies. However, advances in mass spectrometry technology should enable such measurements on a more routine basis. Drug destruction should be evaluated as a potential cause of tumor-mediated resistance in cases in which other mechanisms have been excluded.

NONMUTATIONAL RESISTANCE MECHANISMS

Drug-Tolerant "Persister" Cells

In the earliest days of antimicrobials, it was observed that the killing of a microbial population was rarely complete. A small subpopulation of cells survived but did not have a drug-resistance mutation. When expanded, these cells reverted to antimicrobial sensitivity (Figure 1D). This phenomenon, termed bacterial persistence, is observed in a wide variety of bacterial taxa, including *M. tuberculosis*, *E. coli*, and others (Connolly et al., 2007; Lewis, 2010). It is widely suspected that persister cell populations are responsible for the slow sterilization of many chronic infections and consequent requirement for prolonged antibiotic therapy. There is great interest in identifying molecular determinants of persistence because knowledge of such pathways could lead to more effective antimicrobials that quickly eliminate this reservoir of cells. Such drugs could reduce the chance of late relapse and shorten duration of treatment. In the case of TB, a curative regimen administered over several weeks rather than 6–9 months could have profound consequences on treatment compliance, with obvious public health implications.

We currently have little insight into the molecular basis of persistence. Recent evidence suggests that toxin-antitoxin (TA) modules may be molecular determinants of persistence in some bacteria, presumably by the induction of growth arrest in a subpopulation of cells through toxin-mediated mRNA cleavage (Maisonneuve et al., 2011; Moyed and Bertrand, 1983; Wolfson et al., 1990). But, many questions remain, including the exact mechanisms of TA-mediated persistence, the generalizability of this mechanism to chronic infections, and the therapeutic benefit of targeting TA modules. Similarly, recent evidence from the *M. marinum* system suggests that *M. tuberculosis* drug tolerance in vivo may be mediated by host-inducible drug efflux pumps

(Adams et al., 2011). Although the exact contribution of drug efflux to drug tolerance in vivo remains to be determined, these findings suggest a druggable target that would increase antimicrobial killing by presently available antimycobacterials.

Tumor-Initiating/Cancer Stem Cells

The phenomenon of bacterial persistence shares conceptual similarities with evidence that late relapses of some cancers may be due to tumor-initiating or cancer stem cells that are not eliminated by most treatments (Figure 1D). The existence of cancer stem cells is a topic of considerable controversy (Rossi et al., 2008), but this is not germane to our discussion here. What is relevant is the observation that, even following effective treatment regimens, residual cancer cells persist and are responsible for late relapse. Acute myeloid leukemia (AML) serves as an illustrative example, in which initial treatment with induction chemotherapy is quite effective in inducing remission, but relapse is inevitable unless additional rounds of consolidation chemotherapy are administered.

Several preclinical studies have documented that these residual cells do not have mutations in the drug target or other factors that could explain resistance. Like residual persister cells after antibiotic treatment, these residual cancer cells, when expanded in the absence of drug, give rise to drug-sensitive populations of cancer cells (Sharma et al., 2010). These cancer persister cells tend to express cell-surface antigens typically present on tissue stem cells, raising the possibility that stem cells are inherently drug resistant.

One potential explanation for their resistance is expression of drug efflux pump proteins, such as MDR (multidrug resistance), which are naturally expressed at higher levels in stem cells. However, cancer cells can persist following treatment with drugs that are not MDR substrates. Another potential mechanism is survival signaling mediated through growth factors expressed by adjacent cells in the tumor microenvironment or metastatic niche (Guise, 2010). For example, the CXCR4 receptor plays a critical role in anchoring hematopoietic stem cells (HSCs) and CXCR4-positive AML cells to niches in the bone marrow microenvironment through interaction with the stromal factor SDF-1. CXCR4 inhibitors mobilize HSCs into circulation (DiPersio et al., 2009) and enhance the efficacy and duration of response to induction chemotherapy (Nervi et al., 2009). Similar to antibacterial persister drugs, compounds that selectively eliminate tumor-initiating cells could complement existing cancer therapies (Gupta et al., 2009; Reya et al., 2001).

Feedback due to Signaling Pathway Inhibition

When cancer cells are exposed to drugs that block signaling pathways, such as kinase inhibitors, many tumor cells compensate by activating other signaling pathways. The mechanism of these feedback responses is complex and related to concepts of network robustness and redundancy that have emerged from systems biology research (Lander, 2011). In the context of cancer, inhibition of one signaling pathway initiates compensatory feedback responses that can lead to the activation of another (Carver et al., 2011; Chandarlapaty et al., 2011; O'Reilly et al., 2006; Pratilas et al., 2009). These observations predict that monotherapy with some inhibitors, particularly those targeting the PI3K signaling pathway, is unlikely to be effective unless paired with inhibitors of the compensatory pathway. Interestingly, there is currently no evidence for feedback as a mechanism of resistance to anti-infectious agents, likely because these drugs tend to target essential steps in the replication of the organism (e.g., cell wall synthesis, viral replication, etc.) rather than signaling pathways. However, similar mechanisms of resistance may arise with antimicrobials that target virulence pathways that are not essential to cell viability (Clatworthy et al., 2007).

Lineage Switching

Studies of acquired resistance to EGFR inhibitors in lung cancer reveal that recurrent tumors in some patients undergo a lineage switch from non-small cell to small cell carcinoma (Sequist et al., 2011). Clonality studies document that both tumors arose from a common EGFR mutant tumor rather than from two independent cancers, indicating that the tumor has shifted to a different cellular differentiation pathway. Switching to the small cell lineage (by an unknown mechanism that presumably involves cellular reprogramming) relieves cells of their dependence on EGFR, which was critical for survival of the non-small cell tumor lineage. A somewhat analogous phenomenon has been observed with antibiotics, in which resistance to drugs targeting the bacterial cell wall can occur through the cell wall-deficient bacteria, called L-forms (Allan et al., 2009).

OVERCOMING RESISTANCE

Although our knowledge is still incomplete, progress in deciphering mechanisms of resistance to targeted cancer therapy is already guiding solutions to overcome it. The recognition that point mutations in the BCR-ABL kinase domain confer resistance to the first-generation inhibitor for the ABL kinase inhibitor imatinib led rapidly to efforts to find other ABL inhibitors effective against the mutant BCR-ABL alleles. Within 5 years, two next-generation ABL inhibitors, dasatinib and nilotinib, were approved for the treatment of imatinib-resistant CML (Kantarjian et al., 2007; Talpaz et al., 2006). Dasatinib is notable because its activity against imatinib-resistant BCR-ABL mutants is explained by its ability to bind multiple distinct conformations of BCR-ABL, thereby restricting the potential for escape mutants (Burgess et al., 2005; Shah et al., 2002). A second example has emerged from the discovery that prostate cancers resistant to standard androgen deprivation therapy remain dependent on AR signaling (Chen et al., 2004). Two new drugs, abiraterone and MDV3100, impair AR signaling in this drug-resistant setting and are effective in men with metastatic prostate cancer resistant to standard hormone and chemotherapy (de Bono et al., 2011; Scher et al., 2010; Tran et al., 2009).

These examples underscore the importance of understanding resistance mechanisms and the rapid progress that can be made by leveraging these insights. But they also represent partial solutions because sequential treatment with these different targeted agents is not curative. This current scenario in cancer contrasts strikingly with the current treatment of two previously lethal infections, TB and HIV: combinations of antibiotics cure TB, whereas combinations of antivirals can indefinitely suppress the HIV virus. Physicians treating these diseases in the past faced a situation remarkably similar to that faced by oncologists today. Successful combination regimens arose only after attempts to treat these infections with monotherapy failed.

HIV: Importance of Maximal Suppression of Viral Load at Treatment Initiation

HIV cannot be cured with current antiretroviral therapy, but mortality is dramatically reduced through combination antiretroviral drug therapy. The first major success in HIV therapy came from nucleoside analogs, such as zidovudine (AZT). These analogs inhibit viral replication by causing chain termination when RT incorporates the analogs into the viral cDNA. Although testing of serum

red line). Monotherapy of HIV produces a transient reduction in viral load, which becomes more dramatic and sustained with combination therapy. In contrast, monotherapy for hepatitis B (with entecavir or tenofovir) produces sustained virologic suppression. See text for specific references.

(B) Monotherapy and combination therapy for tuberculosis. The y axis schematically represents clinical response (reduction in bacterial load, clinical improvement, radiographic improvement). Monotherapy and combination therapy have similar efficacy early in treatment, but the benefit of monotherapy is not sustained. (C) Effect of combination therapy on emergence of resistance. The y axis represents the % of patients with resistant bacteria or viruses according to week of treatment. HIV and TB resistance emerges rapidly during monotherapy, leading to the loss of therapeutic effect depicted in panel A (HIV) and panel B (TB). Combination therapy for HIV and TB suppresses the emergence of resistant organisms and allows the sustained therapeutic benefit depicted in (A) and (B). In contrast, monotherapy for hepatitis B with entecavir or tenofovir is not associated with the emergence of resistant viruses, allowing sustained therapeutic benefit with monotherapy.

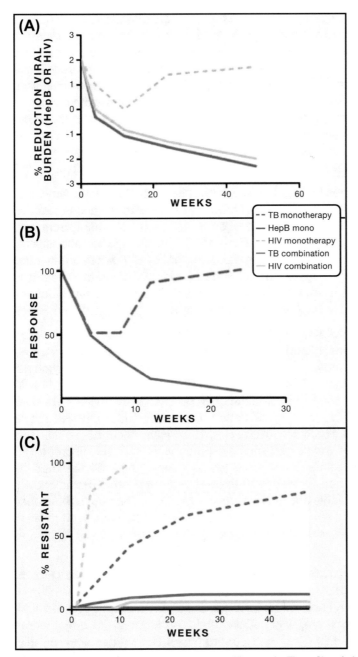

FIGURE 2 Relative Efficacy of Mono- versus Combination Therapy for Three Chronic Infections and Effect on Resistance
(A) The three curves schematically indicate the reduction in viral burden during monotherapy of HIV (dashed green line), combination therapy for HIV (solid green line), and monotherapy for hepatitis B (solid

viral loads was not available at that time, the clinical benefit of AZT was evident after 12 weeks in patients with AIDS. However, the efficacy of the therapy was short lived: CD4 T cell counts returned to the levels of placebo-treated patients in 24 weeks, indicating a significant but transient clinical benefit (Fischl et al., 1987). Similar results were observed with the nucleoside RT inhibitor lamivudine (3TC): nearly 100% of patients had lamivudine resistance by week 12 of monotherapy (Schuurman et al., 1995). Monotherapy with the protease inhibitor saquinavir also revealed substantial short-term clinical benefit, which was lost over the course of the 24 week observation period (Schapiro et al., 1996). This failure of HIV antiviral monotherapy is schematically depicted in Figure 2A.

In all of these examples (and other attempts of monotherapy), the cause of clinical failure was the emergence of virions with mutations that confer resistance to the administered agent (Richman et al., 1994; Schuurman et al., 1995; Zolopa et al., 1999) (Figure 2C). The biologic basis for this rapid emergence of resistance is the massive pool of HIV virions in an infected host, estimated to be 10^{10} new virions per day (Perelson et al., 1996). This pool generates a large number of possible mutant virions, which have a selective growth advantage during monotherapy.

These results rapidly led to the testing of combination regimens, first with dual-nucleoside analogs (Eron et al., 1995) and then with triple-drug regimens, including two nucleosides and a protease inhibitor, such as saquinavir or indinavir. Trials comparing AZT/3TC and AZT/3TC/indinavir revealed that the three-drug combination suppressed HIV-1 viral load more potently than the two-drug combination and substantially lowered the number of HIV-related deaths (Gulick et al., 1997; Hammer et al., 1997). These results have been replicated multiple times with multiple different combination therapy regimens and have led to a highly evolved standard of care for HIV, infection consisting of three-drug combination therapy now known as HAART (highly active antiretroviral therapy). The key to this success is the more rapid and sustained decline in HIV burden achieved through HAART, thereby preventing the emergence of resistant virions (Figures 2A and 2C).

TB: Importance of Preventing Outgrowth of Drug-Resistant Subclones

TB is remarkable for its insidious onset and slow progression, which, in the preantibiotic era, was usually fatal. The first anti-TB drugs were paraminosalicylic acid (PAS) and streptomycin. Both drugs were initially used as monotherapy with significant clinical improvement over 1–2 months of therapy. This success was quickly followed by loss of clinical efficacy due to the emergence of streptomycin- or PAS-resistant TB strains (British Medical

Research Council, 1947, 1948, 1950; Lehmann, 1946). Next, a series of trials were performed comparing streptomycin and PAS monotherapy to streptomycin plus PAS combination therapy. Ultimately, the combination therapy proved superior over monotherapy. All three treatment strategies showed similar improvement at 6 months in terms of the resolution of fever, radiographic improvement, and weight gain, but the combination therapy was ultimately superior because it prevented the emergence of streptomycin resistance (Figures 2B and 2C) (British Medical Research Council, 1950).

Similar results emerged from trials with the current standard-of-care drug isoniazid (INH). INH monotherapy was initially as effective as streptomycin/PAS combination therapy but ultimately failed due to the emergence of INH-resistant strains (British Medical Research Council, 1950, 1952). These insights led to the current, curative three- to four-drug, 6–12 month regimens that are now the standard of care (Fox et al., 1999). The key concept underlying the success of antimicrobial TB therapy is that combination therapy is necessary to prevent the emergence of resistance. In contrast, the success of HIV combination therapy is partially based on a more rapid, synergistic suppression of the infectious agent and partially due to the suppression of resistance (Figure 2).

EXCEPTIONS: EXAMPLES OF SUCCESSFUL MONOTHERAPY

Acute Bacterial Infections

Antibiotic monotherapy for most acute bacterial infections is curative in the absence of pre-existing antimicrobial resistance. Despite substantial evidence for synergistic killing by antibiotic combinations in vitro, definitive evidence supporting the clinical benefit of combination therapy for acute-onset infections, such as bacterial pneumonia or urinary tract infection, is generally lacking (Del Favero et al., 2001). It is not known why antibiotic monotherapy is so effective for most acute infections, but we speculate that the magnitude and kinetics of decline in bacterial burden are sufficiently steep to prevent the emergence of resistance and therefore obviate the need for combination therapy. In contrast, more slowly progressive bacterial infections, such as enterococcal endocarditis, require multiple antibiotics for elimination. It is also likely that the host immune system plays a role in eliminating minimal residual disease because patients with reduced neutrophil counts (e.g., from chemotherapy treatment) often require more prolonged antibiotic therapy and generally recover quickly once the neutropenia resolves.

Hepatitis B

Hepatitis B is a chronic viral infection of hepatocytes that eventually can lead to cirrhosis and hepatocellular carcinoma. Nucleoside analogs, such as entecavir, lamivudine, and tenofovir (the latter two are also used as antiretroviral therapy for HIV), have similar activities against the hepatitis B DNA virus due to the RT activity of the viral polymerase (which copies the pregenomic RNA into DNA). Monotherapy with each of these drugs suppresses viral burden substantially but, surprisingly, is not accompanied by rapid emergence of drug-resistant hepatitis B virions. The incidence of lamivudine-resistant hepatitis B is 20% after 1 year of therapy, and almost no resistance is observed with the newer agents tenofovir or entecavir, with 1 and 5 years of follow up, respectively (Chang et al., 2006; Lok and McMahon, 2009; Marcellin et al., 2008). Notably, the decline in viral burden with hepatitis B monotherapy occurs over months, at a pace comparable to monotherapy for HIV or TB (Figure 2A), indicating that the clinical efficacy of hepatitis B monotherapy is based on the fact that drug-resistant virions rarely emerge. The biologic basis for this serendipitous lack of resistance in hepatitis B is unknown because the same drugs targeting a similar enzyme (i.e., RT) rapidly generate resistant HIV when used as monotherapy (Figure 2C).

Chronic Myeloid Leukemia

As discussed earlier, resistance to monotherapy with ABL kinase inhibitors occurs, but the relapse rate is quite low (~4%–5% in the first year) if treatment is initiated in early chronic phase. By comparison, more than half of patients with most other types of tumors (e.g., lung cancer and melanoma) treated with kinase inhibitors relapse after 1 year. Longer follow up of CML patients receiving imatinib monotherapy has shown that the risk of relapse declines after 5 years and may plateau (Hochhaus et al., 2009), suggesting that many patients with CML can be successfully managed indefinitely with monotherapy. The rate of decline in CML disease burden over the initial 3–6 months of treatment, as measured by PCR for BCR-ABL mRNA transcripts in the blood, is a critical determinant of the durability of response. Specifically, CML patients who obtain a three log reduction in tumor burden on imatinib have a low risk of subsequent relapse (Hughes et al., 2010). Early data with the second-generation inhibitors dasatinib and nilotinib indicate that an even greater proportion of CML patients reach this endpoint (Kantarjian et al., 2010; Saglio et al., 2010). Similar to the importance of achieving a rapid and substantial decline in viral load with HIV therapy, the success of CML monotherapy is most likely due to the dramatic reduction in tumor burden. Extrapolating to lung cancer and melanoma, it may be possible to achieve similar long-term success if we focus our efforts on obtaining deeper reductions in tumor burden using more potent agents and relevant combinations.

PRECISION DIAGNOSTICS: ESSENTIAL FOR TARGETED CANCER DRUGS, MISSED OPPORTUNITY FOR ANTIBACTERIALS

The advent of targeted cancer therapies that are active against a subset of tumors with specific mutations mandated the parallel development of diagnostic tests that can determine the presence of these mutations in clinically relevant timeframes. Examples include the assessment of Her2/Neu gene amplification in breast cancer to identify women who would benefit from trastuzumab; EGFR mutation in lung cancer for sensitivity to erlotinib or gefitinib; BRAF mutation in melanoma for response to vemurafenib (PLX4720); and others. This priority represents a move away from broadly targeted cytotoxic therapies in which tumor/patient-specific molecular diagnostics have not been used to make treatment decisions.

A similar logic could easily guide precision diagnostics of suspected infections to determine the presence of a specific pathogen, rapidly define its drug susceptibility profile, and guide therapy. Examples of this strategy that have gained some traction are (1) CCR5 tropic HIV and the use of maraviroc, a CCR5 antagonist entry inhibitor (Gulick et al., 2008); (2) PCR-based determination of rifampin resistance in TB (Boehme et al., 2010); and (3) HIV genotypic resistance testing, in which the susceptibility of a patient's HIV virus can be predicted based on protease, RT, or integrase mutations present.

Despite these advances, similar diagnostics are either not available or not in widespread use for the most common infectious diseases, largely because of the false comfort in giving broad-spectrum antimicrobials that are relatively nontoxic and safe for any individual patient (Casadevall, 2009). This has led to complacency and blunted the push for specific diagnostics that would allow (1) administration of narrow-spectrum antimicrobials; (2) early recognition of antibiotic resistance; and (iii) differentiation of infections from noninfectious (or viral) diseases with similar clinical presentations. Consequently, we have a situation of antibiotic overuse, which compromises public health by selection for drug-resistant organisms (particularly in hospitals), increases the frequency of life-threatening side effects, such as antibiotic-associated colitis, and results in inefficient and costly heath care expenditures. We note that the problem of resistance as a public health hazard is unique to infectious diseases because the resistant organism is transmissible, and therefore, resistance in one patient can affect the population as a whole. In contrast, although tumor cell resistance is detrimental to the individual patient, it does not harm others who may be afflicted with the same cancer.

CONCLUSIONS AND CHALLENGES

Except in rare examples, such as CML, there is little doubt that combination therapy will be required for sustained clinical benefit of targeted agents against cancer. The experience with infectious diseases highlights the importance of combinations that achieve rapid, efficient cancer suppression (as in HIV) and prevent the emergence of resistance (as in HIV and TB). Current paradigms of oncology drug development are not aligned in ways that allow these goals to be pursued efficiently. Instead, business and regulatory incentives drive commercial drug developers to seek approval for monotherapy indications before embarking on combination studies. Based on the short duration of response to many recently approved anticancer agents, coupled with the precedent of failed monotherapy in chronic infectious diseases, one might ask whether the obligatory phase of monotherapy approval should be bypassed. The FDA recently drafted new guidelines for developing two or more investigational drugs in combination to encourage drug makers to adopt a combination rather than monotherapy strategy for registration (FDA, 2010), but it is too early to assess the impact of these changes. Our perspective as academic physician-scientists with experience in drug development is that additional measures are required to incentivize commercial sponsors to move to combination therapy trials early in drug development. One powerful motivation might be patent life extension, as has been successfully implemented for pediatric indications, particularly when the combination requires collaboration between two different drug makers. Activism on the part of physicians and patients may also be essential. Clinical investigators might collectively "demand" early combination studies from sponsors by refusing to conduct single-agent studies until firm commitments to combination trials are in place. Each of these strategies requires extreme coordination of goals among relevant stakeholders. Early efforts in the TB community to streamline the path of new drugs into combination TB regimens could provide a template for similar efforts in anticancer therapy (Critical Path Institute, 2010).

Another variable that could delay development of combination cancer therapy regimens is the paucity of data on coadministration of two or more targeted agents. In contrast to infectious diseases, in which most drugs are specific for the pathogen, targeted cancer agents typically impact normal and tumor cells, which could lead to unacceptable side effects. The fact that most targeted cancer therapies have highly favorable toxicity profiles relative to cytotoxic chemotherapy is cause for optimism. However, the current focus on monotherapy could complicate the investigation of tolerable combinations because dose and schedule are selected exclusively based on tolerability of monotherapy. Combination studies of targeted therapies typically begin by evaluating tolerability of each monotherapy regimen given simultaneously. If additive toxicities are observed (as might be expected), the combination might be abandoned prematurely. If early development decisions were, instead, driven by strategies that plan for combination therapy rather than monotherapy, clinical safety and efficacy could be assessed more quickly.

Another challenge is the need for better technologies to quantify disease burden over a wide range, as is possible for viral load assessment in HIV and hepatitis B. The rapid approval of ABL kinase inhibitors in CML was enabled, in part, by the availability of highly sensitive and quantitative assays of tumor burden (cytogenetics, PCR) that detect disease well below the threshold of traditional clinical response. Similar assays for solid tumors do not exist today, but proof of concept for quantitative assessment of circulating tumor DNA levels in blood has been established for several types of tumors (Leary et al., 2010). Once in place, these assays might also serve as early endpoints for drug approval, as they did for HIV, hepatitis B, and CML.

Curative regimens must also overcome the problem of subclinical reservoirs of persistent disease. Experimental strategies to define the molecular basis for persistence of cancer cells or microbes deserve more focused attention because the insights gained have the potential to eliminate drug-tolerant cells and thereby avoid prolonged treatment regimens, which are inevitably compromised by chronic toxicity and noncompliance. Recent data in cancer suggest that the immune system could be harnessed to eliminate minimal residual disease. Stimulation of host T lymphocytes by the anti-CTLA4 antibody ipilimumab induces tumor regression and prolongs survival in patients with metastatic melanoma (Robert et al., 2011). Remarkably, some patients remain in remission for years after therapy ceases and may be cured. This long-term success could be explained by the elimination of drug-tolerant persister cells by immune effector cells or by ongoing antitumor immunity that prevents expansion of tumor cells that persist indefinitely. There is clinical precedent for both of these mechanisms from allogeneic marrow transplantation data in CML in which depletion or infusion of donor T lymphocytes can profoundly impact treatment response (Mackinnon, 1997).

Finally, with more and more evidence supporting the potential value of genome-based medicine, the escalating problem of antimicrobial resistance should motivate the infectious diseases community to strive for more precise diagnostics that would allow more specific or limited use of antimicrobial agents. The advent of these technologies in targeted cancer therapy provides a template that the infectious diseases community should leverage.

ACKNOWLEDGMENTS

The authors apologize to those whose work could not be cited due to space limitations. Work in the Sawyers laboratory is supported by the Howard Hughes Medical Institute and grants from NCI and Stand Up to Cancer. Work in the Glickman laboratory is supported by the Geoffrey Beene Cancer Research Center, the Starr Cancer Consortium, and NIAID.

REFERENCES

Adams, K.N., Takaki, K., Connolly, L.E., Wiedenhoft, H., Winglee, K., Humbert, O., Edelstein, P.H., Cosma, C.L., and Ramakrishnan, L. (2011). Drug tolerance in replicating mycobacteria mediated by a macrophage-induced efflux mechanism. Cell *145*, 39–53.

Allan, E.J., Hoischen, C., and Gumpert, J. (2009). Bacterial L-forms. Adv. Appl. Microbiol. *68*, 1–39.

Altamirano, M., Marostenmaki, J., Wong, A., FitzGerald, M., Black, W.A., and Smith, J.A. (1994). Mutations in the catalase-peroxidase gene from isoniazid-resistant *Mycobacterium tuberculosis* isolates. J. Infect. Dis. *169*, 1162–1165.

Antonescu, C.R., Besmer, P., Guo, T., Arkun, K., Hom, G., Koryotowski, B., Leversha, M.A., Jeffrey, P.D., Desantis, D., Singer, S., et al. (2005). Acquired resistance to imatinib in gastrointestinal stromal tumor occurs through secondary gene mutation. Clin. Cancer Res. *11*, 4182–4190.

Bean, J., Brennan, C., Shih, J.Y., Riely, G., Viale, A., Wang, L., Chitale, D., Motoi, N., Szoke, J., Broderick, S., et al. (2007). MET amplification occurs with or without T790M mutations in EGFR mutant lung tumors with acquired resistance to gefitinib or erlotinib. Proc. Natl. Acad. Sci. USA *104*, 20932–20937.

Blanco, J.L., Varghese, V., Rhee, S.Y., Gatell, J.M., and Shafer, R.W. (2011). HIV-1 integrase inhibitor resistance and its clinical implications. J. Infect. Dis. *203*, 1204–1214.

Boehme, C.C., Nabeta, P., Hillemann, D., Nicol, M.P., Shenai, S., Krapp, F., Allen, J., Tahirli, R., Blakemore, R., Rustomjee, R., et al. (2010). Rapid molecular detection of tuberculosis and rifampin resistance. N. Engl. J. Med. *363*, 1005–1015.

Borges, S., Desta, Z., Li, L., Skaar, T.C., Ward, B.A., Nguyen, A., Jin, Y., Storniolo, A.M., Nikoloff, D.M., Wu, L., et al. (2006). Quantitative effect of CYP2D6 genotype and inhibitors on tamoxifen metabolism: implication for optimization of breast cancer treatment. Clin. Pharmacol. Ther. *80*, 61–74.

British Medical Research Council. (1947). EFFECT of streptomycin upon pulmonary tuberculosis; preliminary report of a cooperative study of 223 patients by the Army, Navy and Veterans Administration. Am. Rev. Tuberc. *56*, 485–507.

British Medical Research Council. (1948). Streptomycin treatment of pulmonary tuberculosis. BMJ *2*, 769–782.

British Medical Research Council. (1950). Treatment of pulmonary tuberculosis with streptomycin and para-aminosalicylic acid; a Medical Research Council investigation. BMJ *2*, 1073–1085.

British Medical Research Council. (1952). Treatment of pulmonary tuberculosis with isoniazid; an interim report to the Medical Research Council by their Tuberculosis Chemotherapy Trials Committee. BMJ *2*, 735–746.

Burgess, M.R., Skaggs, B.J., Shah, N.P., Lee, F.Y., and Sawyers, C.L. (2005). Comparative analysis of two clinically active BCR-ABL kinase inhibitors reveals the role of conformation-specific binding in resistance. Proc. Natl. Acad. Sci. USA *102*, 3395–3400.

Campbell, E.A., Korzheva, N., Mustaev, A., Murakami, K., Nair, S., Goldfarb, A., and Darst, S.A. (2001). Structural mechanism for rifampicin inhibition of bacterial rna polymerase. Cell *104*, 901–912.

Carver, B.S., Chapinski, C., Wongvipat, J., Hieronymus, H., Chen, Y., Chandarlapaty, S., Arora, V.K., Le, C., Koutcher, J., Scher, H., et al. (2011). Reciprocal feedback regulation of PI3K and androgen receptor signaling in PTEN-deficient prostate cancer. Cancer Cell 19, 575–586.

Casadevall, A. (2009). The case for pathogen-specific therapy. Expert Opin. Pharmacother. 10, 1699–1703.

Chandarlapaty, S., Sawai, A., Scaltriti, M., Rodrik-Outmezguine, V., Grbovic-Huezo, O., Serra, V., Majumder, P.K., Baselga, J., and Rosen, N. (2011). AKT inhibition relieves feedback suppression of receptor tyrosine kinase expression and activity. Cancer Cell 19, 58–71.

Chang, T.T., Gish, R.G., de Man, R., Gadano, A., Sollano, J., Chao, Y.C., Lok, A.S., Han, K.H., Goodman, Z., Zhu, J.BEHoLD AI463022 Study Group. (, et al. (2006). A comparison of entecavir and lamivudine for HBeAg-positive chronic hepatitis B. N. Engl. J. Med. 354, 1001–1010.

Chen, C.D., Welsbie, D.S., Tran, C., Baek, S.H., Chen, R., Vessella, R., Rosenfeld, M.G., and Sawyers, C.L. (2004). Molecular determinants of resistance to antiandrogen therapy. Nat. Med. 10, 33–39.

Choi, Y.L., Soda, M., Yamashita, Y., Ueno, T., Takashima, J., Nakajima, T., Yatabe, Y., Takeuchi, K., Hamada, T., Haruta, H.ALK Lung Cancer Study Group. (, et al. (2010). EML4-ALK mutations in lung cancer that confer resistance to ALK inhibitors. N. Engl. J. Med. 363, 1734–1739.

Clatworthy, A.E., Pierson, E., and Hung, D.T. (2007). Targeting virulence: a new paradigm for antimicrobial therapy. Nat. Chem. Biol. 3, 541–548.

Connolly, L.E., Edelstein, P.H., and Ramakrishnan, L. (2007). Why is long-term therapy required to cure tuberculosis? PLoS Med. 4, e120.

Cools, J., DeAngelo, D.J., Gotlib, J., Stover, E.H., Legare, R.D., Cortes, J., Kutok, J., Clark, J., Galinsky, I., Griffin, J.D., et al. (2003). A tyrosine kinase created by fusion of the PDGFRA and FIP1L1 genes as a therapeutic target of imatinib in idiopathic hypereosinophilic syndrome. N. Engl. J. Med. 348, 1201–1214.

Cornaglia, G., Giamarellou, H., and Rossolini, G.M. (2011). Metallo-β-lactamases: a last frontier for β-lactams? Lancet Infect. Dis. 11, 381–393.

Critical Path Institute. (2010). Critical path to TB drug regimens initiative. http://c-path.org/pdf/CPTRWorkScope2010.pdf.

de Bono, J.S., Logothetis, C.J., Molina, A., Fizazi, K., North, S., Chu, L., Chi, K.N., Jones, R.J., Goodman, O.B., Jr., Saad, F.COU-AA-301 Investigators. (, et al. (2011). Abiraterone and increased survival in metastatic prostate cancer. N. Engl. J. Med. 364, 1995–2005.

Del Favero, A., Menichetti, F., Martino, P., Bucaneve, G., Micozzi, A., Gentile, G., Furno, P., Russo, D., D'Antonio, D., Ricci, P.Gruppo Italiano Malattie Ematologiche dell'Adulto (GIMEMA) Infection Program. (, et al. (2001). A multicenter, double-blind, placebo-controlled trial comparing piperacillin-tazobactam with and without amikacin as empiric therapy for febrile neutropenia. Clin. Infect. Dis. 33, 1295–1301.

DiPersio, J.F., Micallef, I.N., Stiff, P.J., Bolwell, B.J., Maziarz, R.T., Jacobsen, E., Nademanee, A., McCarty, J., Bridger, G., Calandra, G.3101 Investigators. (2009). Phase III prospective randomized double-blind placebo-controlled trial of plerixafor plus granulocyte colony-stimulating factor compared with placebo plus granulocyte colony-stimulating factor for autologous stem-cell mobilization and transplantation for patients with non-Hodgkin's lymphoma. J. Clin. Oncol. 27, 4767–4773.

Engelman, J.A., Zejnullahu, K., Mitsudomi, T., Song, Y., Hyland, C., Park, J.O., Lindeman, N., Gale, C.M., Zhao, X., Christensen, J., et al. (2007). MET amplification leads to gefitinib resistance in lung cancer by activating ERBB3 signaling. Science 316, 1039–1043.

Eron, J.J., Benoit, S.L., Jemsek, J., MacArthur, R.D., Santana, J., Quinn, J.B., Kuritzkes, D.R., Fallon, M.A., Rubin, M.North American HIV Working Party. (1995). Treatment with lamivudine, zidovudine, or both in HIV-positive patients with 200 to 500 CD4+ cells per cubic millimeter. N. Engl. J. Med. *333*, 1662–1669.

FDA. (2010). Guidance for Industry: Codevelopment of Two or More Unmarketed Investigational Drugs for Use in Combination, U.S.D.o.H.a.H. Services, F.a.D. Administration, and C.f.D.E.a.R. (CDER), eds. http://www.fda.gov/downloads/Drugs/Guidance ComplianceRegulatoryInformation/Guidances/UCM236669.pdf.

Fischl, M.A., Richman, D.D., Grieco, M.H., Gottlieb, M.S., Volberding, P.A., Laskin, O.L., Leedom, J.M., Groopman, J.E., Mildvan, D., Schooley, R.T., et al. (1987). The efficacy of azidothymidine (AZT) in the treatment of patients with AIDS and AIDS-related complex. A double-blind, placebo-controlled trial. N. Engl. J. Med. *317*, 185–191.

Fox, W., Ellard, G.A., and Mitchison, D.A. (1999). Studies on the treatment of tuberculosis undertaken by the British Medical Research Council tuberculosis units, 1946-1986, with relevant subsequent publications. Int. J. Tuberc. Lung Dis. *3*(10, Suppl 2), S231–S279.

Godin-Heymann, N., Bryant, I., Rivera, M.N., Ulkus, L., Bell, D.W., Riese, D.J., 2nd, Settleman, J., and Haber, D.A. (2007). Oncogenic activity of epidermal growth factor receptor kinase mutant alleles is enhanced by the T790M drug resistance mutation. Cancer Res. *67*, 7319–7326.

Gorre, M.E., Mohammed, M., Ellwood, K., Hsu, N., Paquette, R., Rao, P.N., and Sawyers, C.L. (2001). Clinical resistance to STI-571 cancer therapy caused by BCR-ABL gene mutation or amplification. Science *293*, 876–880.

Guise, T. (2010). Examining the metastatic niche: targeting the microenvironment. Semin. Oncol. *37*(Suppl 2), S2–S14.

Gulick, R.M., Mellors, J.W., Havlir, D., Eron, J.J., Gonzalez, C., McMahon, D., Richman, D.D., Valentine, F.T., Jonas, L., Meibohm, A., et al. (1997). Treatment with indinavir, zidovudine, and lamivudine in adults with human immunodeficiency virus infection and prior antiretroviral therapy. N. Engl. J. Med. *337*, 734–739.

Gulick, R.M., Lalezari, J., Goodrich, J., Clumeck, N., DeJesus, E., Horban, A., Nadler, J., Clotet, B., Karlsson, A., Wohlfeiler, M.MOTIVATE Study Teams. (, et al. (2008). Maraviroc for previously treated patients with R5 HIV-1 infection. N. Engl. J. Med. *359*, 1429–1441.

Gupta, P.B., Onder, T.T., Jiang, G., Tao, K., Kuperwasser, C., Weinberg, R.A., and Lander, E.S. (2009). Identification of selective inhibitors of cancer stem cells by high-throughput screening. Cell *138*, 645–659.

Hammer, S.M., Squires, K.E., Hughes, M.D., Grimes, J.M., Demeter, L.M., Currier, J.S., Eron, J.J., Jr., Feinberg, J.E., Balfour, H.H., Jr., Deyton, L.R., et al. (1997). A controlled trial of two nucleoside analogues plus indinavir in persons with human immunodeficiency virus infection and CD4 cell counts of 200 per cubic millimeter or less. AIDS Clinical Trials Group 320 Study Team. N. Engl. J. Med. *337*, 725–733.

Hochhaus, A., O'Brien, S.G., Guilhot, F., Druker, B.J., Branford, S., Foroni, L., Goldman, J.M., Müller, M.C., Radich, J.P., Rudoltz, M.IRIS Investigators. (, et al. (2009). Six-year follow-up of patients receiving imatinib for the first-line treatment of chronic myeloid leukemia. Leukemia *23*, 1054–1061.

Hughes, T.P., Hochhaus, A., Branford, S., Müller, M.C., Kaeda, J.S., Foroni, L., Druker, B.J., Guilhot, F., Larson, R.A., O'Brien, S.G.IRIS investigators. (, et al. (2010). Long-term prognostic significance of early molecular response to imatinib in newly diagnosed chronic myeloid leukemia: an analysis from the International Randomized Study of Interferon and STI571 (IRIS). Blood *116*, 3758–3765.

Kantarjian, H., Shah, N.P., Hochhaus, A., Cortes, J., Shah, S., Ayala, M., Moiraghi, B., Shen, Z., Mayer, J., Pasquini, R., et al. (2010). Dasatinib versus imatinib in newly diagnosed chronic-phase chronic myeloid leukemia. N. Engl. J. Med. *362*, 2260–2270.

Kantarjian, H.M., Giles, F., Gattermann, N., Bhalla, K., Alimena, G., Palandri, F., Ossenkoppele, G.J., Nicolini, F.E., O'Brien, S.G., Litzow, M., et al. (2007). Nilotinib (formerly AMN107), a highly selective BCR-ABL tyrosine kinase inhibitor, is effective in patients with Philadelphia chromosome-positive chronic myelogenous leukemia in chronic phase following imatinib resistance and intolerance. Blood *110*, 3540–3546.

Lander, A.D. (2011). Pattern, growth, and control. Cell *144*, 955–969.

Leary, R.J., Kinde, I., Diehl, F., Schmidt, K., Clouser, C., Duncan, C., Antipova, A., Lee, C., McKernan, K., De La Vega, F.M., et al. (2010). Development of personalized tumor biomarkers using massively parallel sequencing. Sci. Transl. Med. *2*, 20ra14.

Lehmann, J. (1946). Para-aminosalicylic acid in the treatment of tuberculosis. Lancet *1*, 15–16.

Lewis, K. (2010). Persister cells. Annu. Rev. Microbiol. *64*, 357–372.

Lok, A.S., and McMahon, B.J. (2009). Chronic hepatitis B: update 2009. Hepatology *50*, 661–662.

Mackinnon, S. (1997). Donor leukocyte infusions. Baillieres Clin. Haematol. *10*, 357–367.

Maheswaran, S., Sequist, L.V., Nagrath, S., Ulkus, L., Brannigan, B., Collura, C.V., Inserra, E., Diederichs, S., Iafrate, A.J., Bell, D.W., et al. (2008). Detection of mutations in EGFR in circulating lung-cancer cells. N. Engl. J. Med. *359*, 366–377.

Maisonneuve, E., Shakespeare, L.J., Jørgensen, M.G., and Gerdes, K. (2011). Bacterial persistence by RNA endonucleases. Proc. Natl. Acad. Sci. USA *108*, 13206–13211.

Marcellin, P., Heathcote, E.J., Buti, M., Gane, E., de Man, R.A., Krastev, Z., Germanidis, G., Lee, S.S., Flisiak, R., Kaita, K., et al. (2008). Tenofovir disoproxil fumarate versus adefovir dipivoxil for chronic hepatitis B. N. Engl. J. Med. *359*, 2442–2455.

Martinez-Picado, J., and Martínez, M.A. (2008). HIV-1 reverse transcriptase inhibitor resistance mutations and fitness: a view from the clinic and ex vivo. Virus Res. *134*, 104–123.

Maskell, R., Okubadejo, O.A., Payne, R.H., and Pead, L. (1978). Human infections with thymine-requiring bacteria. J. Med. Microbiol. *11*, 33–45.

Moyed, H.S., and Bertrand, K.P. (1983). *hipA*, a newly recognized gene of *Escherichia coli* K-12 that affects frequency of persistence after inhibition of murein synthesis. J. Bacteriol. *155*, 768–775.

Nervi, B., Ramirez, P., Rettig, M.P., Uy, G.L., Holt, M.S., Ritchey, J.K., Prior, J.L., Piwnica-Worms, D., Bridger, G., Ley, T.J., and DiPersio, J.F. (2009). Chemosensitization of acute myeloid leukemia (AML) following mobilization by the CXCR4 antagonist AMD3100. Blood *113*, 6206–6214.

O'Reilly, K.E., Rojo, F., She, Q.B., Solit, D., Mills, G.B., Smith, D., Lane, H., Hofmann, F., Hicklin, D.J., Ludwig, D.L., et al. (2006). mTOR inhibition induces upstream receptor tyrosine kinase signaling and activates Akt. Cancer Res. *66*, 1500–1508.

Pao, W., Miller, V.A., Politi, K.A., Riely, G.J., Somwar, R., Zakowski, M.F., Kris, M.G., and Varmus, H. (2005). Acquired resistance of lung adenocarcinomas to gefitinib or erlotinib is associated with a second mutation in the EGFR kinase domain. PLoS Med. *2*, e73.

Perelson, A.S., Neumann, A.U., Markowitz, M., Leonard, J.M., and Ho, D.D. (1996). HIV-1 dynamics in vivo: virion clearance rate, infected cell life-span, and viral generation time. Science *271*, 1582–1586.

Poulikakos, P.I., Persaud, Y., Janakiraman, M., Kong, X., Ng, C., Moriceau, G., Shi, H., Atefi, M., Titz, B., Gabay, M.T., et al. (2011). RAF inhibitor resistance is mediated by dimerization of aberrantly spliced BRAF(V600E). Nature *480*, 387–390.

Pratilas, C.A., Taylor, B.S., Ye, Q., Viale, A., Sander, C., Solit, D.B., and Rosen, N. (2009). (V600E)BRAF is associated with disabled feedback inhibition of RAF-MEK signaling and elevated transcriptional output of the pathway. Proc. Natl. Acad. Sci. USA *106*, 4519–4524.

Reya, T., Morrison, S.J., Clarke, M.F., and Weissman, I.L. (2001). Stem cells, cancer, and cancer stem cells. Nature *414*, 105–111.

Richman, D.D., Havlir, D., Corbeil, J., Looney, D., Ignacio, C., Spector, S.A., Sullivan, J., Cheeseman, S., Barringer, K., Pauletti, D., et al. (1994). Nevirapine resistance mutations of human immunodeficiency virus type 1 selected during therapy. J. Virol. *68*, 1660–1666.

Robert, C., Thomas, L., Bondarenko, I., O'Day, S., M D, J.W., Garbe, C., Lebbe, C., Baurain, J.F., Testori, A., Grob, J.J., et al. (2011). Ipilimumab plus dacarbazine for previously untreated metastatic melanoma. N. Engl. J. Med. *364*, 2517–2526.

Rossi, D.J., Jamieson, C.H., and Weissman, I.L. (2008). Stems cells and the pathways to aging and cancer. Cell *132*, 681–696.

Saglio, G., Kim, D.W., Issaragrisil, S., le Coutre, P., Etienne, G., Lobo, C., Pasquini, R., Clark, R.E., Hochhaus, A., Hughes, T.P.ENESTnd Investigators. (, et al. (2010). Nilotinib versus imatinib for newly diagnosed chronic myeloid leukemia. N. Engl. J. Med. *362*, 2251–2259.

Schapiro, J.M., Winters, M.A., Stewart, F., Efron, B., Norris, J., Kozal, M.J., and Merigan, T.C. (1996). The effect of high-dose saquinavir on viral load and CD4+ T-cell counts in HIV-infected patients. Ann. Intern. Med. *124*, 1039–1050.

Scher, H.I., and Sawyers, C.L. (2005). Biology of progressive, castration-resistant prostate cancer: directed therapies targeting the androgen-receptor signaling axis. J. Clin. Oncol. *23*, 8253–8261.

Scher, H.I., Beer, T.M., Higano, C.S., Anand, A., Taplin, M.E., Efstathiou, E., Rathkopf, D., Shelkey, J., Yu, E.Y., Alumkal, J.Prostate Cancer Foundation/Department of Defense Prostate Cancer Clinical Trials Consortium. (, et al. (2010). Antitumour activity of MDV3100 in castration-resistant prostate cancer: a phase 1-2 study. Lancet *375*, 1437–1446.

Schimke, R.T., Kaufman, R.J., Alt, F.W., and Kellems, R.F. (1978). Gene amplification and drug resistance in cultured murine cells. Science *202*, 1051–1055.

Schuurman, R., Nijhuis, M., van Leeuwen, R., Schipper, P., de Jong, D., Collis, P., Danner, S.A., Mulder, J., Loveday, C., Christopherson, C., et al. (1995). Rapid changes in human immunodeficiency virus type 1 RNA load and appearance of drug-resistant virus populations in persons treated with lamivudine (3TC). J. Infect. Dis. *171*, 1411–1419.

Scorpio, A., and Zhang, Y. (1996). Mutations in *pncA*, a gene encoding pyrazinamidase/nicotinamidase, cause resistance to the antituberculous drug pyrazinamide in tubercle bacillus. Nat. Med. *2*, 662–667.

Sequist, L.V., Waltman, B.A., Dias-Santagata, D., Digumarthy, S., Turke, A.B., Fidias, P., Bergethon, K., Shaw, A.T., Gettinger, S., Cosper, A.K., et al. (2011). Genotypic and histological evolution of lung cancers acquiring resistance to EGFR inhibitors. Sci. Transl. Med. *3*, 75ra26.

Shah, N.P., Nicoll, J.M., Nagar, B., Gorre, M.E., Paquette, R.L., Kuriyan, J., and Sawyers, C.L. (2002). Multiple BCR-ABL kinase domain mutations confer polyclonal resistance to the tyrosine kinase inhibitor imatinib (STI571) in chronic phase and blast crisis chronic myeloid leukemia. Cancer Cell *2*, 117–125.

Sharma, S.V., Lee, D.Y., Li, B., Quinlan, M.P., Takahashi, F., Maheswaran, S., McDermott, U., Azizian, N., Zou, L., Fischbach, M.A., et al. (2010). A chromatin-mediated reversible drug-tolerant state in cancer cell subpopulations. Cell *141*, 69–80.

Skaggs, B.J., Gorre, M.E., Ryvkin, A., Burgess, M.R., Xie, Y., Han, Y., Komisopoulou, E., Brown, L.M., Loo, J.A., Landaw, E.M., et al. (2006). Phosphorylation of the ATP-binding loop directs oncogenicity of drug-resistant BCR-ABL mutants. Proc. Natl. Acad. Sci. USA *103*, 19466–19471.

Steen, R., and Sköld, O. (1985). Plasmid-borne or chromosomally mediated resistance by Tn7 is the most common response to ubiquitous use of trimethoprim. Antimicrob. Agents Chemother. *27*, 933–937.

Talpaz, M., Shah, N.P., Kantarjian, H., Donato, N., Nicoll, J., Paquette, R., Cortes, J., O'Brien, S., Nicaise, C., Bleickardt, E., et al. (2006). Dasatinib in imatinib-resistant Philadelphia chromosome-positive leukemias. N. Engl. J. Med. *354*, 2531–2541.

Tran, C., Ouk, S., Clegg, N.J., Chen, Y., Watson, P.A., Arora, V., Wongvipat, J., Smith-Jones, P.M., Yoo, D., Kwon, A., et al. (2009). Development of a second-generation antiandrogen for treatment of advanced prostate cancer. Science *324*, 787–790.

Wolfson, J.S., Hooper, D.C., McHugh, G.L., Bozza, M.A., and Swartz, M.N. (1990). Mutants of *Escherichia coli* K-12 exhibiting reduced killing by both quinolone and beta-lactam antimicrobial agents. Antimicrob. Agents Chemother. *34*, 1938–1943.

Zapun, A., Contreras-Martel, C., and Vernet, T. (2008). Penicillin-binding proteins and beta-lactam resistance. FEMS Microbiol. Rev. *32*, 361–385.

Zhang, Y., Heym, B., Allen, B., Young, D., and Cole, S. (1992). The catalase-peroxidase gene and isoniazid resistance of *Mycobacterium tuberculosis*. Nature *358*, 591–593.

Zolopa, A.R., Shafer, R.W., Warford, A., Montoya, J.G., Hsu, P., Katzenstein, D., Merigan, T.C., and Efron, B. (1999). HIV-1 genotypic resistance patterns predict response to saquinavir-ritonavir therapy in patients in whom previous protease inhibitor therapy had failed. Ann. Intern. Med. *131*, 813–821.

ell

Antibody-Based Immunotherapy of Cancer

Louis M. Weiner[1],[*], Joseph C. Murray[2], Casey W. Shuptrine[2]

[1]Department of Oncology, Lombardi Comprehensive Cancer Center, Georgetown University Medical Center, 3800 Reservoir Road NW, Washington, DC 20007, USA, [2]Tumor Biology Training Program, Lombardi Comprehensive Cancer Center, Georgetown University Medical Center, 3800 Reservoir Road NW, Washington, DC 20007, USA

*Correspondence: weinerl@georgetown.edu

Cell, Vol. 148, No. 6, March 16, 2012 © 2012 Elsevier Inc.
http://dx.doi.org/10.1016/j.cell.2012.02.034

SUMMARY

By targeting surface antigens expressed on tumor cells, monoclonal antibodies have demonstrated efficacy as cancer therapeutics. Recent successful antibody-based strategies have focused on enhancing antitumor immune responses by targeting immune cells, irrespective of tumor antigens. We discuss these innovative strategies and propose how they will impact the future of antibody-based cancer therapy.

INTRODUCTION

Specific recognition and elimination of pathological organisms or malignant cells by antibodies were proposed over a century ago by Paul Ehrlich, who is credited for conceptualizing the "magic bullet" theory of targeted therapy. Over the past 30 years, antibody cancer therapeutics have been developed and used clinically in an effort to realize the potential of targeted therapy. The diversity of these targeted approaches reflects the versatility of antibodies as platforms for therapeutic development (Weiner et al., 2010).

Antibodies may target tumor cells by engaging surface antigens differentially expressed in cancers. For example, rituximab targets CD20 in non-Hodgkin B cell lymphoma, trastuzumab targets HER2 in breast cancer, and cetuximab targets EGFR in colorectal cancer (Table S1 available online). The antibodies can invoke tumor cell death by blocking ligand-receptor growth and survival pathways. In addition, innate immune effector mechanisms that engage the

107

CellPress

Fc portion of antibodies (Figure S1) via Fc receptors (FcR) are emerging as equally important (Jiang et al., 2011). The mechanisms include antibody-dependent cellular cytotoxicity (ADCC) and complement-mediated cytotoxicity (CMC); antibody-dependent cellular phagocytosis (ADCP) is likely relevant as well (Figure 1).

Although unconjugated antibodies have had efficacy, molecular genetics and chemical modifications to monoclonal antibodies (mAbs) have advanced their clinical utility. For example, modification of immune effector engagement has improved pharmacokinetic profiles, and conjugating cytotoxic agents to mAbs has enhanced targeted therapeutic delivery to tumors. The increasing facility of antibody modifications has made it possible to construct diverse and efficacious mAb-based therapeutics (Figure S1).

Structural engineering and alternative targets have also expanded the ability of mAbs to stimulate adaptive immune effectors, such as T cells, that can induce antitumor activity. Antibodies directly targeting receptors involved in checkpoint regulation of immune cells have exhibited preclinical and clinical successes. Ongoing studies also suggest that antibodies can indirectly elicit adaptive immunity through antibody-dependent engagement of immune effector mechanisms (Figure 1). Overall, the diverse effects of antibodies and their putative mechanisms of action suggest several exciting directions for developing therapeutic strategies. Some that have achieved recent success are discussed below.

MANIPULATING ANTIBODY STRUCTURE

The natural properties of antibodies that enable specific antigen engagement can be leveraged and improved upon by engineering approaches that increase antitumor activity. One example is the creation of bispecific antibodies (bsAbs) with dual affinities for a tumor antigen and either another tumor antigen or a target in the tumor microenvironment. As the Fc domain of mAbs does not directly activate T cells, the activating receptor for T cells, CD3, is a common target of bsAbs. Catumaxomab is a bsAb that binds the tumor antigen EpCAM, CD3, and innate effector cells through an intact Fc portion (Ruf and Lindhofer, 2001). This bsAb, termed a TriomAb, effectively kills tumor cells in vitro and in vivo and induces protective immunity, most likely through the induction of memory T cells. Catumaxomab's success in a phase II/III clinical trial led to its approval by the European Commission in 2009 for the treatment of malignant ascites. This success spurred the development of other TriomAbs targeted against the tumor antigens HER2/neu (ertumaxomab), CD20 (Bi20/FBTA05; NCT01138579 [see Web Resources section below for information on full urls]), GD2, and GD3 (Ektomun).

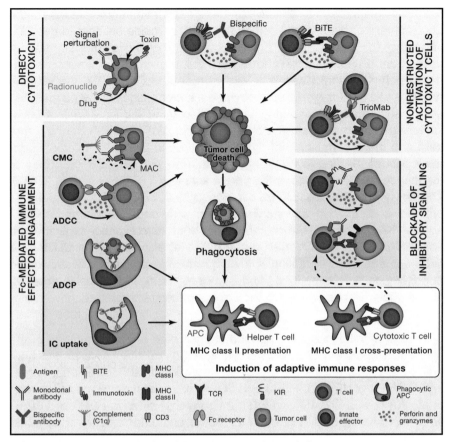

FIGURE 1 Mechanisms of Action of Antibody Immunotherapy in Cancer

Mechanisms of anticancer antibody therapies are diverse and represent the versatility of antibody-based approaches. Here, four different strategies are depicted. Upper left: direct cytotoxicity, in which mAbs can induce direct cytotoxicity in tumor cells by perturbing oncogenic signaling pathways or in which immunoconjugates can carry cytotoxic agents to targeted cells. Lower left: FcR-mediated immune effector engagement, in which the Fc portion of mAbs can engage immune effector functions, including soluble CMC (through the membrane attack complex MAC) as well as NK cells, macrophages, and dendritic cells, through FcRs, allowing for ADCC, ADCP, and IC uptake. Upper right: Nonrestricted activation of cytotoxic T cells, in which tumor-infiltrating CTLs can be activated against tumor cells—independent of T cell receptor (TCR) specificity—by engaging coreceptors on the T cells and tumor antigens. Lower right: blockade of inhibitory signaling, in which cytotoxic lymphocytes, including NK cells and CTLs, express inhibitory receptors for various ligands that may be expressed by tumor cells. Antagonistic antibodies that target these inhibitory receptors can block ligand-receptor interactions so that targeted cytotoxicity can ensue. These four strategies enhance tumor cell death, which can promote phagocytosis of tumor cell antigens, and induction of adaptive immune responses (bottom right) in two ways: MHC class I cross-presentation and priming of cytotoxic T cells and MHC class II presentation and priming of helper T cells. These adaptive immune responses can lead to enhanced—and possibly persistent—antitumor immunity.

A promising approach to directly stimulate T cell immunity with mAbs is the development of bispecific T cell engager (BiTE) molecules that target CD3 and either CD19, EpCAM, or EGFR. Low doses of BiTEs induce antitumor activity, and BiTEs have the added potential to overcome mutations in signaling pathways that classically lead to resistance. In BiTEs, the variable domains of a CD3-targeted antibody and a tumor antigen-targeted antibody are genetically linked, rendering it possible to activate a T cell when it physically engages a tumor cell (Lutterbuese et al., 2010). Lysis of bound tumor cells and the accumulation of cytotoxic T cells in the tumor microenvironment ensues, leading to tumor regression at in vivo doses three orders of magnitude less than those of the parent antibody (Lutterbuese et al., 2010). The newly characterized BiTEs directed against EGFR utilize the parental antibodies cetuximab and panitumumab, with potent antitumor abilities against KRAS- and BRAF-mutated cells that demonstrate resistance to conventional EGFR antibodies (Lutterbuese et al., 2010). The CD19-CD3 BiTE demonstrates significant clinical promise in patients with advanced non-Hodgkin lymphoma (NHL) and is currently being tested in six phase I/II clinical trials. The EpCAM-CD3 BiTE is in a phase I clinical trial.

An alternative method of creating bsAbs relies on the systematic analysis of binding affinities toward a second antigen after random mutation of the light-chain complementarity-determining regions (CDRs) of a parent antibody. Using this technique, bsAbs with two identical Fab regions, targeting VEGFA and HER2 or HER3 and EGFR, have been developed (Schaefer et al., 2011). MEHD7945A, an IgG1 antibody that binds to HER3 and EGFR with high affinity, exhibited equal or better antitumor efficacy than either parent antibody in 12 xenograft models (Schaefer et al., 2011). Although this method has theoretical utility for the development of bsAbs against any combination of two or more antigens, its potential for systematic applicability remains to be fully demonstrated.

The CovX-Body method is another recent technique for the rapid creation of bispecific antibodies (Doppalapudi et al., 2010). By fusing two peptide pharmacophores together and linking this complex to a universal scaffold antibody, a bispecific antibody with known Fc functions can be created. This structure, classified as a bispecific CovX-Body, is reproducible and specific and has the potential for widespread adoption. CVX-241, the first CovX-Body to enter clinical trials (NCT00911898), targets the angiogenesis ligands VEGFA and Ang2. Preclinically, CVX-241 exhibited moderate antitumor effects but, when combined with the chemotherapy agent irinotecan, significantly inhibited tumor growth (Doppalapudi et al., 2010). Another bsAb, MM-111, represents an alternative approach to bispecific engineering, based on linking the variable regions from two different antibodies. MM-111 targets HER2 and HER3 and is in phase I clinical trials. As the repertoire of cancer targets increases, dual targeting techniques may enhance clinical efficacy compared to traditional single-antigen targeting approaches.

STIMULATING PERSISTENT IMMUNITY

Generation of a persistent antitumor immune response is a prevailing goal of cancer immunotherapy. Antibody therapy can indirectly stimulate persistent responses against tumor-associated antigens through induction of adaptive immunity (Figure 1). Hence, therapeutic antibodies can act to promote vaccine-like antitumor effects. Tumor cell death can modulate antigen uptake, maturation, and presentation in antigen-presenting cells (APCs), which are critical for initiating adaptive immunity (Sauter et al., 2000). Beyond inducing tumor cell death by blocking survival pathways, therapeutic antibodies can also coat tumor cells and mark them for recognition by immune cells. APCs, such as dendritic cells (DCs) or macrophages, can phagocytose antibody-coated tumor cells. Antibodies bound to soluble antigens in immune complexes (ICs) can also induce uptake by APCs. Through these various mechanisms of tumor antigen uptake, tumor contents—not only the antibody-targeted antigen—can be processed and presented by major histocompatibility complexes (MHCs) to activate different adaptive immune responses. Antigens presented via MHC class II can prime helper (CD4+) T cell responses important for endogenous antibody (humoral) immunity. In addition, cross-presentation of antigen and MHC class I-restricted priming of cytotoxic (CD8+) T cell, or cytotoxic T lymphocyte (CTL), responses can occur (Dhodapkar et al., 2002). The induction of these T cell responses can enable immunological memory to the presented antigens, which is critical for long-term immunity.

The capacity of mAbs to induce tumor-directed CTL responses is intriguing. Intratumoral CTL composition and distribution have been associated with clinical outcomes (Galon et al., 2006), suggesting the relevance of T cells in antitumor immunity. Moreover, CTLs can target intracellular antigens that are thought to be inaccessible to antibody therapies. Therefore, antibody-initiated cross-presentation of tumor antigens can be exploited to induce adaptive immunity that may extend beyond the targeted antigen. This strategy has been described as the "vaccinal effect" in rituximab therapy of lymphoma (Hilchey et al., 2009) and has been shown to be relevant in antibody therapy of solid tumors. The antibody-dependent promotion of adaptive immunity remains an active and very promising area of research, as the induction of adaptive immunity can be accompanied by efforts to expand, shape, and prolong the host immune response. Because tumors may establish local and systemic immunosuppressive environments, concomitant efforts to neutralize immunosuppressive mechanisms may also amplify the vaccinal effect of mAbs.

MODULATING THE AMPLITUDE OF IMMUNE RESPONSES

Following activation, T cells upregulate the expression of inhibitory receptors, which protects against deleterious autoimmunity. This host-protective mechanism permits tumors to evade persistent immune control due to localized immune tolerance. This control is further manipulated by tumors through downregulation of surface immunogens or through the activation of diverse immune-suppressive mechanisms. This interplay between the immune system and tumor cells, termed immunoediting, allows tumors to escape immune elimination even when tumor-specific immunity is present (Schreiber et al., 2011).

Ipilimumab, an IgG1 mAb, antagonizes the inhibitory receptor CTLA-4, which is expressed on activated T cells. Treatment with this antibody in combination with dacarbazine correlates with a marked increase in overall survival and progression-free survival of previously untreated melanoma patients compared to treatment with dacarbazine alone (Robert et al., 2011). It is clear from the success of this landmark phase III trial that harnessing the activity of T cells will have a therapeutic antitumor benefit, even in the absence of a tumor antigen-targeted strategy. This clinical efficacy highlights the inherent ability of the immune system to recognize and eliminate abnormal self without the aid of tumor-specific therapies.

The PD-1 axis represents another promising immune checkpoint pathway to manipulate. PD-1 is expressed on activated T and B cells and provides a potent inhibitory signal when bound by ligand. Antibodies inhibiting the PD-1 checkpoint may reactivate T cells by blocking APC inhibition of T cells or by stopping tumor cells, which often overexpress PD-1 ligands, from inactivating T cells in the tumor microenvironment. In a phase I trial for refractory solid tumors, Brahmer et al. evaluated an IgG4 mAb, MDX-1106, that recognizes the extracellular domain of PD-1 with high affinity. The antibody was well tolerated in 38 of the 39 patients treated, resulting in three objective responses (two complete responses) and prompting its further evaluation in a phase II trial for clear-cell renal cell carcinoma (NCT01354431) (Brahmer et al., 2010).

Other approaches to enhancing T cell-specific immunity against tumor cells aim to activate stimulatory receptors, including 4-1BB, OX40, CD27, CD40, and DR3. Preclinical models have shown that the stimulation of 4-1BB, OX40, and CD27 leads to proliferation and cytokine production in T cells (Croft, 2009). Interestingly, activation of 4-1BB and OX40 can also inhibit the differentiation and proliferation of regulatory T cells, which contribute to tumor-derived immunosuppression (Croft, 2009). Therefore, agonistic antibodies to these TNF receptors potentially have dual and synergistic immune-promoting roles. Currently, two 4-1BB agonist antibodies are in phase I and II clinical trials for NHL and melanoma, respectively (NCT01307267, NCT00612664), and one agonistic OX40 antibody is being studied in a phase I/II and phase II trial for prostate cancer and melanoma (NCT01303705, NCT01416844).

Although the manipulation of T cells is currently a primary focus, other approaches may leverage the innate immune system. By blocking the function of inhibitory killer cell immunoglobulin-like receptors (KIRs), natural killer (NK) cells could elicit a tumor-specific cytotoxic response without harming normal self cells, as evidenced by the anti-KIR (KIR2DL1/L2/L3 and KIR2DS1/S2) antagonist antibody IPH2101, which is currently in numerous early phase clinical trials (NCT00999830). As a clinical validation of the immunoediting hypothesis, the manipulation of the immune system to elicit an antitumor response has the potential to serve as an efficacious treatment modality across all cancers.

IMMUNOCONJUGATES: TARGETING CYTOTOXIC AGENTS

Early efforts to enhance the antitumor effects of mAbs focused on boosting their direct cytotoxic effects on targeted cells. Conjugation of radionuclides (radioimmunotherapies, or RITs), drugs (antibody-drug conjugates, or ADCs), toxins (immunotoxins), and enzymes (antibody-directed enzyme prodrug therapy, or ADEPT) yielded a multitude of antibodies—or antibody-like molecules—with varying clinical efficacy. Three conjugated antibodies have translated into FDA-approved therapies, all for hematological malignancies. Two are RIT agents targeting CD20 and are indicated for treatment of relapsed and/or rituximab-refractory follicular or low-grade lymphomas: ^{90}Y-ibritumomab tiuxetan and ^{131}I-tositumomab. At least a dozen other RIT agents are in active development, including ten that target solid tumors (Steiner and Neri, 2011). The third approved immunoconjugate, brentuximab vedotin, is an ADC targeting CD30 and carrying the antimitotic drug monomethyl auristatin E. Brentuximab vedotin was recently approved for treatment of anaplastic large cell lymphoma (NCT00866047) and Hodgkin lymphoma (NCT00848926).

The limited translational success of immunoconjugates reflects the challenges of developing a highly cytotoxic agent with acceptable pharmacokinetics and toxicity. RITs can cause systemic toxicity, and many tumors, particularly solid tumors, are inaccessible or insensitive to deliverable doses of radiation. Although a premise of antibody immunotherapy is low toxicity imparted by specificity, "tumor-specific" antigens are often more selective than specific. Additionally, generation of neutralizing antibodies to recombinant immunoconjugates, such as immunotoxins, may limit their clinical utility, similar to the limitations associated with murine antibodies (Kreitman et al., 2009).

These challenges have not thwarted attempts to develop immunoconjugates. Another ADC, trastuzumab-MCC-DM1 (trastuzumab-DM1 or T-DM1), has shown promise in patients with HER2-positive metastatic breast cancer. A phase III trial comparing T-DM1 against capecitabine, a prodrug of 5-fluorouracil, plus lapatinib, a tyrosine kinase inhibitor, is underway (NCT00829166). The most clinically advanced immunotoxin, BL22, contains an anti-CD22 Fv bound to a modified *Pseudomonas* exotoxin. BL22 has shown significant promise in phase II trials for the treatment of hairy cell leukemia (Kreitman et al., 2009). Beyond direct tumor cell cytotoxicity, it is possible that cytotoxic immunoconjugates could induce antigen-targeted adaptive immune responses, though this requires additional study.

A LONGER-TERM VIEW

Activating FcRs involved in immune effector activities are important for antitumor effects (Clynes et al., 2000). However, two recent, independent reports have demonstrated that engagement of either activating and inhibitory FcRs (Wilson et al., 2011) or inhibitory FcR alone (FcgammaRIIB) (Li and Ravetch, 2011) can drive antitumor immune effects of agonistic antibodies targeting death receptor superfamily members. The compelling observation that inhibitory FcRs alone drive productive immune responses is unexpected. Thus, efforts to modify mAb structure to balance engagement of FcRs will remain a critical component of antibody development (Jiang et al., 2011). These findings highlight how antibody therapy can elucidate novel mechanisms of immune effector activity that may differ from fundamental understanding in immunology.

An even more basic tenet of antibody therapy is the concept of targetable antigens. The dogma has been that only soluble extracellular or cell-surface antigens are accessible targets for antibodies. Targeting intracellular antigens—particularly oncogenic or mutated cytosolic proteins specific to tumor cells—has been left to other membrane-permeable treatment modalities. However, cellular immunotherapy targeting endogenous intracellular antigens that are processed and presented at the tumor cell surface, as with the prostate cancer therapy Sipuleucel-T, has already demonstrated its worth.

Remarkably, recent findings suggest that we may have underestimated the capacity for antibodies to target intracellular antigens. Guo and colleagues have demonstrated that mAbs can effectively target intracellular antigens and inhibit tumor growth in mouse models (Guo et al., 2011). The capacity for antibodies to be internalized by tumor cells, thereby allowing for access to intracellular antigens, may explain this provocative observation. Targeting intracellular antigens would profoundly broaden antibody immunotherapy to include tumor-specific mutated intracellular proteins and other intracellular mediators of cell survival and proliferation. As this is a nascent area of research, it is only speculative that intracellular antigen targeting would provide sufficient antitumor effect to translate into clinical efficacy. In concert with targeted strategies that enhance antitumor immunity, even in the face of immune evasion, tolerance, and suppression, it is possible to envision a future where combinatorial antibody approaches transition into cancer immunotherapeutic strategies.

SUPPLEMENTAL INFORMATION

Supplemental Information includes one figure and one table and can be found with this article online at http://www.sciencedirect.com/science/MiamiMultiMediaURL/1-s2.0-S0092867412002346/1-s2.0-S0092867412002346-mmc1.pdf/272196/FULL/S0092867412002346/8512663b8350c3164fd7ad20c19b6f6f/mmc1.pdf.

WEB RESOURCES

Information about the referenced clinical trials can be found at the following url: http://clinicaltrials.gov/ct2/show/xxx, where xxx is the specific number/letter code referenced in the text.

ACKNOWLEDGMENTS

Supported by grants CA51008 and CA50633 from the National Cancer Institute (L.M.W.) and by T32-CA009686 (C.W.S.) from the National Cancer Institute.

REFERENCES

Brahmer, J.R., Drake, C.G., Wollner, I., Powderly, J.D., Picus, J., Sharfman, W.H., Stankevich, E., Pons, A., Salay, T.M., McMiller, T.L., et al. (2010). J. Clin. Oncol. 28, 3167–3175.

Clynes, R.A., Towers, T.L., Presta, L.G., and Ravetch, J.V. (2000). Nat. Med. 6, 443–446.

Croft, M. (2009). Nat. Rev. Immunol. 9, 271–285.

Dhodapkar, K.M., Krasovsky, J., Williamson, B., and Dhodapkar, M.V. (2002). J. Exp. Med. 195, 125–133.

Doppalapudi, V.R., Huang, J., Liu, D., Jin, P., Liu, B., Li, L., Desharnais, J., Hagen, C., Levin, N.J., Shields, M.J., et al. (2010). Proc. Natl. Acad. Sci. USA 107, 22611–22616.

Galon, J., Costes, A., Sanchez-Cabo, F., Kirilovsky, A., Mlecnik, B., Lagorce-Pagès, C., Tosolini, M., Camus, M., Berger, A., Wind, P., et al. (2006). Science 313, 1960–1964.

Guo, K., Li, J., Tang, J.P., Tan, C.P.B., Hong, C.W., Al-Aidaroos, A.Q.O., Varghese, L., Huang, C., and Zeng, Q. (2011). Sci. Transl. Med. 3, ra85.

Hilchey, S.P., Hyrien, O., Mosmann, T.R., Livingstone, A.M., Friedberg, J.W., Young, F., Fisher, R.I., Kelleher, R.J., Jr., Bankert, R.B., and Bernstein, S.H. (2009). Blood 113, 3809–3812.

Jiang, X.-R., Song, A., Bergelson, S., Arroll, T., Parekh, B., May, K., Chung, S., Strouse, R., Mire-Sluis, A., and Schenerman, M. (2011). Nat. Rev. Drug Discov. 10, 101–111.

Kreitman, R.J., Stetler-Stevenson, M., Margulies, I., Noel, P., Fitzgerald, D.J.P., Wilson, W.H., and Pastan, I. (2009). J. Clin. Oncol. 27, 2983–2990.

Li, F., and Ravetch, J.V. (2011). Science 333, 1030–1034.

Lutterbuese, R., Raum, T., Kischel, R., Hoffmann, P., Mangold, S., Rattel, B., Friedrich, M., Thomas, O., Lorenczewski, G., Rau, D., et al. (2010). Proc. Natl. Acad. Sci. USA 107, 12605–12610.

Robert, C., Thomas, L., Bondarenko, I., O'Day, S., M D, J.W., Garbe, C., Lebbe, C., Baurain, J.F., Testori, A., Grob, J.J., et al. (2011). N. Engl. J. Med. 364, 2517–2526.

Ruf, P., and Lindhofer, H. (2001). Blood 98, 2526–2534.

Sauter, B., Albert, M.L., Francisco, L., Larsson, M., Somersan, S., and Bhardwaj, N. (2000). J. Exp. Med. 191, 423–434.

Schaefer, G., Haber, L., Crocker, L.M., Shia, S., Shao, L., Dowbenko, D., Totpal, K., Wong, A., Lee, C.V., Stawicki, S., et al. (2011). Cancer Cell 20, 472–486.

Schreiber, R.D., Old, L.J., and Smyth, M.J. (2011). Science 331, 1565–1570.

Steiner, M., and Neri, D. (2011). Clin. Cancer Res. 17, 6406–6416.

Weiner, L.M., Surana, R., and Wang, S. (2010). Nat. Rev. Immunol. 10, 317–327.

Wilson, N.S., Yang, B., Yang, A., Loeser, S., Marsters, S., Lawrence, D., Li, Y., Pitti, R., Totpal, K., Yee, S., et al. (2011). Cancer Cell 19, 101–113.

ancer Cell

Macrophage Regulation of Tumor Responses to Anticancer Therapies

Michele De Palma[1,*], Claire E. Lewis[2,*]

[1]The Swiss Institute for Experimental Cancer Research (ISREC), School of Life Sciences, Swiss Federal Institute of Technology Lausanne (EPFL), CH-1015 Lausanne, Switzerland, [2]Department of Oncology, Sheffield Cancer Research Centre, University of Sheffield Medical School, Sheffield, S10 2RX, UK
*Correspondence: claire.lewis@sheffield.ac.uk (C.E.L.), michele.depalma@epfl.ch (M.D.P.)

Cancer Cell, Vol. 23, No. 3, March 18, 2013 © 2013 Elsevier Inc.
http://dx.doi.org/10.1016/j.ccr.2013.02.013

SUMMARY

Tumor-associated macrophages (TAMs) promote key processes in tumor progression, like angiogenesis, immunosuppression, invasion, and metastasis. Increasing studies have also shown that TAMs can either enhance or antagonize the antitumor efficacy of cytotoxic chemotherapy, cancer-cell targeting antibodies, and immunotherapeutic agents—depending on the type of treatment and tumor model. TAMs also drive reparative mechanisms in tumors after radiotherapy or treatment with vascular-targeting agents. Here, we discuss the biological significance and clinical implications of these findings, with an emphasis on novel approaches that effectively target TAMs to increase the efficacy of such therapies.

INTRODUCTION

Macrophages phagocytose microbes and present antigens to T cells, therefore constituting a first line of defense against invading pathogens. They also regulate tissue growth, homeostasis, repair, and remodeling via their expression of numerous cytokines, chemokines, growth factors, proteolytic enzymes, and scavenger receptors (Gordon and Martinez, 2010; Murray and Wynn, 2011). As such, macrophages play a central role in developmental processes, such as tissue morphogenesis and vascular and neuronal patterning, but also in pathophysiological responses, like inflammation and organ healing/regeneration (Mantovani et al., 2013; Nucera et al., 2011; Pollard, 2009).

117

CellPress

In selected organs of the adult mouse, the origin of tissue macrophages can be traced back to fetal macrophages that appear before the onset of definitive hematopoiesis (Schulz et al., 2012). In inflamed and remodeling tissues, elevated macrophage turnover is sustained largely from hematopoietic progenitor cells (HPCs), which proliferate and differentiate into promonocytes in the bone marrow (BM) before they are shed into the circulation as monocytes. These then undergo final differentiation into macrophages as they extravasate in the target tissues (Shi and Pamer, 2011). During inflammation and tumor growth, BM-derived HPCs may also accumulate at extramedullary sites, such as the spleen, which can become an important site of monocyte production (Cortez-Retamozo et al., 2012).

Once resident in tissues, macrophages acquire a distinct, tissue-specific phenotype in response to signals present within individual microenvironments. The exact combination of such tissue-specific cues dictates both the differentiation and activation status of these cells. Two extreme forms of the latter are generally referred to as "classical" (or M1) and "alternative" (or M2) activation, which parallel Th1/Th2 programming of adaptive immune cells (Biswas and Mantovani, 2010; Mantovani et al., 2002). During acute inflammation, macrophages are M1-activated by toll-like receptor (TLR) agonists and Th1 cytokines (e.g., interferon [IFN]-γ). This enhances their ability to kill and phagocytose pathogens, upregulate proinflammatory cytokines (e.g., interleukin [IL]-1β, IL-12, and tumor necrosis factor-α [TNF-α]) and reactive molecular species, and present antigens via major histocompatibility complex (MHC) class II molecules (Biswas and Mantovani, 2010; Mantovani et al., 2002). Alternatively, Th2 cytokines, like IL-4 and 13, stimulate monocytes/macrophages to express an M2 activation state. This is characterized by higher production of the anti-inflammatory cytokine, IL-10; lower expression of proinflammatory cytokines; amplification of metabolic pathways that can suppress adaptive immune responses; and the upregulation of cell-surface scavenger receptors, such as mannose receptor (MRC1/CD206) and hemoglobin/aptoglobin scavenger receptor (CD163). As such, M2 macrophage activation may facilitate the resolution of inflammation and promote tissue repair (including angiogenesis) after the acute inflammatory phase (Biswas and Mantovani, 2010; Gordon and Martinez, 2010). In healthy tissues, macrophages often express a mixed M1/M2 phenotype; hence "M1" and "M2" polarization should be regarded as extreme ends of a continuum of activation states, with their exact point on the scale depending on the precise mix of local signals present in a given microenvironment (Biswas and Mantovani, 2010; Lawrence and Natoli, 2011; Sica and Mantovani, 2012).

TUMOR-ASSOCIATED MACROPHAGES

Macrophages are a major cellular component of murine and human tumors, where they are commonly termed tumor-associated macrophages

(TAMs). In this article, we specifically review the role of these cells and their monocyte precursors in tumor responses to anticancer therapies. Other tumor-infiltrating myeloid cells not discussed here include neutrophils, eosinophils, and activated dendritic cells (DCs) (de Visser et al., 2006). Tumors also recruit a variety of immature myeloid cells, often referred to as myeloid-derived suppressor cells (MDSCs), which comprise precursors of both the monocyte-DC (mononuclear) and neutrophil (granulocytic) lineages and are commonly identified by their expression of Gr1 (Ly6C/G) and immunosuppressive activity. Mononuclear MDSCs can further mature into TAMs (Coffelt et al., 2010; Gabrilovich et al., 2012). Finally, there is also evidence for hematopoietic and myeloid progenitor cells homing to tumors and modulating tumor progression (Shaked and Voest, 2009).

Various mouse studies have shown that monocytes are recruited into tumors in large numbers by chemokines secreted by both malignant and stromal cells. These include chemokine (C-C motif) ligand 2 (CCL2, or MCP1), colony-stimulating factor-1 (CSF1), and chemokine (C-X-C motif) ligand 12 (CXCL12, or SDF1) (Murdoch et al., 2008). Upon monocyte differentiation into TAMs, these cells act as a source of local and systemic cues to support the proliferation, survival, and motility of the cancer cells; tumor vascularization (angiogenesis); suppression of antitumor immunity; and intravasation of cancer cells at the primary tumor site and extravasation/growth at distant metastatic sites (Bingle et al., 2006; De Palma et al., 2003; DeNardo et al., 2009; Lewis and Pollard, 2006; Lin et al., 2001; Qian et al., 2011; Qian and Pollard, 2010; Ruffell et al., 2012a; Squadrito and De Palma, 2011; Wyckoff et al., 2004). This impressive array of tumor-promoting functions is consistent with clinical studies showing high macrophage density in many human cancer types to be associated with increased tumor angiogenesis and metastasis, and/or a poor prognosis (Bingle et al., 2002; Clear et al., 2010; Heusinkveld and van der Burg, 2011; Leek et al., 1996). Furthermore, enrichment of a macrophage-related gene signature correlates with reduced survival in some types of human cancer (Engler et al., 2012; Steidl et al., 2010).

A decade ago, it was proposed that TAMs are predominantly polarized in the tumor microenvironment toward an M2-like phenotype and that this underlies their ability to promote the growth and vascularization of tumors (Mantovani et al., 2002). This is also supported by clinical studies showing the predictive value of M2-macrophage associated markers, like CD163 (Heusinkveld and van der Burg, 2011). Flow cytometry and gene expression profiling of mouse and human TAMs has shown that distinct macrophage subpopulations with a variably skewed M2-like phenotype coexist in tumors and that their relative abundance varies with the tumor type (Movahedi et al., 2010; Pucci et al., 2009; Ruffell et al., 2012b). Such complexity likely indicates diverse TAM programming in different microenvironments within individual tumors (Lewis and Pollard, 2006; Qian and Pollard, 2010; Ruffell et al., 2012a;

Squadrito and De Palma, 2011). For example, M2-like TAMs reside in both perivascular and hypoxic regions of different mouse and human tumors (Mazzieri et al., 2011; Movahedi et al., 2010; Pucci et al., 2009). A population of vessel-associated TAMs—also referred to as TIE2-expressing monocytes/ macrophages (TEMs)—is required for tumor angiogenesis (De Palma et al., 2005) and displays a profoundly M2-skewed phenotype characterized by enhanced expression of scavenger receptors (e.g., MRC1 and CD163) and relatively low levels of MHCII molecules and proinflammatory cytokines (Pucci et al., 2009; Squadrito et al., 2012). Interestingly, vascular endothelial cells (ECs) may induce HPCs to directly differentiate into TIE2$^+$MRC1$^+$ macrophages in the perivascular microenvironment, a process that appears to depend on EC-derived CSF1 (He et al., 2012). Also attesting to the complexity of TAM subtypes, recent studies have shown that both the origin and phenotype of TAMs may differ in primary versus metastatic tumors (Qian et al., 2011).

TAMs with a relatively M1-skewed phenotype may be found in incipient or regressing tumors as well as necrotic areas of progressing tumors (Prada et al., 2013; Wang et al., 2011). Gene expression profiling of M1- and M2-like TAMs, however, suggests that such TAM "subtypes" express both canonical M1 and M2 markers, albeit at significantly different levels (Movahedi et al., 2010; Pucci et al., 2009; Squadrito et al., 2012).

MACROPHAGE INVOLVEMENT IN TUMOR RESPONSES TO THERAPY

As will be seen below, TAMs not only enhance tumor growth and progression, but also modulate the efficacy of various forms of anticancer therapy. In some circumstances, they also facilitate tumor regrowth, revascularization, and spread after the treatment.

Chemotherapy

A complex picture has emerged over the past 30 years of the role of TAMs in modulating the antitumor efficacy of chemotherapeutic agents (Figure 1).

---

protumoral monocytes/TAMs via caspase-8 activation (bottom); HRG downregulates PlGF in TAMs, reprogramming them toward an M1-like phenotype, and enhances DOX delivery (left).

(B) Chemoresistance is increased when cytotoxic agents, either directly or indirectly, increase protumoral (M2-like) TAM numbers. The latter cells may also limit chemotherapy delivery by affecting vascular leakage. DOX enhances tumor infiltration by MMP9-expressing monocytes via upregulation of CCL2 (top); PTX enhances tumor infiltration by macrophages via upregulation of CSF1 (right); PTX, GEM, and 5FU enhance tumor infiltration by cathepsin-B/S-expressing monocytes/macrophages, which activate chemoprotective T cells through IL-1β and 17 (bottom); VEGF-expressing TAMs augment vascular leakiness and limit CTX delivery (left).

Abbreviations: DOC, docetaxel; TRAB, trabectedin; CTX, cyclophosphamide; GEM, gemcitabine; 5-FU, 5-fluorouracil; IL-1b, interleukin-1β; IL-17, interleukin-17.

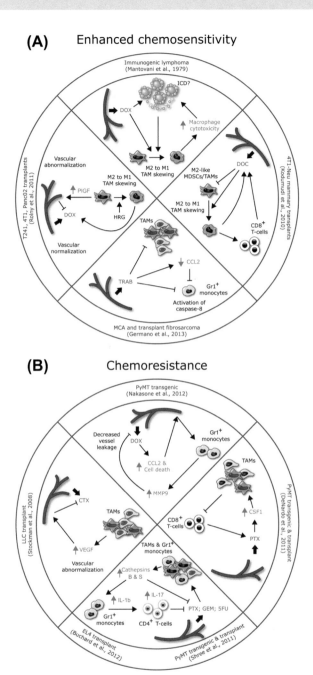

FIGURE 1 TAMs Enhance or Limit the Efficacy of Chemotherapy Depending on the Cytotoxic Agent Applied and/or Mouse Tumor Model Used

(A) Chemosensitivity is increased when cytotoxic agents, either directly or indirectly, increase the cytotoxicity of TAMs or deplete monocytes, TAMs, or M2-like TAMs. The latter cells can also be reprogrammed by agents like HRG, which in turn enhances chemotherapy delivery. DOX enhances the cytotoxicity of macrophages/TAMs, a process possibly involving ICD (top); DOC promotes the expansion of cytotoxic M1-like MDSCs/TAMs, which enhance antitumor T cell responses (right); TRAB depletes

Early studies showed that the antitumor efficacy of doxorubicin (DOX; an anthracycline formerly known as adriamycin) is reduced when mice bearing immunogenic leukemia or lymphoma transplants were given macrophage toxins (Mantovani et al., 1979; Figure 1A). Furthermore, the in vivo administration of DOX enhanced the tumoricidal activity of macrophages ex vivo. Interestingly, macrophages did not enhance the efficacy of DOX against poorly immunogenic lymphomas, suggesting that tumor immunogenicity may influence the ability of macrophages to modulate the antitumor activity of DOX. In contrast, macrophage depletion failed to limit the antitumor activity of another anthracycline, daunorubicin (formerly daunomycin) (Mantovani et al., 1979), possibly because the latter is per se toxic toward macrophages in vivo (Mantovani, 1977). Together, these early reports suggested that some cytotoxic agents are able to foster the antitumor activities of TAMs, at least in leukemia and/or immunogenic (transplant) tumor models. In this regard, innate immune cells, like macrophages and DCs, are known to mediate "immunogenic cell death" (ICD), a process that encompasses chemotherapy-induced cancer cell death and release of "eat-me" signals (e.g., ATP and high-mobility group B1 [HMGB1]); activation of mononuclear phagocytes and enhancement of their antigen-presenting capacity; and promotion of T cell responses against immunogenic tumors. Of note, only a few chemotherapeutics are known to induce ICD, one of which is DOX (Kroemer et al., 2012).

TAMs can also contribute in other ways to the modulation of tumor responses to chemotherapy. Figure 1 shows that this can vary markedly between different cytotoxic agents and tumor models. For example, the antitumor activity of the taxane docetaxel involves the depletion of immunosuppressive (M2-like) TAMs and the concomitant activation or expansion of antitumoral (M1-like) monocytes/MDSCs in 4T1-Neu mammary tumor implants. Indeed, in vitro T cell assays showed that docetaxel-treated monocytes/MDSCs are able to enhance tumor-specific, cytotoxic T cell responses (Kodumudi et al., 2010). Trabectedin, a DNA-damaging agent approved for soft tissue sarcomas, inhibited the growth of mouse fibrosarcomas primarily by depleting mononuclear phagocytes, including monocytes and TAMs (Germano et al., 2013). Mechanistically, it activates caspase 8 and induces apoptosis specifically in monocytes/macrophages via TRAIL-R2, a death receptor not expressed by other leukocytes. Interestingly, trabectedin also depleted circulating monocytes and TAMs in patients with soft-tissue sarcomas. These findings support the notion that the antitumor activity of some cytotoxic agents may depend, at least in part, on their ability to reprogram or deplete protumoral mononuclear phagocytes (Kodumudi et al., 2010; Germano et al., 2013). It remains to be seen whether the mode of action of trabectedin also entails the promotion of adaptive antitumor immune responses, unleashed through the depletion of immunosuppressive TAMs (Figure 1A).

There is also compelling evidence for TAMs limiting the efficacy of chemotherapy (Figure 1B). For example, TAM depletion by anti-CSF1 antibodies enhanced the efficacy of combination chemotherapy (cyclophosphamide, methotrexate, and 5-fluorouracil) in chemoresistant, human breast cancer xenografts grown in immunodeficient mice (Paulus et al., 2006). Similarly, TAM depletion enhanced the efficacy of paclitaxel (PTX, a taxane) in immunocompetent, MMTV-PyMT mouse mammary tumors (DeNardo et al., 2011). At variance with some other cytotoxic drugs (e.g., trabectedin), PTX did not affect tumor growth by depleting TAMs. Rather, it augmented their recruitment to the tumors by upregulating tumor-derived CSF1. Consistent with the known immunosuppressive functions of TAMs, the increased TAM numbers in PTX-treated tumors limited tumor infiltration by CD8+ cytotoxic T cells and possibly reduced their tumoricidal activity. These important findings suggest that TAMs may limit the therapeutic activity of PTX in breast cancer, at least in part, by suppressing specific antitumor immune responses (DeNardo et al., 2011).

TAMs may also release "chemoprotective" factors. Shree et al. (2011) reported increased TAM numbers in PTX-treated MMTV-PyMT tumors and showed that TAM secretion of the lysosomal enzymes, cathepsins B and S, protected cancer cells from PTX-induced cell death and so limited the efficacy of this agent (Shree et al., 2011). Indeed, a pan-cathepsin inhibitor improved the response of MMTV-PyMT tumors to PTX. Interestingly, coculture experiments showed that such macrophage-derived cathepsins protect cancer cells from the direct cytotoxic effects of several chemotherapeutics, including DOX and etoposide (Shree et al., 2011). In this regard, a recent study showed that two broadly used chemotherapeutics, gemcitabine and 5-fluorouracil, induce monocytes/MDSCs to release cathepsin B from lysosomes (Bruchard et al., 2013). This activates the inflammasome and enhances monocyte/MDSC secretion of IL-1β. In turn, IL-1β prompted secretion of IL-17 by CD4+ T cells, which then blunted the anticancer effects of chemotherapy (Figure 1B). These data provide a molecular mechanism linking myeloid cell-derived cathepsins to chemoprotection.

While DOX may stimulate macrophage cytotoxicity toward immunogenic leukemias (Mantovani et al., 1979), its effects on TAMs appear to vary with the tumor type. In the transgenic MMTV-PyMT mammary tumor model, DOX induction of necrotic cell death led to increased tumor infiltration by CCL2 receptor (CCR2)+ monocytes/TAMs, a process that relied on upregulation of CCL2 (Nakasone et al., 2012). Interestingly, the antitumor activity of DOX was enhanced in Ccr2 knockout hosts, which lack CCR2+ monocytes. While the effect of DOX on the cytotoxic activity of TAMs was not examined in this study, the authors showed that matrix-metalloproteinase (MMP)-9 produced by recruited myeloid cells decreased blood vessel leakiness and limited drug delivery to the tumors, suggesting that, at least in MMTV-PyMT tumors, increased vascular permeability is associated with a better response to DOX (Nakasone et al., 2012). It should be noted that, in other tumor models, downregulating the expression of proangiogenic factors, like vascular endothelial growth factor (VEGF) or placental growth factor (PlGF) by TAMs "normalized" the tumor-associated vasculature, decreased vessel leakiness, and enhanced chemotherapy delivery to tumors (Rolny et al., 2011; Stockmann et al., 2008; Figures 1A and 1B). It remains to be seen whether the different effects of DOX in leukemia versus the above mammary tumor model (Nakasone et al., 2012) reflect differences in tumor immunogenicity or more complex aspects of the tumor microenvironment.

Taken together, the above studies show that different chemotherapeutic agents may induce distinct responses in monocytes/macrophages, which can either enhance or antagonize the activity of the anticancer drug, possibly in a tumor-type dependent fashion. Tumor immunogenicity along with the intrinsic sensitivity of TAMs to the drug and their activation state (M1 versus M2-like) may be important determinants of such TAM-mediated responses. Furthermore, cytotoxic drugs often target multiple cell types in tumors, so tumor-type specific stromal cell signatures (Coussens et al., 2013) could influence the ability of TAMs to respond to and modulate the activity of a given chemotherapeutic. Indeed, cytotoxic drugs could have both direct and indirect effects on TAM behavior. For example, taxanes profoundly alter macrophage gene expression in vitro (Javeed et al., 2009) but also induce tumor damage and cancer cell death, which may trigger a reparative, "wound healing" response in TAMs (Mantovani et al., 2013). Further studies are now warranted to distinguish between the role of TAMs in the chemoprotection described above (DeNardo et al., 2011; Nakasone et al., 2012; Shree et al., 2011) and the reparative responses that occur in tumors after therapy.

Finally, TAMs may enhance tumor chemoresistance by providing survival signals to tumor-initiating/cancer stem cells (CSCs). For example, TAMs were found to release milk fat globule-epidermal growth factor 8 protein (MFG-E8) to help protect lung and colon CSCs from cisplatin. This relied, at least in part, on MFG-E8-induced activation of STAT3, which enhanced CSC chemoresistance (Jinushi et al., 2011). Moreover, TAM depletion has been shown to improve antitumor T cell responses and the efficacy of chemotherapy in a pancreatic cancer model, in part by decreasing the frequency, tumor-initiating capacity, and STAT3 activation of CSCs (Mitchem et al., 2013).

Tumor Irradiation

Tumor irradiation is widely used to treat many cancer types. Early studies correlated high TAM numbers in mouse tumors with poor tumor responses to irradiation (Milas et al., 1987). Recent data suggest that radiation-induced DNA damage and activation of the v-abl Abelson murine leukemia viral oncogene homolog 1 (ABL1) kinase promote *Csf1* gene transcription and upregulation of tumor CSF1, which in turn recruits CSF1R-expressing myeloid cells (including TAMs) that enhance posttherapy tumor regrowth. Indeed, a CSF1R inhibitor improved tumor response to radiotherapy in a prostate cancer model (Xu et al., 2013; Figure 2).

Antibody-mediated depletion of CD11b⁺ myeloid cells in human head and neck tumors grown in immunodeficient mice also reduced tumor regrowth after therapy (Ahn et al., 2010). In a model of orthotopic human glioblastoma,

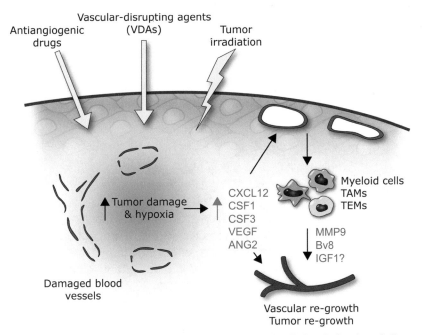

FIGURE 2 TAMs Promote Tumor Regrowth Following Tumor Irradiation, Antiangiogenic Drugs and VDAs

These anticancer therapies cause tumor necrosis, vascular damage, and hypoxia, which together or separately induce the upregulation of several myeloid cell/monocyte chemoattractants, including CXCL12, CSF1, CSF3, VEGF, and ANG2, in the tumor microenvironment. De novo recruitment of myeloid cells drives tumor regrowth via their effects on the tumor blood vessels (mediated, e.g., by MMP9, Bv8, and IGF1) and, possibly, the cancer cells.

Abbreviations: CSF3, granulocyte-colony stimulating factor; Bv8, prokineticin.

See also Figure 1.

local irradiation dramatically enhanced tumor infiltration by CD11b[+] myeloid cells (Kioi et al., 2010). Interestingly, a high proportion of these cells were F4/80[+]TIE2[+] TEMs, and their recruitment was dependent on the hypoxic induction of the chemoattractant, CXCL12, in the irradiated tumors (Figure 2). Upregulation of CXCL12 and increased TEM infiltration were also observed in lung and mammary tumors grown subcutaneously following irradiation (Kozin et al., 2010). In the latter study, TEMs congregated mainly around the remaining blood vessels in treated tumors (Kozin et al., 2010), suggesting that they may stimulate tumor recurrence by promoting EC survival and vascular regrowth through their expression of prosurvival factors like insulin growth factor 1 (IGF1) and fibroblast growth factor 2 (FGF2) (De Palma et al., 2005; Pucci et al., 2009). However, the location and, possibly, the function of M2-like TAMs in irradiated tumors may vary with tumor type. In irradiated orthotopic astrocytomas, arginase-1 (ARG1)[+] M2-like TAMs were found to accumulate mainly in avascular, hypoxic areas rather than at perivascular sites (Chiang et al., 2012). This suggests that the reparative mechanisms employed by M2-like TAMs in postirradiated tumors may be regulated by distinct microenvironmental signals in different tumor types. It is also conceivable that the functions of M2-like TAMs in irradiated tumors are similar to those of M2-like macrophages driving tissue repair in healing organs, such as following acute renal injury and myocardial infarction (Mantovani et al., 2013).

Vascular-Targeted Therapies

VEGF is a proangiogenic cytokine that also functions as a potent monocyte chemoattractant (Barleon et al., 1996). It is, therefore, possible that the antiangiogenic and antitumor effects of VEGF blockade could result, at least in part, from impaired monocyte/TAM recruitment. However, this seems increasingly unlikely, as it is now established that therapeutic interception of VEGF is counteracted by the compensatory induction of other proangiogenic factors, some of which are involved in monocyte/myeloid cell chemoattraction (Bergers and Hanahan, 2008; Ferrara, 2010).

Tumor hypoxia and necrosis dramatically increase after the selective destruction of tumor blood vessels by high-dose antiangiogenic drugs or vascular-disrupting agents (VDAs) (Bergers and Hanahan, 2008). When tumors are treated with VDAs, like combretastatin-A4-phosphate (CA-4-P), the selective disruption of the tumor-associated vasculature results in vessel collapse, reduced blood flow, induction of tumor hypoxia, and secondary tumor cell death. As in irradiated tumors, VDA-induced hypoxia was associated with elevated levels of CXCL12 and increased TEM infiltration in mammary tumor models (Welford et al., 2011; Figure 2). Blocking this CA-4-P-induced TEM recruitment, either using the CXCR4 antagonist, plerixafor (AMD3100), or by genetic

TEM depletion, markedly increased the efficacy of CA-4-P treatment in subcutaneous N202 (Neu[+]) mammary carcinomas (Welford et al., 2011).

Blocking the proangiogenic factor angiopoietin-2 (ANG2) also leads to angiogenesis inhibition and increased tumor hypoxia (Daly et al., 2013; Mazzieri et al., 2011). As seen in CA-4-P-treated tumors (Welford et al., 2011), the latter events were associated with an enhanced recruitment of MRC1[+] TEMs, which may have limited the efficacy of ANG2 blockade (Mazzieri et al., 2011). Sorafenib, which targets several receptor tyrosine kinases (including VEGF receptor 2 [VEGFR2] and platelet-derived growth factor receptor [PDGFR]) and Raf kinases, was also shown to increase CXCL12 levels and TAM infiltration in hepatocellular carcinoma xenografts. Depletion of TAMs by clodronate-loaded liposomes (clodrolip) augmented the inhibitory effects of sorafenib on tumor angiogenesis, growth, and metastasis in this tumor model (Zhang et al., 2010). Moreover, TAM depletion by clodrolip (Zeisberger et al., 2006) or a CSF1R inhibitor (Priceman et al., 2010) increased the antiangiogenic and antitumor effects of VEGF/VEGFR2 antibodies in subcutaneous tumor models. Together, these data support the rationale for combining antiangiogenic drugs with macrophage targeting strategies to increase the efficacy of the former, particularly in tumors that are refractory or develop resistance to anti-VEGF therapy.

Targeted Therapies by Monoclonal Antibodies

Although a role for TAMs in modulating the efficacy of oncogene-targeted, small molecule inhibitors has yet to be elucidated, there is now increasing evidence for TAMs contributing to the cytotoxicity of therapeutic monoclonal antibodies (moAbs). TAMs express surface receptors that bind the Fc fragment of antibodies and enable them to engage in Ab-dependent cellular cytotoxicity/phagocytosis (ADCC/ADCP). Trastuzumab, a moAb against the human epidermal growth factor receptor-2 (HER2), not only interrupts HER2 signaling in breast cancer cells, thereby slowing their proliferation rate, but also induces Fcγ receptor (FcγR)-mediated activation of macrophage cytotoxicity (Clynes et al., 2000) and priming of antigen-specific CD8[+] T cell responses in MMTV-Neu tumors (Park et al., 2010) (Figure 3A). In one study, TAM depletion limited the efficacy of a moAb directed against tissue factor (CD142)-expressing human breast carcinoma cells inoculated in mice (Grugan et al., 2012). Macrophages also enhance lymphoma elimination in mice in response to rituximab, a moAb against CD20, primarily through FcγR-dependent ADCP (Chao et al., 2010; Minard-Colin et al., 2008). The significance of the aforementioned studies is supported by clinical findings suggesting that certain *FCGR* polymorphisms may bear predictive value for the clinical efficacy of trastuzumab or rituximab therapy in breast cancer and lymphoma, respectively (Mellor et al., 2013). Furthermore, high TAM

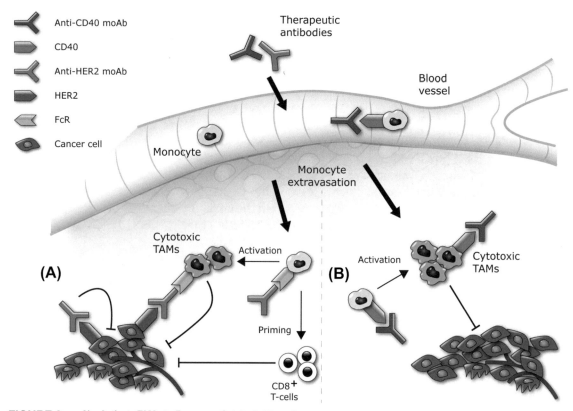

FIGURE 3 moAbs Activate TAMs to Express a Cytotoxic Phenotype
Binding of therapeutic antibodies to monocytes/macrophages may enhance their tumoricidal activity.
(A) Binding of therapeutic (cancer-cell targeted) moAbs (e.g., anti-HER2) to monocytes/TAMs via Fc-receptors (FcRs) induces FcR-mediated activation of macrophage cytotoxicity/phagocytosis (ADCC/ADCP) and priming of adaptive antitumor immunity (e.g., CD8+ T cells).
(B) Binding of immunotherapeutic moAbs (e.g., anti-CD40) to monocytes/TAMs triggers their activation to a cytotoxic (M1-like) phenotype.

numbers correlate with a better prognosis in rituximab-treated patients (Taskinen et al., 2007). Engineered recombinant proteins that can enhance the interactions between FcγR-expressing immune cells and moAbs, like the recently described "grababodies" (Cai et al., 2013), may thus have the potential to increase ADCC/ADCP in tumors. It should be noted, however, that engagement of macrophage-FcγRs by serum or therapeutic antibodies (e.g., the anti-EGFR moAb cetuximab) was shown to enhance the immunosuppressive, proangiogenic, and protumoral functions of TAMs both in experimental tumor models and human cancer (Andreu et al., 2010; Pander et al., 2011).

Immunotherapies

As mentioned previously, TAMs can be potent immunosuppressors that limit the cytotoxic activity of CD8$^+$ cytotoxic T cells in progressing tumors (DeNardo et al., 2011). The analysis of human breast cancer tissues showed that a high stromal TAM density correlates inversely with CD8$^+$ T cell numbers (DeNardo et al., 2011). In a preclinical study, clodrolip-mediated depletion of TAMs enhanced tumor infiltration by HPV16 E7-specific CD8$^+$ T cells in a HPV16 E6$^+$/E7$^+$ mouse model of cervical cancer (Lepique et al., 2009). TAM-mediated immunosuppression is mediated, at least in part, by induction of T cell apoptosis and nitrosylation of T cell receptors via macrophage products, like ARG1, NOS2, and peroxynitrite (Gabrilovich et al., 2012).

It should be noted that the study by DeNardo et al. (2011) analyzed the leukocyte composition of established tumors (DeNardo et al., 2011), in which immunosuppressive, M2-like TAMs likely predominate over tumoricidal (M1-like) macrophages. It is possible that incipient tumors, which are likely to be more immunogenic than established lesions, contain higher proportions of M1-like TAMs, which could initiate and/or potentiate adaptive immune responses (Prada et al., 2013; Wang et al., 2011). In this regard, macrophages were shown to acutely engulf myeloma cells inoculated subcutaneously in syngenic mice and to activate myeloma-specific CD4$^+$ Th1 cells, which then enhanced the tumoricidal activity of macrophages through IFN-γ secretion (Corthay et al., 2005). In certain immunoprivileged organs, such as the eye, macrophages promote the effector functions of CD4$^+$ T cells, and their depletion enhances rather than inhibits intraocular tumor growth (Dace et al., 2008). Thus, the type of macrophage activation—which may correlate with tumor stage (Prada et al., 2013; Wang et al., 2011)—may dictate the magnitude of antitumoral T cell responses in mouse models of cancer.

Based on the above, strategies to deplete TAMs or block cancer-induced M2-like macrophage programming (see below) may have the potential to enhance T cell-mediated antitumor responses and improve the efficacy of immunotherapies (Coussens et al., 2013; Hagemann et al., 2008; Jaiswal et al., 2010). Intriguingly, increasing data suggest that the efficacy of some forms of immunotherapy may also depend on effective reprogramming of TAMs toward an M1-like phenotype. For example, intravesical instillation of *Mycobacterium bovis* bacillus Calmette-Guérin, which is used for the treatment of superficial bladder cancer, reduces tumor recurrence by stimulating the cytotoxic activity of macrophages. Macrophage-mediated killing of bladder cancer cells relies on both direct effector-target cell contacts and the release of soluble cytotoxic factors, such as TNF-α, IFN-γ, and

NO, from the macrophages (Luo and Knudson, 2010). An agonistic antibody to the TNF receptor superfamily member, CD40, was recently reported to bind to circulating monocytes, trigger their recruitment into mouse pancreatic tumors, and activate their tumoricidal functions (Figure 3B). These CD40-activated, cytotoxic (M1-skewed) TAMs were also found to enhance the efficacy of gemcitabine in a small cohort of patients with surgically incurable pancreatic cancer (Beatty et al., 2011). Finally, macrophages and DCs express programmed cell death ligand-1 (PDL1, also known as B7-H1), a major negative regulatory ligand that suppresses T cell activation through its receptor-programmed cell death protein 1. The promising therapeutic activity of anti-PDL1 moAbs in patients with advanced cancer (Brahmer et al., 2012) will no doubt prompt further studies of the possible inhibition of PDL1 expression on TAMs to improve the efficacy of chemo- or antiangiogenic therapies.

CONCLUDING REMARKS: IMPLICATIONS FOR CANCER TREATMENT

In light of this growing body of evidence for TAMs modulating the effects of various anticancer therapies, attempts are now being made to either target key molecules that regulate their recruitment into tumors or re-educate these cells toward a cytotoxic M1-like phenotype. The efficacy of CSF1R inhibitors in blocking the enhanced uptake of monocytes during PTX treatment in preclinical studies (DeNardo et al., 2011) has prompted clinical trials of their use in combination with either PTX or the antiproliferative agent eribulin (http://www.clinicaltrials.gov). Various preclinical studies have also highlighted ways to reprogram TAMs from an M2 to an M1-like phenotype in tumors. These include the use of histidine-rich glycoprotein (HRG), which induces macrophage downregulation of PlGF, promotes the normalization of blood vessels, and increases delivery and efficacy of chemotherapy in mouse tumor models (Rolny et al., 2011; Figure 1A). Other strategies to reprogram TAMs include blockade of nuclear factor-kB signaling (Hagemann et al., 2008) or their exposure to anti-IL-10R antibodies combined with the TLR9 ligand CpG (Guiducci et al., 2005). The latter resulted in hemorrhagic tumor necrosis, activation of DCs and cytotoxic T cells, and clearance of tumor remnants.

However, there is still much to learn about the mechanisms regulating TAM functions during chemotherapy, as well as other forms of therapy discussed in this review. Importantly, a number of key questions need to be addressed before approaches that combine macrophage targeting (or reprogramming) and conventional cancer therapies can be translated into more effective treatments. Why do preclinical studies employing distinct chemotherapeutic drugs and/or tumor models show different and, at times, contradictory roles for TAMs in modulating tumor responses to such agents? Why do TAMs apparently limit the effects of chemotherapy in some tumor types but not others? Are the distinct TAM subtypes present in individual tumors differentially responsive to chemotherapy? If yes, what are the specific features of the TAM subset(s) that either enhance or promote the antitumor activity of cytotoxic agents? And what are the signals in tumors that regulate these TAM responses? Such information might help selectively target the TAMs that limit chemotherapy while leaving antitumoral or tissue-resident macrophages unaffected. Furthermore, most preclinical studies to date have focused on primary, nonmetastatic tumors. So, are we confident that TAMs in metastatic tumors (Qian et al., 2011) behave in the same way during therapy as those in the primary tumor site?

Mouse tumor models, including genetically engineered mouse models (GEMMs), are being used extensively to study mechanisms underlying tumor (and TAM) responses to anticancer therapies. However, even sophisticated GEMMs of cancer cannot simulate the endless variations in TAM abundance, distribution, and phenotypes between and within different types and subtypes of human cancer (Coussens et al., 2013; De Palma and Hanahan, 2012). Nor do they necessarily model the ability of such tissues to recruit monocytes during therapy. Future work should therefore aim to define the identities and molecular profiles of distinct TAM subtypes in human cancer biopsies before, during, and after therapy. Specific TAM signatures could then be used to stratify patients carrying defined genetic lesions in order to explore how such signatures correlate with the response of individual patients to chemo-, radio-, or targeted therapies, and/or the emergence of secondary resistance (DeVita and Costa, 2010). If such studies demonstrate the predictive value of specific TAM subtypes for individual tumor responses, then their further characterization in mouse tumor models could help develop more effective cancer therapies. Undeniably, such clinical approaches should consider the biological complexities on a tumor (sub)type and individual patient basis and harness them to design effective personalized therapies.

ACKNOWLEDGMENTS

We apologize to the authors whose work we could not cite or illustrate in the figures because of space constraints. The authors thank Luisa Iruela-Arispe and Livio Trusolino for critical reading of the manuscript and Bruno Liardon for help with the figures. Work in the authors' laboratories is supported by grants from the European Research Council (ERC; Tie2+monocytes), National Centres of Competence in Research (Oncology program), Anna Fuller Fund (to M.D.P.), and Cancer Research UK (to C.E.L.).

REFERENCES

Ahn, G.O., Tseng, D., Liao, C.H., Dorie, M.J., Czechowicz, A., and Brown, J.M. (2010). Inhibition of Mac-1 (CD11b/CD18) enhances tumor response to radiation by reducing myeloid cell recruitment. Proc. Natl. Acad. Sci. USA 107, 8363–8368.

Andreu, P., Johansson, M., Affara, N.I., Pucci, F., Tan, T., Junankar, S., Korets, L., Lam, J., Tawfik, D., DeNardo, D.G., et al. (2010). FcRgamma activation regulates inflammation-associated squamous carcinogenesis. Cancer Cell 17, 121–134.

Barleon, B., Sozzani, S., Zhou, D., Weich, H.A., Mantovani, A., and Marmé, D. (1996). Migration of human monocytes in response to vascular endothelial growth factor (VEGF) is mediated via the VEGF receptor flt-1. Blood 87, 3336–3343.

Beatty, G.L., Chiorean, E.G., Fishman, M.P., Saboury, B., Teitelbaum, U.R., Sun, W., Huhn, R.D., Song, W., Li, D., Sharp, L.L., et al. (2011). CD40 agonists alter tumor stroma and show efficacy against pancreatic carcinoma in mice and humans. Science 331, 1612–1616.

Bergers, G., and Hanahan, D. (2008). Modes of resistance to anti-angiogenic therapy. Nat. Rev. Cancer 8, 592–603.

Bingle, L., Brown, N.J., and Lewis, C.E. (2002). The role of tumour-associated macrophages in tumour progression: implications for new anticancer therapies. J. Pathol. 196, 254–265.

Bingle, L., Lewis, C.E., Corke, K.P., Reed, M.W., and Brown, N.J. (2006). Macrophages promote angiogenesis in human breast tumour spheroids in vivo. Br. J. Cancer 94, 101–107.

Biswas, S.K., and Mantovani, A. (2010). Macrophage plasticity and interaction with lymphocyte subsets: cancer as a paradigm. Nat. Immunol. 11, 889–896.

Brahmer, J.R., Tykodi, S.S., Chow, L.Q., Hwu, W.J., Topalian, S.L., Hwu, P., Drake, C.G., Camacho, L.H., Kauh, J., Odunsi, K., et al. (2012). Safety and activity of anti-PD-L1 antibody in patients with advanced cancer. N. Engl. J. Med. 366, 2455–2465.

Bruchard, M., Mignot, G., Derangère, V., Chalmin, F., Chevriaux, A., Végran, F., Boireau, W., Simon, B., Ryffel, B., Connat, J.L., et al. (2013). Chemotherapy-triggered cathepsin B release in myeloid-derived suppressor cells activates the Nlrp3 inflammasome and promotes tumor growth. Nat. Med. 19, 57–64.

Cai, Z., Fu, T., Nagai, Y., Lam, L., Yee, M., Zhu, Z., and Zhang, H. (2013). scFv-based "grababody" as a general strategy to improve recruitment of immune effector cells to antibody-targeted tumors. Cancer Res.. http://dx.doi.org/10.1158/0008-5472.CAN-12-3920.

Chao, M.P., Alizadeh, A.A., Tang, C., Myklebust, J.H., Varghese, B., Gill, S., Jan, M., Cha, A.C., Chan, C.K., Tan, B.T., et al. (2010). Anti-CD47 antibody synergizes with rituximab to promote phagocytosis and eradicate non-Hodgkin lymphoma. Cell 142, 699–713.

Chiang, C.S., Fu, S.Y., Wang, S.C., Yu, C.F., Chen, F.H., Lin, C.M., and Hong, J.H. (2012). Irradiation promotes an m2 macrophage phenotype in tumor hypoxia. Front Oncol. 2, 89.

Clear, A.J., Lee, A.M., Calaminici, M., Ramsay, A.G., Morris, K.J., Hallam, S., Kelly, G., Macdougall, F., Lister, T.A., and Gribben, J.G. (2010). Increased angiogenic sprouting in poor prognosis FL is associated with elevated numbers of CD163+ macrophages within the immediate sprouting microenvironment. Blood 115, 5053–5056.

Clynes, R.A., Towers, T.L., Presta, L.G., and Ravetch, J.V. (2000). Inhibitory Fc receptors modulate in vivo cytotoxicity against tumor targets. Nat. Med. 6, 443–446.

Coffelt, S.B., Lewis, C.E., Naldini, L., Brown, J.M., Ferrara, N., and De Palma, M. (2010). Elusive identities and overlapping phenotypes of proangiogenic myeloid cells in tumors. Am. J. Pathol. 176, 1564–1576.

Cortez-Retamozo, V., Etzrodt, M., Newton, A., Rauch, P.J., Chudnovskiy, A., Berger, C., Ryan, R.J., Iwamoto, Y., Marinelli, B., Gorbatov, R., et al. (2012). Origins of tumor-associated macrophages and neutrophils. Proc. Natl. Acad. Sci. USA 109, 2491–2496.

Corthay, A., Skovseth, D.K., Lundin, K.U., Røsjø, E., Omholt, H., Hofgaard, P.O., Haraldsen, G., and Bogen, B. (2005). Primary antitumor immune response mediated by CD4+ T cells. Immunity 22, 371–383.

Coussens, L.M., Zitvogel, L., and Palucka, A.K. (2013). Neutralizing tumor-promoting chronic inflammation: a magic bullet? Science 339, 286–291.

Dace, D.S., Chen, P.W., and Niederkorn, J.Y. (2008). CD4+ T-cell-dependent tumour rejection in an immune-privileged environment requires macrophages. Immunology 123, 367–377.

Daly, C., Eichten, A., Castanaro, C., Pasnikowski, E., Adler, A., Lalani, A.S., Papadopoulos, N., Kyle, A.H., Minchinton, A.I., Yancopoulos, G.D., and Thurston, G. (2013). Angiopoietin-2 functions as a Tie2 agonist in tumor models, where it limits the effects of VEGF inhibition. Cancer Res. 73, 108–118.

De Palma, M., and Hanahan, D. (2012). The biology of personalized cancer medicine: facing individual complexities underlying hallmark capabilities. Mol. Oncol. 6, 111–127.

De Palma, M., Venneri, M.A., Roca, C., and Naldini, L. (2003). Targeting exogenous genes to tumor angiogenesis by transplantation of genetically modified hematopoietic stem cells. Nat. Med. 9, 789–795.

De Palma, M., Venneri, M.A., Galli, R., Sergi Sergi, L., Politi, L.S., Sampaolesi, M., and Naldini, L. (2005). Tie2 identifies a hematopoietic lineage of proangiogenic monocytes required for tumor vessel formation and a mesenchymal population of pericyte progenitors. Cancer Cell 8, 211–226.

de Visser, K.E., Eichten, A., and Coussens, L.M. (2006). Paradoxical roles of the immune system during cancer development. Nat. Rev. Cancer 6, 24–37.

DeNardo, D.G., Barreto, J.B., Andreu, P., Vasquez, L., Tawfik, D., Kolhatkar, N., and Coussens, L.M. (2009). CD4(+) T cells regulate pulmonary metastasis of mammary carcinomas by enhancing protumor properties of macrophages. Cancer Cell 16, 91–102.

DeNardo, D.G., Brennan, D.J., Rexhepaj, E., Ruffell, B., Shiao, S.L., Madden, S.F., Gallagher, W.M., Wadhwani, N., Keil, S.D., Junaid, S.A., et al. (2011). Leukocyte complexity predicts breast cancer survival and functionally regulates response to chemotherapy. Cancer Discov. 1, 54–67.

DeVita, V.T., Jr., and Costa, J. (2010). Toward a personalized treatment of Hodgkin's disease. N. Engl. J. Med. 362, 942–943.

Engler, J.R., Robinson, A.E., Smirnov, I., Hodgson, J.G., Berger, M.S., Gupta, N., James, C.D., Molinaro, A., and Phillips, J.J. (2012). Increased microglia/macrophage gene expression in a subset of adult and pediatric astrocytomas. PLoS ONE 7, e43339.

Ferrara, N. (2010). Role of myeloid cells in vascular endothelial growth factor-independent tumor angiogenesis. Curr. Opin. Hematol. 17, 219–224.

Gabrilovich, D.I., Ostrand-Rosenberg, S., and Bronte, V. (2012). Coordinated regulation of myeloid cells by tumours. Nat. Rev. Immunol. 12, 253–268.

Germano, G., Frapolli, R., Belgiovine, C., Anselmo, A., Pesce, S., Liguori, M., Erba, E., Uboldi, S., Zucchetti, M., Pasqualini, F., et al. (2013). Role of macrophage targeting in the antitumor activity of trabectedin. Cancer Cell 23, 249–262.

Gordon, S., and Martinez, F.O. (2010). Alternative activation of macrophages: mechanism and functions. Immunity 32, 593–604.

Grugan, K.D., McCabe, F.L., Kinder, M., Greenplate, A.R., Harman, B.C., Ekert, J.E., van Rooijen, N., Anderson, G.M., Nemeth, J.A., Strohl, W.R., et al. (2012). Tumor-associated macrophages promote invasion while retaining Fc-dependent anti-tumor function. J. Immunol. 189, 5457–5466.

Guiducci, C., Vicari, A.P., Sangaletti, S., Trinchieri, G., and Colombo, M.P. (2005). Redirecting in vivo elicited tumor infiltrating macrophages and dendritic cells towards tumor rejection. Cancer Res. 65, 3437–3446.

Hagemann, T., Lawrence, T., McNeish, I., Charles, K.A., Kulbe, H., Thompson, R.G., Robinson, S.C., and Balkwill, F.R. (2008). "Re-educating" tumor-associated macrophages by targeting NF-kappaB. J. Exp. Med. *205*, 1261–1268.

He, H., Xu, J., Warren, C.M., Duan, D., Li, X., Wu, L., and Iruela-Arispe, M.L. (2012). Endothelial cells provide an instructive niche for the differentiation and functional polarization of M2-like macrophages. Blood *120*, 3152–3162.

Heusinkveld, M., and van der Burg, S.H. (2011). Identification and manipulation of tumor associated macrophages in human cancers. J. Transl. Med. *9*, 216.

Jaiswal, S., Chao, M.P., Majeti, R., and Weissman, I.L. (2010). Macrophages as mediators of tumor immunosurveillance. Trends Immunol. *31*, 212–219.

Javeed, A., Ashraf, M., Riaz, A., Ghafoor, A., Afzal, S., and Mukhtar, M.M. (2009). Paclitaxel and immune system. Eur. J. Pharm. Sci. *38*, 283–290.

Jinushi, M., Chiba, S., Yoshiyama, H., Masutomi, K., Kinoshita, I., Dosaka-Akita, H., Yagita, H., Takaoka, A., and Tahara, H. (2011). Tumor-associated macrophages regulate tumorigenicity and anticancer drug responses of cancer stem/initiating cells. Proc. Natl. Acad. Sci. USA *108*, 12425–12430.

Kioi, M., Vogel, H., Schultz, G., Hoffman, R.M., Harsh, G.R., and Brown, J.M. (2010). Inhibition of vasculogenesis, but not angiogenesis, prevents the recurrence of glioblastoma after irradiation in mice. J. Clin. Invest. *120*, 694–705.

Kodumudi, K.N., Woan, K., Gilvary, D.L., Sahakian, E., Wei, S., and Djeu, J.Y. (2010). A novel chemoimmunomodulating property of docetaxel: suppression of myeloid-derived suppressor cells in tumor bearers. Clin. Cancer Res. *16*, 4583–4594.

Kozin, S.V., Kamoun, W.S., Huang, Y., Dawson, M.R., Jain, R.K., and Duda, D.G. (2010). Recruitment of myeloid but not endothelial precursor cells facilitates tumor regrowth after local irradiation. Cancer Res. *70*, 5679–5685.

Kroemer, G., Galluzzi, L., Kepp, O., and Zitvogel, L. (2012). Immunogenic Cell Death in Cancer Therapy. Annu. Rev. Immunol.. http://dx.doi.org/10.1146/annurev-immunol-032712-100008.

Lawrence, T., and Natoli, G. (2011). Transcriptional regulation of macrophage polarization: enabling diversity with identity. Nat. Rev. Immunol. *11*, 750–761.

Leek, R.D., Lewis, C.E., Whitehouse, R., Greenall, M., Clarke, J., and Harris, A.L. (1996). Association of macrophage infiltration with angiogenesis and prognosis in invasive breast carcinoma. Cancer Res. *56*, 4625–4629.

Lepique, A.P., Daghastanli, K.R., Cuccovia, I.M., and Villa, L.L. (2009). HPV16 tumor associated macrophages suppress antitumor T cell responses. Clin. Cancer Res. *15*, 4391–4400.

Lewis, C.E., and Pollard, J.W. (2006). Distinct role of macrophages in different tumor microenvironments. Cancer Res. *66*, 605–612.

Lin, E.Y., Nguyen, A.V., Russell, R.G., and Pollard, J.W. (2001). Colony-stimulating factor 1 promotes progression of mammary tumors to malignancy. J. Exp. Med. *193*, 727–740.

Luo, Y., and Knudson, M.J. (2010). Mycobacterium bovis bacillus Calmette-Guérin-induced macrophage cytotoxicity against bladder cancer cells. Clin. Dev. Immunol. *2010*, 357591.

Mantovani, A. (1977). In vitro and in vivo cytotoxicity of adriamycin and daunomycin for murine macrophages. Cancer Res. *37*, 815–820.

Mantovani, A., Polentarutti, N., Luini, W., Peri, G., and Spreafico, F. (1979). Role of host defense mechanisms in the antitumor activity of adriamycin and daunomycin in mice. J. Natl. Cancer Inst. *63*, 61–66.

Mantovani, A., Sozzani, S., Locati, M., Allavena, P., and Sica, A. (2002). Macrophage polarization: tumor-associated macrophages as a paradigm for polarized M2 mononuclear phagocytes. Trends Immunol. *23*, 549–555.

Mantovani, A., Biswas, S.K., Galdiero, M.R., Sica, A., and Locati, M. (2013). Macrophage plasticity and polarization in tissue repair and remodelling. J. Pathol. *229*, 176–185.

Mazzieri, R., Pucci, F., Moi, D., Zonari, E., Ranghetti, A., Berti, A., Politi, L.S., Gentner, B., Brown, J.L., Naldini, L., and De Palma, M. (2011). Targeting the ANG2/TIE2 axis inhibits tumor growth and metastasis by impairing angiogenesis and disabling rebounds of proangiogenic myeloid cells. Cancer Cell *19*, 512–526.

Mellor, J.D., Brown, M.P., Irving, H.R., Zalcberg, J.R., and Dobrovic, A. (2013). A critical review of the role of Fc gamma receptor polymorphisms in the response to monoclonal antibodies in cancer. J. Hematol. Oncol. *6*, 1.

Milas, L., Wike, J., Hunter, N., Volpe, J., and Basic, I. (1987). Macrophage content of murine sarcomas and carcinomas: associations with tumor growth parameters and tumor radiocurability. Cancer Res. *47*, 1069–1075.

Minard-Colin, V., Xiu, Y., Poe, J.C., Horikawa, M., Magro, C.M., Hamaguchi, Y., Haas, K.M., and Tedder, T.F. (2008). Lymphoma depletion during CD20 immunotherapy in mice is mediated by macrophage FcgammaRI, FcgammaRIII, and FcgammaRIV. Blood *112*, 1205–1213.

Mitchem, J.B., Brennan, D.J., Knolhoff, B.L., Belt, B.A., Zhu, Y., Sanford, D.E., Belaygorod, L., Carpenter, D., Collins, L., Piwnica-Worms, D., et al. (2013). Targeting tumor-infiltrating macrophages decreases tumor-initiating cells, relieves immunosuppression, and improves chemotherapeutic responses. Cancer Res. *73*, 1128–1141.

Movahedi, K., Laoui, D., Gysemans, C., Baeten, M., Stangé, G., Van den Bossche, J., Mack, M., Pipeleers, D., In't Veld, P., De Baetselier, P., and Van Ginderachter, J.A. (2010). Different tumor microenvironments contain functionally distinct subsets of macrophages derived from Ly6C(high) monocytes. Cancer Res. *70*, 5728–5739.

Murdoch, C., Muthana, M., Coffelt, S.B., and Lewis, C.E. (2008). The role of myeloid cells in the promotion of tumour angiogenesis. Nat. Rev. Cancer *8*, 618–631.

Murray, P.J., and Wynn, T.A. (2011). Protective and pathogenic functions of macrophage subsets. Nat. Rev. Immunol. *11*, 723–737.

Nakasone, E.S., Askautrud, H.A., Kees, T., Park, J.H., Plaks, V., Ewald, A.J., Fein, M., Rasch, M.G., Tan, Y.X., Qiu, J., et al. (2012). Imaging tumor-stroma interactions during chemotherapy reveals contributions of the microenvironment to resistance. Cancer Cell *21*, 488–503.

Nucera, S., Biziato, D., and De Palma, M. (2011). The interplay between macrophages and angiogenesis in development, tissue injury and regeneration. Int. J. Dev. Biol. *55*, 495–503.

Pander, J., Heusinkveld, M., van der Straaten, T., Jordanova, E.S., Baak-Pablo, R., Gelderblom, H., Morreau, H., van der Burg, S.H., Guchelaar, H.J., and van Hall, T. (2011). Activation of tumor-promoting type 2 macrophages by EGFR-targeting antibody cetuximab. Clin. Cancer Res. *17*, 5668–5673.

Park, S., Jiang, Z., Mortenson, E.D., Deng, L., Radkevich-Brown, O., Yang, X., Sattar, H., Wang, Y., Brown, N.K., Greene, M., et al. (2010). The therapeutic effect of anti-HER2/neu antibody depends on both innate and adaptive immunity. Cancer Cell *18*, 160–170.

Paulus, P., Stanley, E.R., Schäfer, R., Abraham, D., and Aharinejad, S. (2006). Colony-stimulating factor-1 antibody reverses chemoresistance in human MCF-7 breast cancer xenografts. Cancer Res. *66*, 4349–4356.

Pollard, J.W. (2009). Trophic macrophages in development and disease. Nat. Rev. Immunol. *9*, 259–270.

Prada, C.E., Jousma, E., Rizvi, T.A., Wu, J., Dunn, R.S., Mayes, D.A., Cancelas, J.A., Dombi, E., Kim, M.O., West, B.L., et al. (2013). Neurofibroma-associated macrophages play roles in tumor growth and response to pharmacological inhibition. Acta Neuropathol. *125*, 159–168.

Priceman, S.J., Sung, J.L., Shaposhnik, Z., Burton, J.B., Torres-Collado, A.X., Moughon, D.L., Johnson, M., Lusis, A.J., Cohen, D.A., Iruela-Arispe, M.L., and Wu, L. (2010). Targeting distinct tumor-infiltrating myeloid cells by inhibiting CSF-1 receptor: combating tumor evasion of antiangiogenic therapy. Blood *115*, 1461–1471.

Pucci, F., Venneri, M.A., Biziato, D., Nonis, A., Moi, D., Sica, A., Di Serio, C., Naldini, L., and De Palma, M. (2009). A distinguishing gene signature shared by tumor-infiltrating Tie2-expressing monocytes, blood "resident" monocytes, and embryonic macrophages suggests common functions and developmental relationships. Blood *114*, 901–914.

Qian, B.Z., and Pollard, J.W. (2010). Macrophage diversity enhances tumor progression and metastasis. Cell *141*, 39–51.

Qian, B.Z., Li, J., Zhang, H., Kitamura, T., Zhang, J., Campion, L.R., Kaiser, E.A., Snyder, L.A., and Pollard, J.W. (2011). CCL2 recruits inflammatory monocytes to facilitate breast-tumour metastasis. Nature *475*, 222–225.

Rolny, C., Mazzone, M., Tugues, S., Laoui, D., Johansson, I., Coulon, C., Squadrito, M.L., Segura, I., Li, X., Knevels, E., et al. (2011). HRG inhibits tumor growth and metastasis by inducing macrophage polarization and vessel normalization through downregulation of PlGF. Cancer Cell *19*, 31–44.

Ruffell, B., Affara, N.I., and Coussens, L.M. (2012). Differential macrophage programming in the tumor microenvironment. Trends Immunol. *33*, 119–126.

Ruffell, B., Au, A., Rugo, H.S., Esserman, L.J., Hwang, E.S., and Coussens, L.M. (2012). Leukocyte composition of human breast cancer. Proc. Natl. Acad. Sci. USA *109*, 2796–2801.

Schulz, C., Gomez Perdiguero, E., Chorro, L., Szabo-Rogers, H., Cagnard, N., Kierdorf, K., Prinz, M., Wu, B., Jacobsen, S.E., Pollard, J.W., et al. (2012). A lineage of myeloid cells independent of Myb and hematopoietic stem cells. Science *336*, 86–90.

Shaked, Y., and Voest, E.E. (2009). Bone marrow derived cells in tumor angiogenesis and growth: are they the good, the bad or the evil? Biochim. Biophys. Acta *1796*, 1–4.

Shi, C., and Pamer, E.G. (2011). Monocyte recruitment during infection and inflammation. Nat. Rev. Immunol. *11*, 762–774.

Shree, T., Olson, O.C., Elie, B.T., Kester, J.C., Garfall, A.L., Simpson, K., Bell-McGuinn, K.M., Zabor, E.C., Brogi, E., and Joyce, J.A. (2011). Macrophages and cathepsin proteases blunt chemotherapeutic response in breast cancer. Genes Dev. *25*, 2465–2479.

Sica, A., and Mantovani, A. (2012). Macrophage plasticity and polarization: in vivo veritas. J. Clin. Invest. *122*, 787–795.

Squadrito, M.L., and De Palma, M. (2011). Macrophage regulation of tumor angiogenesis: implications for cancer therapy. Mol. Aspects Med. *32*, 123–145.

Squadrito, M.L., Pucci, F., Magri, L., Moi, D., Gilfillan, G.D., Ranghetti, A., Casazza, A., Mazzone, M., Lyle, R., Naldini, L., and De Palma, M. (2012). miR-511-3p modulates genetic programs of tumor-associated macrophages. Cell Rep. *1*, 141–154.

Steidl, C., Lee, T., Shah, S.P., Farinha, P., Han, G., Nayar, T., Delaney, A., Jones, S.J., Iqbal, J., Weisenburger, D.D., et al. (2010). Tumor-associated macrophages and survival in classic Hodgkin's lymphoma. N. Engl. J. Med. *362*, 875–885.

Stockmann, C., Doedens, A., Weidemann, A., Zhang, N., Takeda, N., Greenberg, J.I., Cheresh, D.A., and Johnson, R.S. (2008). Deletion of vascular endothelial growth factor in myeloid cells accelerates tumorigenesis. Nature *456*, 814–818.

Taskinen, M., Karjalainen-Lindsberg, M.L., Nyman, H., Eerola, L.M., and Leppä, S. (2007). A high tumor-associated macrophage content predicts favorable outcome in follicular lymphoma patients treated with rituximab and cyclophosphamide-doxorubicin-vincristine-prednisone. Clin. Cancer Res. *13*, 5784–5789.

Wang, B., Li, Q., Qin, L., Zhao, S., Wang, J., and Chen, X. (2011). Transition of tumor-associated macrophages from MHC class II(hi) to MHC class II(low) mediates tumor progression in mice. BMC Immunol. *12*, 43.

Welford, A.F., Biziato, D., Coffelt, S.B., Nucera, S., Fisher, M., Pucci, F., Di Serio, C., Naldini, L., De Palma, M., Tozer, G.M., and Lewis, C.E. (2011). TIE2-expressing macrophages limit the therapeutic efficacy of the vascular-disrupting agent combretastatin A4 phosphate in mice. J. Clin. Invest. *121*, 1969–1973.

Wyckoff, J., Wang, W., Lin, E.Y., Wang, Y., Pixley, F., Stanley, E.R., Graf, T., Pollard, J.W., Segall, J., and Condeelis, J. (2004). A paracrine loop between tumor cells and macrophages is required for tumor cell migration in mammary tumors. Cancer Res. *64*, 7022–7029.

Xu, J., Escamilla, J., Mok, S., David, J., Priceman, S.J., West, B.L., Bollag, G., McBride, W.H., and Wu, L. (2013). Abrogating the protumorigenic influences of tumor-infiltrating myeloid cells by CSF1R signaling blockade improves the efficacy of radiotherapy in prostate cancer. Cancer Res.. http://dx.doi.org/10.1158/0008-5472.CAN-12-3981 Published online February 19, 2013.

Zeisberger, S.M., Odermatt, B., Marty, C., Zehnder-Fjällman, A.H., Ballmer-Hofer, K., and Schwendener, R.A. (2006). Clodronate-liposome-mediated depletion of tumour-associated macrophages: a new and highly effective antiangiogenic therapy approach. Br. J. Cancer *95*, 272–281.

Zhang, W., Zhu, X.D., Sun, H.C., Xiong, Y.Q., Zhuang, P.Y., Xu, H.X., Kong, L.Q., Wang, L., Wu, W.Z., and Tang, Z.Y. (2010). Depletion of tumor-associated macrophages enhances the effect of sorafenib in metastatic liver cancer models by antimetastatic and antiangiogenic effects. Clin. Cancer Res. *16*, 3420–3430.

ends in Molecular Medicine

Recombinant Viral Vaccines for Cancer

Ryan Cawood, Thomas Hills, Suet Ling Wong, Aliaa A. Alamoudi,
Storm Beadle, Kerry D. Fisher, Leonard W. Seymour*

Department of Oncology, University of Oxford, Oxford, OX3 7DQ, UK

Correspondence: len.seymour@oncology.ox.ac.uk

Trends in Molecular Medicine, Vol. 18, No. 9, September 2012 © 2012 Elsevier Inc.
http://dx.doi.org/10.1016/j.molmed.2012.07.007

SUMMARY

Cancer arises from 'self' in a series of steps that are all subject to immunoediting. Therefore, therapeutic cancer vaccines must stimulate an immune response against tumour antigens that have already evaded the body's immune defences. Vaccines presenting a tumour antigen in the context of obvious danger signals seem more likely to stimulate a response. This approach can be facilitated by genetic engineering using recombinant viral vectors expressing tumour antigens, cytokines, or both, from an immunogenic virus particle. We overview clinical attempts to use these agents for systemic immunisation and contrast the results with strategies employing direct intratumoural administration. We focus on the challenge of producing an effective response within the immune-suppressive tumour microenvironment, and discuss how the technology can overcome these obstacles.

CANCER IMMUNITY

It has long been recognised that malignant cells are antigenically distinct from normal tissues, although tumour-associated antigens (TAAs) found in the context of a clinically apparent tumour tend not to result in tumour elimination. The genetic and physiological changes associated with neoplasia are therefore often considered poorly immunogenic. However, although TAAs originally arise from 'self', they also often exist in an immunosuppressive microenvironment – highlighting a role for both central (thymic depletion) and peripheral [e.g., regulatory T cell (Treg)-mediated] tolerance that can frustrate attempts at therapeutic vaccination. It is also becoming increasingly clear that when the immune system attempts to limit tumour development through cancer cell destruction, this process can select actively for cells capable of

139

CellPress

surviving in the immunocompetent host. Unlike the simplistic view that cancer cells inherently lack sufficiently immunogenic TAAs, tumours are made up of cells with a range of antigenic profiles that are shaped by a complex and dynamic interplay between the immune system and developing tumour. This process has been termed 'cancer immunoediting' [1].

The acquisition by emerging tumour cells of active mechanisms to avoid immune destruction allows cancer cells to express antigens that might stimulate a response in the context of normal tissue. These 'immune escape' mechanisms include downregulating proteins that can activate the immune system (such as the major histocompatibility complexes and interferon-response pathways) and expressing immunosuppressive mediators such as interleukin (IL)-10 and transforming growth factor β (TGF-β), to create a locally immunosuppressive microenvironment [2]. Successful cancer immunotherapy must therefore overcome diverse immune escape mechanisms, some of which operate to limit antigen presentation and others which operate during the effector phase. In the case of limited antigen presentation it is necessary to find ways to increase the quantity or improve the quality of effector immune cells, whereas modulating the immunosuppressive tumour microenvironment may inhibit cytotoxic activity of cells that can recognise TAAs and change the outcome of the effector phase.

One of the most successful branches of medicine is the development of vaccines for a range of infectious diseases. Attenuated strains or related viruses of reduced pathogenicity have shown particular success in prophylactic vaccination for diseases such as smallpox, polio, measles, mumps, and rubella [3]. The advent of genetic engineering has led to the exploitation of these intrinsically immunogenic pathogens as vectors for delivering and expressing exogenous proteins against which an immune response is desired, but which are not normally sufficiently immunogenic [4]. Viral vectors allow *in situ* antigen expression from a virus particle with intrinsic adjuvant activity, recognised by innate receptors that have evolved to detect pathogen-associated molecular patterns (PAMPs). It is therefore enticing to consider that recombinant viral vectors may make effective therapeutic cancer vaccines, strengthening anticancer responses and overcoming tolerance by manipulating the context in which TAAs are presented to the immune system. Some of the molecular strategies that might be applied are illustrated in Figure 1.

In this review, we describe the different viral vectors that have been used clinically in attempts to stimulate an anticancer immune response. Vaccination strategies using viral vectors expressing TAAs and immunomodulatory agents are first discussed, either alone or in combination. We contrast this concept with intratumoural (IT) vaccination strategies that attempt to

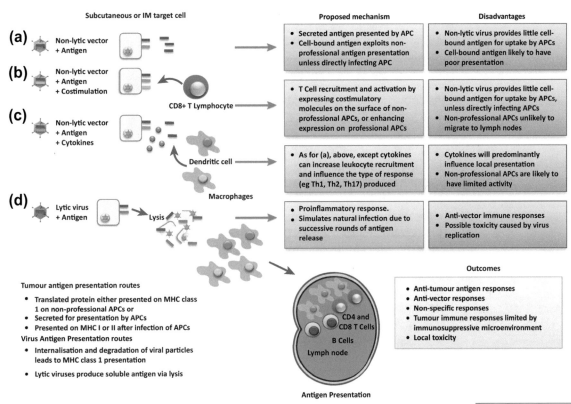

Subcutaneous or IM target cell

(a) Non-lytic vector + Antigen

(b) Non-lytic vector + Antigen + Costimulation

CD8+ T Lymphocyte

(c) Non-lytic vector + Antigen + Cytokines

Dendritic cell

Macrophages

(d) Lytic virus + Antigen

Lysis

Proposed mechanism

- Secreted antigen presented by APC
- Cell-bound antigen exploits non-professional antigen presentation unless directly infecting APC

- T Cell recruitment and activation by expressing costimulatory molecules on the surface of non-professional APCs, or enhancing expression on professional APCs

- As for (a), above, except cytokines can increase leukocyte recruitment and influence the type of response (eg Th1, Th2, Th17) produced

- Proinflammatory response.
- Simulates natural infection due to successive rounds of antigen release

Disadvantages

- Non-lytic virus provides little cell-bound antigen for uptake by APCs
- Cell-bound antigen likely to have poor presentation

- Non-lytic virus provides little cell-bound antigen for uptake by APCs, unless directly infecting APCs
- Non-professional APCs unlikely to migrate to lymph nodes

- Cytokines will predominantly influence local presentation
- Non-professional APCs are likely to have limited activity

- Anti-vector immune responses
- Possible toxicity caused by virus replication

Outcomes

CD4 and CD8 T Cells

B Cells

Lymph node

- Anti-tumour antigen responses
- Anti-vector responses
- Non-specific responses
- Tumour immune responses limited by immunosuppressive microenvironment
- Local toxicity

Tumour antigen presentation routes

- Translated protein either presented on MHC class 1 on non-professional APCs or
- Secreted for presentation by APCs
- Presented on MHC I or II after infection of APCs

Virus Antigen Presentation routes

- Internalisation and degradation of viral particles leads to MHC class 1 presentation
- Lytic viruses produce soluble antigen via lysis

Antigen Presentation

TRENDS in Molecular Medicine

FIGURE 1 **Methods of cancer vaccination at intramuscular and subcutaneous sites.**
The injection of recombinant viruses into non-tumour tissue has been used extensively as a method to induce systemic anticancer responses. **(a)** Encoded tumour-associated antigens (TAAs) can be either secreted, for presentation by antigen-presenting cells (APCs), or expressed on MHC class I at the surface of the infected cell (and MHC class II if infecting an APC). Using non-lytic viral vectors, little cell-bound antigen will be provided to APCs, and non-professional APCs are unlikely to traffic to the lymph nodes. **(b)** The expression of costimulatory molecules alongside TAAs attempts to increase costimulatory signaling by professional APCs, or convert non-professional APCs into T cell activating cells. **(c)** The expression of cytokines can be used to increase leukocyte infiltration to the vaccination site and skew the immune response towards desired outcome. **(d)** Replicating viral vectors are known to induce local inflammation, and their successive rounds of replication can lead to efficient and prolonged antigen release to the immune system, mimicking natural infections.

induce local and systemic anticancer responses using both replicating and non-replicating viruses. Finally, we highlight emerging approaches designed to capitalise on growing insights into cancer immunity, and consider strategies to produce a new generation of more effective recombinant cancer vaccines. Currently at least 40 clinical trials are recruiting patients with neoplastic disease where the treatment regimen includes the use of a recombinant virus expressing an antigen, cytokine or costimulatory molecule.

RECOMBINANT VIRAL VACCINES EXPRESSING TAAS

Many different viruses have been genetically modified to express TAAs for intramuscular (IM) or subcutaneous (SC) injection. Most investigation has focused on members of the adenoviridae, poxviridae, and herpesviridae families.

Adenovirus is a DNA virus that has been used extensively in the fields of vaccination, gene therapy, and oncolytic virotherapy. Its genome is well-characterised and amenable to the insertion of exogenous DNA, and it is an ideal candidate platform for the presentation of antigens to the immune system. Early cancer vaccination studies used replication-incompetent variants of serotypes Ad2 and Ad5 encoding a range of TAAs. However, most humans have pre-existing immune responses against these viruses, the result of lifelong exposure to environmental virus, and although preclinical animal models often respond well to vaccination, humans show variable vaccine responses and little therapeutic benefit [5–7].

Poxviruses have several beneficial features as vectors for cancer vaccines, including broad cell tropism, potent immunogenicity, and pre-existing 'neutralising' immunity only in those patients that have received vaccinia virus (the smallpox vaccine). Some members of this family have a tropism for monocytes and immature myeloid-derived dendritic cells [8], hence these viruses may infect antigen-presenting cells (APCs) directly and thereby improve antigen presentation. Vaccinia virus, fowlpox virus, and canarypox virus have all been investigated clinically for cancer vaccination, delivered by either SC or IM injection. For example, recombinant vaccinia virus (rVV) based on the attenuated Wyeth strain has been tested in a Phase I clinical trial as a vaccine against human carcino-embryonic antigen (CEA) [9]. No CEA specific responses were observed in the first assessment; however, subsequent T cell expansion in vitro produced cell lines capable of lysing CEA-expressing tumour cells, suggesting that the lack of response in patients may reflect inadequate clonal expansion and/or cytotoxicity in vivo. A series of similar studies using recombinant Wyeth strain vaccinia also showed no evidence of anti-CEA cellular responses in vivo, and although seven of 32 patients were found to have elevated levels of anti-CEA antibodies, these were of low titre and low affinity, and were not in sufficient quantity to modulate the circulating levels of CEA [10].

Although rVV virus has a good safety profile, modified vaccinia Ankara (MVA) is often a preferred vaccine platform because of its reduced replication in primary human cells. In one example, MVA has been used to overexpress the TAA oncofoetal protein 5T4 in a vector known as TROVAX. Human 5T4 (h5T4) is expressed on the surface of a range of carcinomas. Preclinical efficacy was attributed to CD4+ T cells and polyclonal h5T4-specific antibodies induced by vaccination, since only weak interferon γ (IFNγ) ELISPOT CD8+ T cell responses were induced [11]. In a Phase I/II trial of metastatic colorectal cancer, 14 of 17 patients that could be evaluated demonstrated a humoral response to 5T4. Periods of disease stabilisation ranging from three to 18 months were observed in five patients, and retrospective statistical analysis revealed an association between antibody responses and either patient survival or time to disease progression [12]. Cellular and antibody responses were readily observed in subsequent trials of prostate cancer [13] and alongside interferon-α treatment for renal cell carcinoma [14,15]. Prostate cancer patients who mounted cellular responses showed increased time to progression, although in renal cancer cellular and antibody responses were not clearly associated with efficacy. A large randomised Phase III trial involving 733 patients with metastatic renal cancer has been conducted using MVA-5T4 in combination with first-line treatment of sunitinib, interleukin-2, or interferon-α. No overall survival benefit was seen in the vaccine arm, however, analysis in this larger trial did reveal a significant correlation between the magnitude of 5T4-specific antibody response and improved patient survival [16].

Early studies of TAA-expressing recombinant viral vaccines often showed significant anticancer activity in animal models, but clinical outcomes from human trials produced marginal benefit at best. This finding could reflect several factors, including tolerance to the TAA in humans and the effects of tumour-associated immune-suppressive mechanisms. Neutralisation of the vaccine virus by the human immune system is also thought to be a key factor limiting successful vaccination, and in other disease settings this has been addressed using recombinant vaccines based on uncommon viral vector serotypes [17]. Finally, the possibility that the immune response may focus on strong viral antigens and ignore the coexpressed TAAs, so-called 'epitope dominance', has led to the 'prime–boost' approach, where two different recombinant viral vaccines expressing the same TAA are used consecutively, aiming to minimise anti-vector responses and focus activity against the TAA. This concept is discussed in detail below.

RECOMBINANT VIRAL VACCINES EXPRESSING TUMOUR-ASSOCIATED ANTIGENS AND IMMUNOMODULATORY MOLECULES

Viral vaccines expressing tumour antigens and costimulatory molecules

In this approach, viral vectors are used to enhance T cell activation by expressing TAAs alongside costimulatory molecules (Table 1). This can be achieved by infecting APCs, or by infecting other cell types and endowing them with the ability to provide the costimulatory signals necessary to activate TAA-specific T cells fully.

Table 1 Viral vaccines encoding TAAs and immunostimulatory agents

Agent	Virus	TAA [malignancy]	Immunostimulatory agent	Immunological and clinical outcomes	Refs
ALVAC–CEA–B7.1	Canary pox	CEA (Unresectable, CEA-positive malignancies)	CD80 (B7.1)	3/18 patients showed evidence of disease stabilisation. All three patients had CEA-specific T cell responses.	[18]
Vaccinia with HLA-A201 restricted epitopes	Vaccinia	HLA-A201 restricted gp100, Melan-A (MART-1), and tyrosinase (Metastatic melanoma)	CD80 (B7.1) and CD86 (B7.2)	Single agent injection directly into lymph node, or given as a prime followed by peptide boosting; both gave Ag-specific CD8+ responses. No overall survival benefit.	[20,21]
rV–B7.1, rV–PSA, and rF–PSA	Vaccinia prime, fowl-pox boost.	PSA (Non-metastatic prostate cancer treated locally, but with subsequently rising PSA levels)	CD80 (B7.1)	Assessed alongside nilutamide in a cross over study. Patients treated in the vaccine arm first showed a trend towards increased time to disease progression. Retrospective analysis showed benefit was more marked in particular patient subsets.	[19]
TRICOM–PSA	Vaccinia prime, fowl-pox boost	PSA (Metastatic castration-resistant prostate cancer)	TRICOM: CD80 (B7.1) CD54 (ICAM-1) CD58 (LFA3)	TRICOM vectors increased TAA-specific CTL responses, particularly with GM-CSF or IL-2. In prostate cancer an increase in progression-free survival was observed.	[24,25]
PANVAC	Vaccinia prime, fowl-pox boost	CEA and MUC-1 (Advanced CEA/MUC1 expressing tumours. Breast and Ovarian cancer)	TRICOM: CD80 (B7.1) CD54 (ICAM-1) CD58 (LFA3)	PANVAC is capable of eliciting TAA-specific CD4 and CD8 responses. A large Phase III trial in pancreatic cancer failed to show an overall survival benefit. However, clinical responses have been observed in other malignancies.	[27,28]

Table 1 Viral vaccines encoding TAAs and immunostimulatory agents *Continued*

Agent	Virus	TAA [malignancy]	Immunostimulatory agent	Immunological and clinical outcomes	Refs
TG4010	Modified vaccinia Ankara (MVA)	MUC-1 (Renal cell carcinoma, prostate cancer, and NSCLC)	IL-2	Limited evidence of improved survival in renal and prostate cancer. TG4010 has been extensively studied in NSCLC, with the most recent trial (using TG4010 as an adjunct to standard chemotherapy) reporting increased PFS for those in the vaccine group. Trials are ongoing.	[29, 31–33]
TG1031	Vaccinia	MUC-1 (Inoperable breast cancer recurrences)	IL-2	Well tolerated but little evidence of clinical efficacy.	[30]

ALVAC–CEA–B7.1 is a canarypox virus expressing the TAA CEA and the costimulatory molecule CD80 (B7.1). Building on promising preclinical data, ALVAC–CEA–B7.1 was injected IM into patients with advanced, unresectable CEA-expressing malignancies. The virus could induce CEA-specific peripheral blood T cells in a proportion of patients, and 3 of 16 patients demonstrated transient disease stabilisation, but no disease regression. All three patients had evidence of CEA-specific T cell responses [18].

In a prime–boost approach, priming with vaccinia virus expressing prostate-specific antigen (PSA) and B7.1, followed by boosting with a recombinant fowlpox–PSA, has been investigated alongside nilutamide chemotherapy in the treatment of prostate cancer [19]. Although there was a hint of increased time to progression in patients randomised to the vaccine arm first (median 5.1 years versus 3.4 years), this was not significant ($P = 0.13$). Retrospective analysis of subgroups suggested particular benefits in patients with a Gleason score (a pathological measure, with higher Gleason score predicting worse prognosis) of ≤ 7 ($P = 0.033$), levels of PSA < 20 ng/dl ($P = 0.013$), or previous treatment with radiotherapy ($P = 0.018$). However, retrospective subgroup analysis of data should be interpreted with caution.

To further strengthen antigen presentation, two costimulatory molecules, CD80 (B7.1) and CD86 (B7.2), have been coexpressed from a recombinant, UV-inactivated vaccinia virus encoding HLA-A201 restricted gp100, Melan-A (MART-1), and tyrosinase epitopes. This has been assessed in metastatic melanoma where priming with the viral vector was followed by peptide boosting, and granulocyte macrophage colony-stimulating factor (GM-CSF) was used as a systemic adjuvant. Cytotoxic T-lymphocyte (CTL) responses were readily induced and disease regression was observed in some individual lesions. However, these responses translated poorly into survival benefit [20]. This agent was subsequently assessed after direct injection into the inguinal lymph nodes and, again, induction of epitope-specific CTL responses was observed, but these did not correlate with improved survival [21].

TRICOM vaccines (triad of costimulatory molecules) are poxviruses (vaccinia and fowlpox) expressing TAAs alongside the three costimulatory molecules CD80 (B7.1), CD54 (ICAM-1), and CD58 (LFA3) [22]. Following promising preclinical development results [23] a vaccinia prime–fowlpox boost 'PROSTVAC-VF' regime was assessed, with both vectors expressing PSA alongside the TRICOM proteins, in a Phase II clinical trial for metastatic prostate cancer. Patients were randomised to receive either PROSTVAC–VF and GM-CSF, or empty viral vector and saline injections. The study did not meet its primary objective of improved progression-free survival, but did find an increased median overall survival compared with control subjects (25.1 months versus 16.6 months; $P = 0.015$) [24]. However, interpretation of these findings is made difficult by methodological issues in the study, and an ongoing Phase III clinical trial should yield more information. Interestingly, the PROSTVAC-VF investigators have gone on to note that Treg suppressive function can be decreased following vaccination in patients with improved survival, and increased in patients with shorter than anticipated survival [25]. This suggests that any vaccine efficacy may be mediated, at least in part, by the modification of the Treg response.

A vaccinia prime–fowlpox boost regime has also been studied in pancreatic cancer using PANVAC, with each vector encoding two TAAs (CEA and MUC1) alongside TRICOM [26]. Phase II results had been promising – increasing median survival in those patients with a pre-trial life expectancy of ≤3 months – but a Phase III trial to characterise efficacy better in these patients with advanced cancer did not demonstrate a survival benefit. More encouragingly, two recent studies enrolling patients with a variety of metastatic, progressive, MUC-1- or CEA-expressing malignancies showed TAA-specific immune responses [27] and suggested survival benefit in particular patient subgroups. Furthermore, a recent study using vectors expressing CEA and MUC1 alongside TRICOM to treat patients with metastatic breast and ovarian cancer showed signs of activity in patients with more limited disease [28]. It therefore seems that prime–boost approaches using vectors expressing TAA alongside costimulatory antigens can deliver clinical benefit, although there is considerable variability between different patients and even between different trials.

Viral vaccines expressing tumour antigens and interleukin-2

Interleukin-2 (IL-2) is a pleiotropic cytokine that is licensed for systemic administration in several malignancies. Poxviruses expressing the TAAs MUC-1 and IL-2 (VV–MUC1–IL-2 and MVA–MUC1–IL-2) have been studied in the context of a variety of malignancies including non-small cell lung cancer [29], breast cancer [30], renal cell cancer [31], and prostate cancer [32]. Overall, results have been varied, with promising immunogenicity findings and some evidence of limited clinical efficacy (Table 1).

TG4010 is a MVA vector encoding the TAAs MUC-1 and IL-2. A report of a randomised, open-label, Phase IIb study of TG4010 as an adjunct to conventional chemotherapy in patients with non-small cell lung cancer (NSCLC) showed that vaccination could increase progression-free survival [33]. Consequently, a confirmatory study is underway.

IT VACCINATION TO ENHANCE IMMUNE ACTIVATION

There is a growing body of evidence supporting the notion that the local environment within solid tumours is at least partially immune-privileged [2,34,35], and this may affect both antigen-presentation efficiency within tumours and the effector phase of cell killing. It follows that even successful vaccination at peripheral sites may not produce responses capable of mediating anticancer effects within the tumour microenvironment. Although there is no firm immunological consensus, it is conceivable that the use of recombinant vaccines directly within tumour nodules may be a superior method for generating effective responses in the context of tumour cell expression, perhaps by focussing immune scrutiny onto a range of TAAs that had previously gone unrecognised. Some of the mechanisms that may be employed are outlined in Figure 2.

Poxviruses have been employed as vectors to express costimulatory molecules after IT administration, ostensibly to endow tumour cells with antigen-presenting capabilities and enhance anticancer CTL responses. Although antigen presentation normally requires migration of the APC to lymph nodes, something unlikely to happen in this setting, in a small pilot study injection of melanoma lesions with a recombinant vaccinia virus vector expressing TRICOM was nevertheless shown to induce clinical responses in more than 30% of patients [36]. Whereas the vaccinia used in this study did not express a TAA, recombinant vaccinia virus expressing both PSA and TRICOM (rV–PSA/TRICOM) has been injected IT in 21 patients with locally recurrent prostate cancer. Early data indicate that vaccination results in significantly higher numbers of tumour-infiltrating CD4$^+$ and CD8$^+$ T cells. Additionally, vaccination reduced local Treg function; 16 patients (76%) had stable or improved serum PSA levels. Further trials to characterise the immunological and clinical endpoints better are planned, although the initial results of IT vaccination using recombinant pox viruses are encouraging [37].

ONCOLYTIC VIRAL VACCINES

Replicating viruses often induce a better immune response than non-replicating vectors [38]. In recent years, 'oncolytic' viruses have been developed that

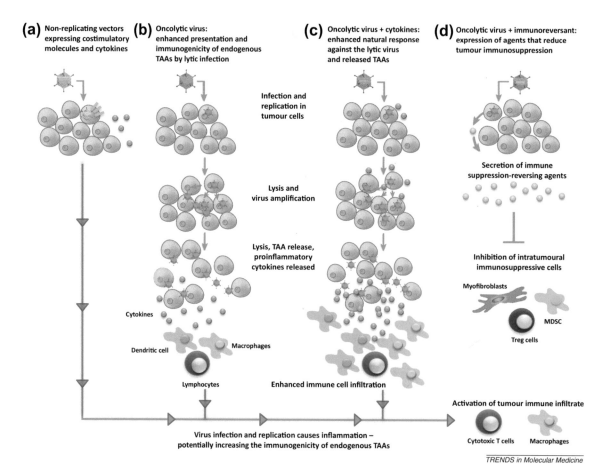

(a) Non-replicating vectors expressing costimulatory molecules and cytokines

(b) Oncolytic virus: enhanced presentation and immunogenicity of endogenous TAAs by lytic infection

(c) Oncolytic virus + cytokines: enhanced natural response against the lytic virus and released TAAs

(d) Oncolytic virus + immunoreversant: expression of agents that reduce tumour immunosuppression

Infection and replication in tumour cells

Lysis and virus amplification

Lysis, TAA release, proinflammatory cytokines released

Cytokines

Dendritic cell   Macrophages

Lymphocytes   Enhanced immune cell infiltration

Virus infection and replication causes inflammation – potentially increasing the immunogenicity of endogenous TAAs

Secretion of immune suppression-reversing agents

Inhibition of intratumoural immunosuppressive cells

Myofibroblasts

MDSC

Treg cells

Activation of tumour immune infiltrate

Cytotoxic T cells   Macrophages

TRENDS in Molecular Medicine

FIGURE 2 Intratumoural (IT) injection of viruses as a strategy to enhance cancer vaccination.
(a) IT injection of non-replicating viruses expressing cytokines and/or costimulatory molecules can be used to influence the immune environment within the tumour to improve antigen processing and presentation. **(b)** Oncolytic viruses are potently immunogenic, many being based on vaccine virus strains. The direct injection of oncolytic viruses into tumours has been used with the aim of stimulating an anticancer immune response while also causing direct cellular cytotoxicity. **(c)** Oncolytic viruses expressing cytokines, often termed oncolytic vaccines, have also been investigated extensively, and have demonstrated activity in a variety of tumours when injected directly into the tumour mass. **(d)** The observation that tumours often exhibit an immunosuppressive phenotype, that limits attack by the immune system, is leading the development of new virus strains that encode inhibitors of key immunosuppressive pathways and networks.

replicate and lyse cancer cells selectively, before spreading to infect adjacent cells and repeating the process. Some oncolytic viruses are showing good clinical results and are moving into advanced phase trials. Virus replication and cytotoxicity within the tumour may also enhance the vaccine effect *in situ* through local recruitment of immune cells, and there is a clear possibility of combining the science of oncolytic viruses and virally vectored vaccines to produce a new paradigm that might be termed 'oncolytic vaccines' (Table 2).

Table 2 'Oncolytic' viruses modified to express immunomodulatory cytokines

Cytokine	Virus	Additional modifications	Route	Immunological and clinical outcomes	Refs
Clinical Studies					
GM-CSF	JX-594 (vaccinia, Jennerex Inc)	Δ Thymidine kinase	IT	Phase I liver cancer: 3/14 patients showed partial response, 6/14 stable disease.	[62]
				Phase I/II melanoma: 6/10 patients exhibited significant increase in white blood cells. Immune stimulation evident by lymphocyte infiltration into tumours.	[63]
				Phase I refractory solid tumours: 12/23 patients, stable disease.	[64]
			IV	Ongoing Phase I/II (colorectal, solid refractory tumours), Phase I (paediatric cancers), Phase II (hepatocellular).	
GM-CSF	JX-963 (vaccinia, Jennerex Inc)	Δ Thymidine kinase, Δ vaccinia growth factor	IV	Rabbit tumour model: induction of tumour-specific cytotoxic lymphocytes; significant efficacy against primary liver tumours and metastases.	[65]
GM-CSF	OncoVex–GMCSF (HSV, BioVex Inc)	Δ ICP34.5 and α47	IT	Phase I breast, head and neck, gastrointestinal cancer.	[66]
				Phase I/II head and neck with cisplatin treatment: 14/17 patients showed response to tumour. 93% patients complete remission (confirmed by neck dissection). Prolonged progression-free survival in two thirds of patients.	[67]
				Phase II melanoma: 13/50 patients responded; most maintained for 7–31 months.	[47]
				Phase II melanoma: induction of local and systemic tumour specific T cell responses, decreased regulatory T cells.	[48]
				Ongoing Phase III (melanoma).	
GM-CSF	Ad5-D24–GMCSF (Ad)	Δ in Rb binding site of E1A	IT	Trialed in 16 patients with various advanced metastatic tumours: three showed objective responses, five had stable disease.	[68]
GM-CSF	Ad5-RGD-D24–GMCSF (Ad)	Δ in Rb binding site of E1A, RGD modified fibre	IT/IP	Trialed in seven patients with various advanced solid tumours: three showed disease stabilisation.	[69]

Continued

Table 2 'Oncolytic' viruses modified to express immunomodulatory cytokines *Continued*

Cytokine	Virus	Additional modifications	Route	Immunological and clinical outcomes	Refs
GM-CSF	CG0070 (Ad5, Cell Genesys Inc)	E2F-regulated E1A promoter	IVS	Phase I bladder cancer: anti-tumour responses noted in patients receiving weekly doses and in 2/5 patients receiving monthly treatment.	[70]
Preclinical Studies					
GM-CSF	Measles virus (Edmonston strain) GM-CSF	Attenuated by *in vitro* serial passage	IT	Murine lymphoid tumour model: significant tumour regression. Predominant anti-tumour cellular response due to influx of neutrophils.	[71]
IL-12	NV1042 (HSV)	Δ α47	IT	Murine models of squamous cell carcinoma, hepatic and prostate cancer: increased immune cell infiltration and decreased angiogenesis; inhibition of growth of distant tumours.	[72]
IL-12	HSVM002 (HSV)	Δ ICP34.5	IT	Murine neuroblastoma model: prolonged survival. Tumours displayed influx of CD4+ and CD8+ T cells, and macrophages.	[73]
IL-12	Ad–DHscIL12 (Ad)	Hypoxia-regulated E1A, Δ Rb binding site, E3 deletions, E4 and E1A were E2F-regulated	IT	Syrian hamster pancreatic cancer model: potent anti-tumour effect compared with parental Ad (minus IL-12). Less toxic than non-replicating AdIL-12 due to specific tumour expression and decreased systemic exposure of cytokine.	[74]
IL-12	rVSV–IL-12 (VSV)	None	IT	Murine squamous cell carcinoma model: significant reduction in tumour size. Prolonged survival.	[75]
IFN-α	KD3-IFN (Ad5)	Overexpression of ADP, mutated E1A	IT	Immunodeficient mice model of hepatocellular carcinoma: growth suppression of tumours and prolonged survival; decreased toxicity.	[76]
IFN-β	JX-795 (vaccinia)	Δ Thymidine kinase, Δ B18R	IT/IV	Murine tumour model: 100% complete response after local delivery, fewer with systemic administration.	[77]
IL-24/TRAIL	ZD55–IL24 (Ad5)/ZD55–TRAIL (Ad5)	Δ in E1B	IT	Murine colorectal carcinoma model: inhibition of tumour growth	[78]

Table 2 'Oncolytic' viruses modified to express immunomodulatory cytokines *Continued*

Cytokine	Virus	Additional modifications	Route	Immunological and clinical outcomes	Refs
IL-24	Ad.PEG–E1A–mda-7 (Ad5)	PEG-3-regulated E1A	IV	Immunodeficient mice model of prostate cancer: eradication of targeted and non-targeted (distant) tumours, suggesting innate immune role.	[79]
IL-24	CRAdRGD-flt-IL24 (Ad5)	VEGFR-1-regulated E1A, RGD modified fibre	IT	Murine glioma model: increased median survival times. Treatment in combination with temozolomide (TMZ).	[80]
IL-4 & IL4/IL10	R8306 (IL-4) and R8308 (IL-1-) (HSV-1)	Wild-type thymidine kinase	IT	Murine glioma model: HSV-IL4 showed prolonged survival compared when compared with saline or HSV-IL-10. HSV expressing IL-10 showed no efficacy.	[81]
IL-4/CD40 ligand/6CK	rHSVQ1–mIL-4 (HSV)	Δ ICP34.5 and ICP6	IT	Murine glioma model: slowed tumour growth.	[82]
IL-4	HYPR-Ad–IL-4 (Ad5)	Hypoxia (HIF)-regulated E1A, E3 deletion	IT	Immunodeficient mouse model of glioma: 70% decrease in tumour volume, tumour infiltration by leukocytes. Six-fold greater efficacy compared with wild-type Ad5.	[83]
IL-4	VSV–IL-4	None	IT	Murine models of breast cancer and melanoma: complete tumour regression, increased granulocyte infiltration and anti-tumour cytotoxic T cell responses.	[84]
IL-18/IL-12/B7-1	Triple armed HSV	Δ ICP34.5, ICP6, α47	IT	Murine neuroblastoma model: decreased tumour volume in both treated and non-treated tumours.	[85]

Abbreviations: ADP, adenovirus death protein; Ad, adenovirus; 6CK, secondary lymphoid tissue chemokine; FLT1, Fms tyrosine kinase related 1; GM-CSF, granulocyte macrophage colony stimulating factor; IFN-α, interferon-α; IFN-β, interferon-β; IL-12, interleukin-12; IV, intravenous; IVS, intravesical; IT, intratumoural; HSV, herpes simplex virus; MDA-7, melanoma differentiation associated-7; PEG3, progression-elevated gene-3; Rb, retinoblastoma protein; VEGFR-1, vascular endothelial growth factor receptor-1; VSV, vesicular stomatitis virus.

One approach to engineering oncolytic vaccines is to encode a TAA. This has received relatively little clinical attention to date, although an attenuated measles virus has been engineered to express CEA and injected intraperitoneally (IP) to patients with ovarian cancer [39]. Patients in this study were preselected for powerful anti-measles neutralising activity and for safety; although the treatment proved safe, anti-CEA responses were not stimulated. Clearly greater success might be observed by applying the virus in settings where it might not encounter a strong neutralising environment. An alternative type of oncolytic vaccine is one where the virus has been engineered to express immunomodulatory proteins and cytokines. Such viruses could combine direct cytotoxicity, adjuvanticity, and cytokine expression, and may represent platforms for potent immune stimulation.

GRANULOCYTE MACROPHAGE COLONY STIMULATING FACTOR

The cytokine GM-CSF has been extensively investigated for immunotherapy [40,41]. It plays a role in the maturation and activation of macrophages and dendritic cells, which can lead to the induction of antigen-specific T cells and the recruitment of natural killer (NK) cells [42]. The oncolytic vaccines that have advanced furthest in human clinical trials encode GM-CSF (Table 2), although the benefits of GM-CSF expression in this context are not clear and it is known that high levels of GM-CSF attract myeloid-derived suppressor cells that can impair immune responses [43].

JX-594 is an oncolytic vaccinia virus expressing GM-CSF that contains a deletion in the thymidine kinase (TK) locus to restrict its replication to cycling cells, and it has been evaluated as a treatment for several malignancies. A dose-escalating study by direct injection of the virus into primary and secondary liver cancers showed responses in both injected and non-injected tumours, and four of six patients treated at the maximum tolerated dose showed grade I–III increases in neutrophil counts, thought to reflect expression of GM-CSF [44]. Similarly, a Phase I/II trial of JX-594 to treat malignant melanoma in eight patients found that the number of GM-CSF responsive cells in the circulation increased after IT injection of the virus [45].

Herpes simplex virus type 1 (HSV-1) with deletions in the ICP34.5 and α47 loci [46], also expressing GM-CSF, has been explored in several Phase I and II trials, including studies of head and neck cancer and malignant melanoma [47]. Immunological analysis demonstrated increased melanoma-specific T cells and decreased numbers of Treg, CD8+ T-suppressor cells, and myeloid-derived suppressor cells [48]. OncoVex GM-CSF is currently undergoing Phase III trial by direct IT injection in melanoma patients. In this trial the control arm will receive subcutaneous GMCSF protein. This should make it possible to determine the therapeutic benefit of exploiting a viral vector to generate the GMCSF *in vivo*.

An oncolytic adenovirus expressing GM-CSF has also been developed, called Ad5-Δ24-GMCSF. This virus is incapable of binding the tumour-suppressor retinoblastoma protein (pRb), which allows the virus to replicate only in cells that are cycling and/or pRb defective. Following IT injection of 16 patients with various advanced metastatic tumours [49], three patients showed objective responses and five patients showed stable disease. Low levels of GM-CSF could be detected systemically, but there was no increase in leukocyte counts. However, T cell responses to adenovirus were noted. An integrin re-targeted form of Ad5-Δ24–GMCSF has also been tested in seven patients with different advanced solid tumours, including breast and colorectal cancers [50]. A total of three patients demonstrated disease stabilisation.

Overall the clinical results observed testing oncolytic viruses (which may be viewed as oncolytic vaccines) are encouraging and justify further study.

MODULATING THE TUMOUR IMMUNE MICROENVIRONMENT

Tumours often promote an immune-suppressive microenvironment that may well be a barrier to natural or vaccine-induced anti-tumour immunity. This may occur through the induction of immune-suppressor cells such as Tregs and myeloid-derived suppressor cells (MDSC), as well as through the secretion of soluble immune-suppressive factors such as TGFβ, IL-10, prostaglandin E2 (PGE2), and FASL [35]. These insights have led to new approaches in cancer immunotherapy that aim to reverse local tumour immune suppression, either as an independent therapeutic approach or together with cancer vaccination [51]. Recent licensing of ipilimumab, a monoclonal antibody that antagonises the inhibitory CTL-associated antigen 4 (CTLA-4) to reverse T cell attenuation and promote effective anticancer immune responses, as a treatment for metastatic melanoma, provides clinical validation of this approach [52]. Interestingly, a Phase I trial combining PSA–TRICOM with ipilimumab therapy in metastatic, hormone-insensitive prostate cancer has recently been completed. Investigators noted no increased incidence in immune-associated adverse effects and encouraging reductions in PSA levels – particularly in those individuals not previously treated with chemotherapy [53]. Further clinical trials of this combination are being planned.

Oncolytic viral vaccines show particular promise in addressing local immune suppression by inducing inflammatory cytokine expression that can enhance innate and adaptive anti-tumour immune responses and reduce tumour-induced immune suppression. For example, oncolytic vaccinia virus has been reported to reduce immune-suppressive cells such as MDSC, Tregs, and tumour-associated macrophages in a murine tumour model [54]. TGFβ inhibits the activation, maturation, and differentiation of innate and adaptive immune cells [55,56], and, indeed, anti-TGFβ treatment has also been shown to enhance cancer vaccine efficacy in various tumour models [57,58]. The therapeutic benefits of blocking TGFβ signalling prompted the design of viruses armed with the soluble TGFβ receptor II (STGFβRII) to antagonise TGFβ function in tumours. Adenovirus armed with STGFβRII fused to the Fc portion of human IgG could inhibit tumour growth and metastasis [59,60]; however, these studies were performed in nude mice and the effects of reversing TGFβ immune suppression in the context of adaptive immunity remain to be addressed.

Oncolytic vaccines armed with immune suppression-reversing agents are an attractive approach to enhance tumour immunity and vaccine efficacy, although there remains a significant challenge in identifying approaches that can target the wide range of immune-suppressive mechanisms in human tumours.

IMPLICATIONS FOR THE FUTURE OF CANCER VACCINES

Many different approaches are being explored for using recombinant viruses to produce cancer vaccine responses. The main approaches, together with their strengths and weaknesses, are outlined in Figures 1 and 2. The simplest approaches simply use the virus itself as a proinflammatory 'adjuvant' to enhance responses against encoded TAAs, however, viral vector versatility means that immune modulators may also be encoded for local expression. In a further step, genetically programming the viral vector to replicate and spread through tumours not only increases immunogenicity but also increases the amount of TAA or immune-modulator being produced. Coupled with the diversity of available viral vectors, with their different tropisms, life cycles, and packaging capacities that allow coexpression of several agents simultaneously, we clearly have platform technologies to explore and exploit a broad swath of proposed vaccine mechanisms. As many of the viruses described in this review progress towards and through advanced clinical trials, using control arms of either the virus or recombinant protein alone, the therapeutic benefit of delivering transgenes using viruses will become clearer and easier to define.

Increasingly, we are limited more by our understanding of the pathways of cancer immune evasion rather than by our ability to develop recombinant vaccines and immunotherapies capable of addressing them. Notably, it is clear that preclinical model systems can be poorly predictive of clinical challenges, with promising preclinical results often not translating into patient survival benefits. This may be due, at least in part, to the genetic diversity of the outbred human population, presenting with highly heterogeneous malignancies, but may also reflect qualitative differences between slow-growing clinical disease and the rapidly growing transplantable tumours that are often used in murine models. It may also reflect the presence of pre-existing vector-neutralising immunity in patients but not in animal models. Nonetheless, many recombinant vaccines can successfully induce detectable anti-TAA responses in clinical trials, although the correlation between this immune stimulation measured in the laboratory and the practical benefits for treated patients has been somewhat disappointing. It may well be that more sophisticated interrogation of clinical immune responses (e.g., measuring local responses in the tumour rather than systemic responses, measuring T cell binding avidity rather than T cell frequency, and assessing T cell subsets and natural killer cells) may correlate more closely with clinical outcomes. The measurement of local responses deserves particular attention because we are increasingly realising that the quantity, quality, and distribution of tumour-infiltrating lymphocytes are highly significant prognostic indicators. Furthermore, it is becoming clear that clinical trials of vaccines (and other immunotherapies) must be designed carefully to ensure that the endpoints measure outcomes relevant to the mechanism of action of the agent. For example, the size of a tumour on radiological examination may reasonably be expected to temporarily increase as immune cells infiltrate the tumour, and overall survival is likely to be a more important measure of efficacy than time to progression, given the kinetics of vaccine-based therapies.

Vaccine-based immunotherapy often assumes that, if a systemic immune response of sufficient magnitude can be achieved in a human cancer patient, this will result in clinical benefit. Although we have noted some evidence for this in some studies we have discussed, it may be that using vaccination to increase the quantity and quality of anticancer effector cells present systemically is insufficient to mediate clinically meaningful local effects. In fact, to shift favourably the balance between anticancer effector cells and tumour-induced immune suppression, we may need to generate a local immune response or combine vaccination with strategies that abolish or reduce local immune suppression. In this vein, an upcoming clinical trial will examine the benefit of combining systemic TROVAX vaccination with cyclophosphamide treatment in an effort to abrogate Treg effects [61]. Strategies that combine vaccination with chemotherapeutics are establishing themselves in preclinical and clinical trials, and it seems likely that cancer vaccines will be best used as part of a multi-modal treatment regimen.

If manipulating the local tumour immunosuppressive microenvironment is crucial to a successful immunotherapy strategy, it is unsurprising that we see some promising results with IT vaccination and oncolytic virotherapy. Viruses represent agents with a degree of functional versatility that is not available with other classes of therapeutics. As we understand more about the mechanisms underlying tumour immune suppression, the engineering potential of viral vectors should be harnessed in therapeutic efforts to define, and then overcome, this barrier. Similarly, increased knowledge of cancer biology will provide a new range of targets for vaccine design, and the synergy between increased knowledge and engineering potential provides us with an ever-improving arsenal of immunotherapies for cancer.

ACKNOWLEDGMENTS

We gratefully acknowledge the financial support of Cancer Research UK (R.C., K.D.F.), the Rhodes Trust (T.H.), Medical Research Council (S.L.W.), and Scholarship programme, King Abdulaziz University, Saudi Arabia (A.A.).

REFERENCES

1 Schreiber, R.D. et al. (2011) Cancer immunoediting: integrating immunity's roles in cancer suppression and promotion. Science 331, 1565–1570

2 Vesely, M.D. et al. (2011) Natural innate and adaptive immunity to cancer. Annu. Rev. Immunol. 29, 235–271

3 Francis, D.P. (2010) Successes and failures: worldwide vaccine development and application. Biologicals 38, 523–528

4 Rollier, C.S. et al. (2011) Viral vectors as vaccine platforms: deployment in sight. Curr. Opin. Immunol. 23, 377–382

5 Zhai, Y.F. et al. (1996) Antigen-specific tumor vaccines – development and characterization of recombinant adenoviruses encoding MART1 or gp100 for cancer therapy. J. Immunol. 156, 700–710

6 Rosenberg, S.A. et al. (1998) Immunizing patients with metastatic melanoma using recombinant adenoviruses encoding MART-1 or gp100 melanoma antigens. J. Natl. Cancer Inst. 90, 1894–1900

7 Lubaroff, D.M. et al. (2009) Phase I clinical trial of an adenovirus/prostate-specific antigen vaccine for prostate cancer: safety and immunologic results. Clin. Cancer Res. 15, 7375–7380

8 Yu, Q. et al. (2009) Poxvirus tropism for primary human leukocytes and hematopoietic cells. Methods Mol. Biol. 515, 309–328

9 Tsang, K.Y. et al. (1995) Generation of human cytotoxic T-cells specific for human carcinoembryonic antigen epitopes from patients immunized with recombinant vaccinia–CEA vaccine. J. Natl. Cancer Inst. 87, 982–990

10 Conry, R.M. et al. (2000) Human autoantibodies to carcinoembryonic antigen (CEA) induced by a vaccinia–CEA vaccine. Clin. Cancer Res. 6, 34–41

11 Harrop, R. et al. (2006) Active treatment of murine tumors with a highly attenuated vaccinia virus expressing the tumor associated antigen 5T4 (TroVax) is CD4+ T cell dependent and antibody mediated. Cancer Immunol. Immunother. 55, 1081–1090

12 Harrop, R. et al. (2006) Vaccination of colorectal cancer patients with modified vaccinia Ankara delivering the tumor antigen 5T4 (TroVax) induces immune responses which correlate with disease control: a phase I/II trial. Clin. Cancer Res. 12, 3416–3424

13 Amato, R.J. et al. (2008) Vaccination of prostate cancer patients with modified vaccinia Ankara delivering the tumor antigen 5T4 (TroVax) – a phase 2 trial. J. Immunother. 31, 577–585

14 Amato, R.J. et al. (2009) Vaccination of renal cell cancer patients with modified vaccinia Ankara delivering the tumor antigen 5T4 (TroVax) alone or administered in combination with interferon-alpha (IFN-alpha). A Phase 2 Trial. *J. Immunother.* 32, 765–772

15 Hawkins, R.E. et al. (2009) Vaccination of patients with metastatic renal cancer with modified vaccinia Ankara encoding the tumor antigen 5T4 (TroVax) given alongside interferon-alpha. *J. Immunother.* 32, 424–429

16 Amato, R.J. et al. (2010) Vaccination of metastatic renal cancer patients with MVA-5T4: A randomized, double-blind, placebo-controlled phase III study. *Clin. Cancer Res.* 16, 5539–5547

17 Sheehy, S.H. et al. (2012) Phase Ia clinical evaluation of the safety and immunogenicity of the Plasmodium falciparum blood-stage antigen AMA1 in ChAd63 and MVA vaccine vectors. *PLoS ONE* 7, e31208

18 Horig, H. et al. (2000) Phase I clinical trial of a recombinant canarypoxvirus (ALVAC) vaccine expressing human carcinoembryonic antigen and the B7.1 co-stimulatory molecule. *Cancer Immunol. Immunother.* 49, 504–514

19 Arlen, P.M. et al. (2005) Antiandrogen, vaccine and combination therapy in patients with nonmetastatic hormone refractory prostate cancer. *J. Urol.* 174, 539–546

20 Zajac, P. et al. (2003) Phase I/II clinical trial of a nonreplicative vaccinia virus expressing multiple HLA-A0201-restricted tumor-associated epitopes and costimulatory molecules in metastatic melanoma patients. *Hum. Gene Ther.* 14, 1497–1510

21 Adamina, M. et al. (2010) Intranodal immunization with a vaccinia virus encoding multiple antigenic epitopes and costimulatory molecules in metastatic melanoma. *Mol. Ther.* 18, 651–659

22 Hodge, J.W. et al. (1999) A triad of costimulatory molecules synergize to amplify T-cell activation. *Cancer Res.* 59, 5800–5807

23 Aarts, W.M. et al. (2002) Vector-based vaccine/cytokine combination therapy to enhance induction of immune responses to a self-antigen and antitumor activity. *Cancer Res.* 62, 5770–5777

24 Kantoff, P.W. et al. (2010) Overall survival analysis of a phase II randomized controlled trial of a poxviral-based PSA-targeted immunotherapy in metastatic castration-resistant prostate cancer. *J. Clin. Oncol.* 28, 1099–1105

25 Gulley, J.L. et al. (2010) Immunologic and prognostic factors associated with overall survival employing a poxviral-based PSA vaccine in metastatic castrate-resistant prostate cancer. *Cancer Immunol. Immunother.* 59, 663–674

26 Schuetz, T. et al. (2005) Extended survival in second-line pancreatic cancer after therapeutic vaccination. *J. Clin. Oncol.* 23, 2576

27 Gulley, J.L. et al. (2008) Pilot study of vaccination with recombinant CEA–MUC-1–TRICOM poxviral-based vaccines in patients with metastatic carcinoma. *Clin. Cancer Res.* 14, 3060–3069

28 Mohebtash, M. et al. (2011) A pilot study of MUC-1/CEA/TRICOM poxviral-based vaccine in patients with metastatic breast and ovarian cancer. *Clin. Cancer Res.* 17, 7164–7173

29 Ramlau, R. et al. (2008) A phase II study of Tg4010 (MVA–MUC1-IL2) in association with chemotherapy in patients with stage III/IV non-small cell lung cancer. *J. Thorac. Oncol.* 3, 735–744

30 Scholl, S.M. et al. (2000) Recombinant vaccinia virus encoding human MUC1 and IL2 as immunotherapy in patients with breast cancer. *J. Immunother.* 23, 570–580

31 Oudard, S. et al. (2011) A phase II study of the cancer vaccine TG4010 alone and in combination with cytokines in patients with metastatic renal clear-cell carcinoma: clinical and immunological findings. *Cancer Immunol. Immunother.* 60, 261–271

32 Dreicer, R. *et al*. (2009) MVA–MUC1-IL2 vaccine immunotherapy (TG4010) improves PSA doubling time in patients with prostate cancer with biochemical failure. *Invest. New Drugs* 27, 379–386

33 Quoix, E. *et al*. (2011) Therapeutic vaccination with TG4010 and first-line chemotherapy in advanced non-small-cell lung cancer: a controlled phase 2B trial. *Lancet Oncol.* 12, 1125–1133

34 Schlom, J. (2012) Therapeutic cancer vaccines: current status and moving forward. *J. Natl. Cancer Inst.* 104, 599–613

35 Mellor, A.L. and Munn, D.H. (2008) Creating immune privilege: active local suppression that benefits friends, but protects foes. *Nat. Rev. Immunol.* 8, 74–80

36 Kaufman, H.L. *et al*. (2006) Local delivery of vaccinia virus expressing multiple costimulatory molecules for the treatment of established tumors. *Hum. Gene Ther.* 17, 239–244

37 Heery, C.P. *et al*. (2011) Intraprostatic vaccine administration in patients with locally recurrent prostate cancer. *J. Clin. Oncol.* 29(Suppl. 7), abstract 141

38 Robert-Guroff, M. (2007) Replicating and non-replicating viral vectors for vaccine development. *Curr. Opin. Biotechnol.* 18, 546–556

39 Galanis, E. *et al*. (2010) Phase I trial of intraperitoneal administration of an oncolytic measles virus strain engineered to express carcinoembryonic antigen for recurrent ovarian cancer. *Cancer Res.* 70, 875

40 Kaur, B. *et al*. (2009) 'Buy one get one free': armed viruses for the treatment of cancer cells and their microenvironment. *Curr. Gene Ther.* 9, 341–355

41 Melcher, A. *et al*. (2011) Thunder and lightning: immunotherapy and oncolytic viruses collide. *Mol. Ther.* 19, 1008–1016

42 Shi, Y. *et al*. (2006) Granulocyte-macrophage colony-stimulating factor (GM-CSF) and T-cell responses: what we do and don't know. *Cell Res.* 16, 126–133

43 Parmiani, G. *et al*. (2007) Opposite immune functions of GM-CSF administered as vaccine adjuvant in cancer patients. *Ann. Oncol.* 18, 226–232

44 Park, B.H. *et al*. (2008) Use of a targeted oncolytic poxvirus, JX-594, in patients with refractory primary or metastatic liver cancer: a phase I trial. *Lancet Oncol.* 9, 533–542

45 Hwang, T.H. *et al*. (2011) A mechanistic proof-of-concept clinical trial with JX-594, a targeted multi-mechanistic oncolytic poxvirus, in patients with metastatic melanoma. *Mol. Ther.* 19, 1913–1922

46 Hernandez-Alcoceba, R. (2011) Recent advances in oncolytic virus design. *Clin. Transl. Oncol.* 13, 229–239

47 Senzer, N.N. *et al*. (2009) Phase II clinical trial of a granulocyte-macrophage colony-stimulating factor-encoding, second-generation oncolytic herpesvirus in patients with unresectable metastatic melanoma. *J. Clin. Oncol.* 27, 5763–5771

48 Kaufman, H.L. *et al*. (2010) Local and distant immunity induced by intralesional vaccination with an oncolytic herpes virus encoding GM-CSF in patients with stage IIIc and IV melanoma. *Ann. Surg. Oncol.* 17, 718–730

49 Cerullo, V. *et al*. (2010) Oncolytic adenovirus coding for granulocyte macrophage colony-stimulating factor induces antitumoral immunity in cancer patients. *Cancer Res.* 70, 4297

50 Pesonen, S. *et al*. (2012) Integrin targeted oncolytic adenoviruses Ad5-D24-RGD and Ad5-RGD-D24-GMCSF for treatment of patients with advanced chemotherapy refractory solid tumors. *Int. J. Cancer* 130, 1937–1947

51 Stewart, T.J. and Smyth, M.J. (2011) Improving cancer immunotherapy by targeting tumor-induced immune suppression. *Cancer Metastasis Rev.* 30, 125–140

52 Lesterhuis, W.J. *et al*. (2011) Cancer immunotherapy – revisited. *Nat. Rev. Drug Discov.* 10, 591–600

53 Madan, R.A. *et al.* (2012) Ipilimumab and a poxviral vaccine targeting prostate-specific antigen in metastatic castration-resistant prostate cancer: a phase 1 dose-escalation trial. *Lancet Oncol.* 15, 501–508

54 Thorne, S.H. *et al.* (2010) Targeting localized immune suppression within the tumor through repeat cycles of immune cell-oncolytic virus combination therapy. *Mol. Ther.* 18, 1698–1705

55 Li Yang, Y.P. and Moses, Harold L. (2010) TGF-β and immune cells: an important regulatory axis in the tumor microenvironment and progression. *Trends Immunol.* 31, 220–227

56 Miyazono, H.I.a.K. (2010) TGFbeta signalling: a complex web in cancer progression. *Nat. Rev. Cancer* 10, 415–424

57 Terabe, M. *et al.* (2009) Synergistic enhancement of CD8+ T cell-mediated tumor vaccine efficacy by an anti-transforming growth factor-monoclonal antibody. *Clin. Cancer Res.* 15, 6560–6569

58 Ueda, R. *et al.* (2009) Systemic inhibition of transforming growth factor in glioma-bearing mice improves the therapeutic efficacy of glioma-associated antigen peptide vaccines. *Clin. Cancer Res.* 15, 6551–6559

59 Seth, P. *et al.* (2006) Development of oncolytic adenovirus armed with a fusion of soluble transforming growth factor receptor II and human immunoglobulin Fc for breast cancer therapy. *Hum. Gene Ther.* 17, 1152–1161

60 Hu, Z. *et al.* (2010) Systemic delivery of an oncolytic adenovirus expressing soluble transforming growth factor receptor Fc fusion protein can inhibit breast cancer bone metastasis in a mouse model. *Hum. Gene Ther.* 21, 1623

61 Godkin, A. (2011) TaCTiCC (TroVax® And Cyclophosphamide Treatment In Colorectal Cancer). A pilot study to assess the effect of regulatory T cell depletion on 5T4-containing MVA (TROVAX®) vaccination in patients with INOPERABLE metastatic colorectal cancer. http://www.controlled-trials.com/ISRCTN54669986

62 Park, B-H. *et al.* (2008) Use of a targeted oncolytic poxvirus, JX-594, in patients with refractory primary or metastatic liver cancer: a phase I trial. *Lancet Oncol.* 9, 533–542

63 Hwang, T-H. *et al.* (2011) A mechanistic proof-of-concept clinical trial with JX-594, a targeted multi-mechanistic oncolytic poxvirus, in patients with metastatic melanoma. *Mol. Ther.* 10, 1913–1922

64 Breitbach, C.J. *et al.* (2011) Intravenous delivery of a multi-mechanistic cancer-targeted oncolytic poxvirus in humans. *Nature* 477, 99–102

65 Lee, J.H. *et al.* (2010) Oncolytic and immunostimulatory efficacy of a targeted oncolytic poxvirus expressing human GM-CSF following intravenous administration in a rabbit tumor model. *Cancer Gene Ther.* 17, 73–79

66 Hu, J.C.C. *et al.* (2006) A phase I study of OncoVEX(GM-CSF), a second-generation oncolytic herpes simplex virus expressing granulocyte macrophage colony-stimulating factor. *Clin. Cancer Res.* 12, 6737–6747

67 Harrington, K.J. *et al.* (2010) Phase I/II study of oncolytic HSV(GM-CSF) in combination with radiotherapy and cisplatin in untreated stage III/IV squamous cell cancer of the head and neck. *Clin. Cancer Res.* 16, 4005–4015

68 Cerullo, V. *et al.* (2010) Oncolytic adenovirus coding for granulocyte macrophage colony-stimulating factor induces antitumoral immunity in cancer patients. *Cancer Res.* 70, 4297–4309

69 Pesonen, S. *et al.* (2011) Integrin targeted oncolytic adenoviruses Ad5-D24-RGD and Ad5-RGD-D24-GMCSF for treatment of patients with advanced chemotherapy refractory solid tumors. *Int. J. Cancer* 130, 1937–1947

70 McKiernan, J.M. *et al.* (2008) Phase 1 study of multi-dose administration of intravesical CG0070 in patients with non-muscle invasive bladder cancer (NMIBC). *J. Urol.* 179, 616

71 Grote, D. *et al*. (2003) Neutrophils contribute to the measles virus-induced antitumor effect: Enhancement by granulocyte macrophage colony-stimulating factor expression. *Cancer Res.* 63, 6463–6468

72 Varghese, S. *et al*. (2006) Enhanced therapeutic efficacy of IL-12, but not GM-CSF, expressing oncolytic herpes simplex virus for transgenic mouse derived prostate cancers. *Cancer Gene Ther.* 13, 253–265

73 Parker, J.N. *et al*. (2000) Engineered herpes simplex virus expressing IL-12 in the treatment of experimental murine brain tumors. *Proc. Natl. Acad. Sci. U.S.A.* 97, 2208–2213

74 Bortolanza, S. *et al*. (2009) Treatment of pancreatic cancer with an oncolytic adenovirus expressing interleukin-12 in syrian hamsters. *Mol. Ther.* 17, 614–622

75 Shin, E.J. *et al*. (2007) Interleukin-12 expression enhances vesicular stomatitis virus oncolytic therapy in murine squamous cell carcinoma. *Laryngoscope* 117, 210–214

76 Shashkova, E.V. *et al*. (2007) Targeting interferon-alpha increases antitumor efficacy and reduces hepatotoxicity of E1A-mutated spread-enhanced oncolytic adenovirus. *Mol. Ther.* 15, 598–607

77 Kirn, D.H. *et al*. (2007) Targeting of interferon-beta to produce a specific, multi-mechanistic oncolytic vaccinia virus. *PLoS Med.* 4, 2001–2012

78 Zhao, L. *et al*. (2006) The antitumor activity of TRAIL and IL-24 with replicating oncolytic adenovirus in colorectal cancer. *Cancer Gene Ther.* 13, 1011–1022

79 Greco, A. *et al*. (2010) eradication of therapy-resistant human prostate tumors using an ultrasound-guided site-specific cancer terminator virus delivery approach. *Mol. Ther.* 18, 295–306

80 Kaliberova, L.N. *et al*. (2009) CRAdRGDflt–IL24 virotherapy in combination with chemotherapy of experimental glioma. *Cancer Gene Ther.* 16, 794–805

81 Andreansky, S. *et al*. (1998) Treatment of intracranial gliomas in immunocompetent mice using herpes simplex viruses that express murine interleukins. *Gene Ther.* 5, 121

82 Terada, K. *et al*. (2006) Development of a rapid method to generate multiple oncolytic HSV vectors and their *in vivo* evaluation using syngeneic mouse tumor models. *Gene Ther.* 13, 705–714

83 Cherry, T. *et al*. (2010) Second-generation HIF-activated oncolytic adenoviruses with improved replication, oncolytic, and antitumor efficacy. *Gene Ther.* 17, 1430–1441

84 Fernandez, M. *et al*. (2002) Genetically engineered vesicular stomatitis virus in gene therapy: application for treatment of malignant disease. *J. Virol.* 76, 895–904

85 Ino, Y. *et al*. (2006) Triple combination of oncolytic herpes simplex virus-1 vectors armed with interleukin-12, interleukin-18, or soluble B7-1 results in enhanced antitumor efficacy. *Clin. Cancer Res.* 12, 643–652

ends in Molecular Medicine

Dynamics of Targeted Cancer Therapy

Ivana Bozic[1,2], Benjamin Allen[1], Martin A. Nowak[1,2,3],*

[1]*Program for Evolutionary Dynamics, Harvard University, Cambridge, MA 02138, USA,* [2]*Department of Mathematics, Harvard University, Cambridge, MA 02138, USA,* [3]*Department of Organismic and Evolutionary Biology, Harvard University, Cambridge, MA 02138, USA*

Correspondence: martin_nowak@harvard.edu

Trends in Molecular Medicine, Vol. 18, No. 6, June 2012 © 2012 Elsevier Inc.

http://dx.doi.org/10.1016/j.molmed.2012.04.006

SUMMARY

Targeted cancer therapies offer renewed hope for an eventual 'cure for cancer'. At present, however, their success is often compromised by the emergence of resistant tumor cells. In many cancers, patients initially respond to single therapy treatment but relapse within months. Mathematical models of somatic evolution can predict and explain patterns in the success or failure of anticancer drugs. These models take into account the rate of cell division and death, the mutation rate, the size of the tumor, and the dynamics of tumor growth including density limitations caused by geometric and metabolic constraints. As more targeted therapies become available, mathematical modeling will provide an essential tool to inform the design of combination therapies that minimize the evolution of resistance.

TARGETED CANCER THERAPY

Targeted cancer therapies are drugs that interfere with specific molecular structures implicated in tumor development [1]. In contrast to chemotherapy, which acts by killing both cancer cells as well as normal cells that divide rapidly, targeted therapies are a much sharper instrument and offer the prospect of more effective cancer treatment, with fewer side effects. Most targeted therapies are either small-molecule drugs that act on targets found inside the cell (usually protein tyrosine kinases) or monoclonal antibodies directed against tumor-specific proteins on the cell surface [2].

The first drug that was rationally developed to block a known oncogene was imatinib, a small-molecule drug that effectively blocks the activity of the

161

CellPress

BCR-ABL kinase protein in chronic myeloid leukemia (CML) [3]. The success of imatinib for treating CML is striking: the response rate to imatinib treatment is 90% compared with 35% that can be achieved with conventional chemotherapy [4]. Moreover, most patients taking imatinib achieve complete cytogenetic remission and those who do have an overall survival rate similar to the general population [5,6]. Unfortunately, many of the more recent targeted therapies are not as successful over time. An example is the EGFR tyrosine kinase inhibitor gefitinib, used to treat the 10% of patients with non-small cell lung cancer (NSCLC) who have EGFR-activating mutations. Patients taking gefitinib have a higher response rate and longer progression-free survival (75% and 11 months, respectively) compared with those treated with standard chemotherapy (30% and 5 months); however, after 2 years, disease progresses in more than 90% of patients who initially responded to gefitinib treatment [7].

The failures of targeted therapies in patients who initially respond to treatment are usually due to acquired resistance. This resistance is often caused by a single genetic alteration in tumor cells, arising either before or during treatment [8,9]. In the case of CML, several mutations in the *BCR-ABL* kinase domain have been shown to cause resistance to imatinib [10]. In the case of NSCLC, a mutation in *EGFR* is observed in approximately 50% of patients [11,12]. The mutation that confers resistance to targeted therapy does not necessarily arise in the gene that is targeted. For example, resistance to BRAF inhibitor PLX4032 (vemurafinib), used in the treatment of melanomas, does not occur via mutations in the *BRAF* gene [13].

The current situation has interesting parallels to the treatment of HIV with AZT (coincidentally, a failed cancer drug) in the 1990s. AZT impedes HIV progression, but during prolonged treatment the virus usually develops resistance. It was only after the introduction of combination therapies with several HIV inhibitors that the disease became controllable in most patients. The hope for cancer is that similarly, as more targeted therapies become available, combination targeted therapies will be able to achieve indefinite remission in most cancer patients. However, the situation in cancer is more complicated than in HIV: because every cancer is genetically unique, many targeted therapies are needed for effective combination therapies to be available for all cancers.

To understand why some targeted therapies succeed while others ultimately fail, it is important to study the evolutionary process by which resistance arises. Mathematical evolutionary models have previously provided great insight into the gradual escape of HIV from the immune system [14–18] and the response of HIV to treatment [19–21], and similar models can be applied to the evolution of tumors.

MODELING THE EVOLUTION OF RESISTANCE TO CANCER THERAPY

Evolutionary modeling of cancer has a rich history dating to the 1950s, when Nordling [22] and Armitage and Doll [23,24] showed how patterns in the age incidence of cancer could be explained by somatic evolutionary processes involving multiple mutations. Mathematical evolutionary models have elucidated important patterns in the genetic and clinical progression of cancer [25–32] and its response to treatment [33–36]. Attolini and Michor [37] provide a comprehensive review of the history and development of this field.

Evolutionary modeling is particularly useful for understanding the emergence of acquired resistance to treatment, either conventional chemotherapy or targeted therapy (Table 1). Investigations of this question usually model tumor growth and evolution as a branching process – a stochastic process in which cells divide and die at random. Mutations that confer resistance appear at random during cell divisions. In most models, the tumor and its clonal subpopulations (including those resistant to treatment) grow exponentially on average. However, many clones that arise subsequently disappear due to stochastic drift – fluctuations caused by randomness in cell division and death.

Goldie and Coldman [38–44] were the first to mathematically investigate the evolution of resistance to cancer therapy (chemotherapy, in their case).

Table 1 Models of the Evolution of Resistance to Cancer Therapy

Event of Interest	Model of Tumor Dynamics[a]	Refs
Resistant cells exist when tumor reaches detectable size	Exponential growth	[33,36,38,40,41,43,45]
Treatment fails due to resistance acquired during treatment	Exponential decay	[47,48]
Treatment fails due to resistance acquired before or during treatment	Exponential growth, then decay during treatment	[35,50]
Treatment fails due to resistance acquired before or during treatment	Density-dependent growth, then decay during treatment	This work

[a]Blue curves show the dynamics of sensitive cells, whereas red curves show a possible (stochastic) trajectory of the resistant cell population. In some cases, resistant cells may arise and subsequently disappear due to stochastic drift.

Specifically, they calculated the probability that resistant cells exist in a tumor that has grown exponentially to a certain size. One assumption made in their models is that resistance mutations are neutral (that is, they have no effect on fitness in the absence of treatment). Later work by Iwasa *et al.* [36], Haeno *et al.* [45], and Durrett and Moseley [46] relaxed this assumption by including the possibility that resistance mutations also confer a fitness advantage or disadvantage. A common feature of these investigations is their focus on the question of whether resistant cells exist in a tumor of detectable size. Although this is a valuable question, it does not fully address whether treatment will eradicate the tumor because resistant cells may disappear during treatment due to stochastic drift, especially if they are only present in small numbers when treatment begins.

More recent work [47–49] has addressed the probability that a treatment will eradicate a tumor, if a given number of resistant cells are present at the start of treatment. In these models, the number of sensitive cells declines exponentially due to treatment. The number of resistant cells is expected to grow exponentially on average, but they may be eliminated due to stochastic drift. In these studies, the probability of eradication was calculated in a variety of situations, including cases in which multiple mutations are required for resistance (e.g., when combination therapies are used). The formulas derived there provide an important component for calculating the overall probability of tumor eradication.

Komarova and Wodarz [35,50] derived an overall formula for the probability of tumor eradication in a fully stochastic model. They considered a tumor cell population that grows exponentially up to a certain size until treatment begins. During treatment, the number of sensitive cells declines exponentially, as in previous models. In their model, resistance can arise either before or during treatment. The authors calculated the probability of tumor eradication, based on the size of the tumor at the start of treatment and the rate at which resistance mutations appear. They found that resistance is more likely to arise during tumor growth rather than treatment. This effect is magnified if resistance requires multiple mutations (e.g., in the case of several drugs). A limiting assumption in this model is that tumors grow exponentially until treatment is initiated. Although tumors are believed to initially grow exponentially, their growth can slow as they expand, due to nutrient shortages or geometric constraints [51–53]. Because of these restrictions, tumors often reach a steady state, with little or no tumor growth until further mutations arise [54–56].

EFFECTS OF DENSITY DEPENDENCE ON THE EVOLUTION OF RESISTANCE

We present a method for quantifying the evolution of resistance in tumors that grow subject to density limitations. We assume that tumor growth is initially exponential, but this growth slows as the tumor size increases, and the tumor eventually reaches a steady state. In this steady state, density constraints prevent further growth, unless new mutations arise that allow the tumor to overcome these constraints. The key parameters of our model are the number N of tumor cells at steady state; the time T that the tumor remains at steady state before treatment; the initial rates of division (r) and death (d) of tumor cells; the rate u at which resistance mutations are produced; and the division and death rates (r' and d', respectively) of sensitive cells under treatment, in the absence of density constraints.

Mathematically, this method is based on a density-dependent branching process model of tumor growth. In this model, tumor evolution starts with a single sensitive cell. Sensitive cells divide at rate $r/(1+\eta X)$ and die at rate d, where X is the current total number of cells in the tumor and $\eta=(r-d)/(Nd)$. From these formulas we can see that tumor growth is initially exponential with rate $r-d$, but that the division rate decreases as the tumor approaches size N. The net growth rate, $r/(1+\eta X)-d$, is positive for $X<N$ and negative for $X>N$, thus the tumor will remain at approximately size N (steady state), with small fluctuations, until treatment starts. At every division, one of the daughter cells will, with probability u, receive a mutation conferring resistance to treatment. We initially assume that resistance mutations are selectively neutral before treatment. After the tumor has been at steady state for time T, treatment is initiated. We assume that treatment affects only sensitive cells, reducing their division rate to $r'/(1+\eta X)$ with $r' \le r$, and increasing the death rate to $d' \ge d$. We assume that $r' < d'$, so that the sensitive cell population declines approximately exponentially during treatment (otherwise the treatment is ineffective).

The dynamics of tumor size in this model can be approximately described by three phases: (i) expansion (up to size N); (ii) steady state (for time T); and (iii) treatment (Figure 1).

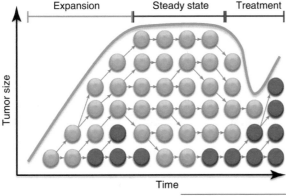

FIGURE 1 Density-dependent model of the evolution of acquired resistance.
Sensitive cells (blue) initially grow exponentially, but this growth slows due to density constraints. Resistant cells (red) arise through mutation. When treatment begins the sensitive cells decline, leaving room for resistant cells to grow.

Resistance mutations can arise during any of the three phases. However, the majority of resistance mutations will die out shortly after being produced, due to stochastic drift. For example, during the expansion and treatment phases, new resistance mutations disappear with probability approximately d/r, and those resistance mutations that do not survive drift have a median lifetime of $\log(2 - d/r)/(r - d)$ days [57]. For the parameter values $r = 0.25/$day, $d = 0.24/$day [31,54], only one out of every 25 resistance mutations survives stochastic drift, and the majority disappear within 5 days. Even resistant clones that grow to 20 cells still have a 44% $[=(d/r)^{20}]$ chance of disappearing due to drift. Resistance mutations arising during steady state have even slimmer chances: such a mutation has probability $1/(1 + dt)$ of surviving for t days after being produced. This probability decreases to zero as t increases.

Treatment will eradicate the tumor as long as no resistance mutations survive long enough to cause treatment failure. Considering that resistance mutations can arise during any phase, we write the overall probability P of tumor eradication as:

$$P = P_1 P_2 P_3.$$

Here P_1, P_2, and P_3 are the probabilities that no resistance mutations leading to treatment failure arise during expansion, steady state, and treatment, respectively.

P_1, the probability that no resistance mutation leading to treatment failure arises during expansion, is calculated in the supplementary material as:

$$P_1 = \exp\left(-Nu\frac{1-\xi}{d/r-\xi}\log\left(\frac{1-\xi}{1-d/r}\right)\right), \xi = \frac{d(r-d)T+d}{d(r-d)T+r}.$$

P_2, the probability that no resistance mutation leading to treatment failure arises during steady state, is also calculated in the supplementary material as:

$$P_2 = \left(1 + \frac{d}{r}(r-d)T\right)^{-Nu}.$$

P_3, the probability that no resistance mutation leading to treatment failure arises during treatment, was calculated in previous work [47–49]:

$$P_3 = \exp\left(-Nu\frac{r-d}{r}\frac{r'}{d'-r'}\right).$$

The accuracy of our formula for the probability of tumor eradication, using the above expressions for P_1, P_2, and P_3, is verified by simulations in the supplementary material (Figure S1).

The formulas above apply to the case that resistance mutations are selectively neutral. In the supplementary material, we also consider resistance mutations that carry a fitness cost, so that resistant cells divide at a reduced rate $\hat{r}/(1 + \eta X)$, with $\hat{r} < r$. The analysis is similar in this case, but the formulas for P_1, P_2, and P_3 are more complicated.

These formulas allow us to compare the relative importance of the three phases to the overall probability of eradication (Figure 2). Suppose, for example, that a tumor remains in steady state for a long period of time ($T \to \infty$). In this case, if resistance mutations are neutral (before treatment), P_1 increases to 1 and P_2 decreases to zero, while P_3 remains constant. Thus, treatment failure is probable in this case, due to resistance acquired during steady

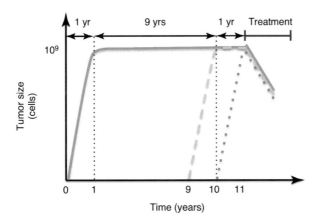

Time spent at steady state (years)	Probability that treatment fails due to resistance acquired during...			Probability of treatment success $(P = P_1 P_2 P_3)$
	Growth $(1-P_1)$	Steady state $(1-P_2)$	Treatment $(1-P_3)$	
0	0.63	0	0.03	0.36
1	0.35	0.78	0.03	0.14
10	0.10	0.97	0.03	0.02

TRENDS in Molecular Medicine

FIGURE 2 Treatment failure due to resistance acquired during three phases of tumor dynamics.
Top: sample growth trajectories of three tumors, which spend different amounts of time at steady state ($N=10^9$) before treatment. Bottom: probabilities of treatment failure due to resistance acquired during each of the three phases, and overall probability of treatment success, for these three trajectories. As more time is spent in steady state, treatment failure becomes increasingly probable, due to resistance acquired during this phase.

state. The outcome is similar if resistance mutations are costly, except that P_2 decreases not to zero but to an intermediate value:

$$P_2 \rightarrow \left(\frac{r-d}{\hat{r}-d}\right)^{-Nu r/\hat{r}}$$

In the opposite scenario, if treatment begins while the tumor is expanding (i.e., $T = 0$ and there is no steady state), our method reproduces results obtained by Komarova and Wodarz [35] using their biphasic (expansion then treatment) model. Their results (and ours) indicate that acquired resistance leading to treatment failure is more likely to arise during expansion than during treatment (in symbols, $P_3 > P_1$). This is true as long as the decline of sensitive tumor cells during treatment is faster than their growth during the expansion phase – a reasonable assumption for most targeted cancer therapies.

A common feature of all our results is that the probability of tumor eradication can be expressed as M^{-Nu}, where M is a positive quantity that depends on T, r, r', and d, but not on N or u. From this insight, we reason that (a) for $Nu \ll 1$, tumor eradication is almost certain; (b) for $Nu \gg 1$, treatment failure is almost certain; and (c) in between these two regimes, the probability of eradication declines sharply as Nu increases. A useful rule of thumb is that doubling the value of Nu has the effect of squaring the probability of treatment success. (For example, a 60% success probability would become 36% if Nu were doubled.) This exponential dependence on the product of tumor size and mutation rate was first noticed by Goldie and Coldman [38], who considered only the more limited question of whether resistant cells exist after a tumor

Table 2 The Probability of Treatment Success, Depending on Tumor Size and Time Spent at Steady State[a]

	$T=0$	$T=1$ year	$T=10$ years
Neutral			
$N=10^7$	0.990	0.981	0.964
$N=10^8$	0.902	0.822	0.690
$N=10^9$	0.357	0.140	0.024
$N=10^{10}$	0.0	0.0	0.0
Deleterious			
$N=10^7$	0.992	0.987	0.986
$N=10^8$	0.925	0.881	0.867
$N=10^9$	0.459	0.283	0.241
$N=10^{10}$	0.0	0.0	0.0

[a]Parameter values: $r=0.25$, $r'=0.1$, $d=d'=0.24$, $u=10^{-9}$. A deleterious resistant cell has a fitness disadvantage of 1% before treatment compared with sensitive cells ($\hat{r} = 0.99\,r$).

grows exponentially up to a certain size. Our findings extend this principle to the entire process of density-dependent tumor growth and treatment.

In Table 2 we show numerical results for the probability of treatment success as a function of the number of cells at steady state, N, and time spent there, T. These results illustrate points (a), (b), and (c) above. Additionally, they show that when N and $1/u$ are of similar orders of magnitude, the time spent at steady state has a significant effect on probability of treatment success. For example, when $N = 10^9$ (which corresponds to a tumor of approximately $1\,cm^3$) and resistance mutations arise at rate $u = 10^{-9}$, waiting to treat for a year after the tumor reaches the carrying capacity decreases the probability of treatment success from 36% to 14%. Waiting for 9 more years further decreases this chance to only 2%. This result reveals that treatment success depends critically not only on the size of the tumor but also its age, underscoring the importance of early detection and treatment.

Another important question is how long the treatment should last in order to eradicate all sensitive cells in the tumor. In the supplementary material we calculate the time until there is probability p that all sensitive cells have been eradicated:

$$t = \frac{1}{d' - r'} \log \left(\frac{-d' + r' p^{1/N}}{-d' + d' p^{1/N}} \right).$$

For example, in a tumor with $N = 10^9$ cells, using parameter values $r' = 0.22$, $d' = 0.24$, it will take 3.1 years of treatment to achieve a 99% probability that all sensitive cells have been eradicated. If treatment effectiveness is increased so that $r' = 0.1$, $d' = 0.24$, it will only take 0.5 years to achieve a 99% probability of eradication of sensitive cells. We caution, however, that eradication may take significantly longer if there are latent tumor cells unaffected by treatment.

APPLICATIONS AND EXTENSIONS

We believe our model may be useful in understanding resistance to many targeted therapies, and provides an important correction to models that assume exponential growth. The parameter values, including the division rate, death rate, rate of resistance mutation, and tumor size at steady state, may vary significantly among different types of cancer. Additionally, it may be appropriate to vary the functional form of the density limitation depending on the type of cancer (e.g., density limitations may apply differently in liquid versus solid tumors). However, our results remain applicable to other forms of density dependence as long as the three-phase approximation (exponential growth, steady state, treatment) is reasonably accurate.

Another important consideration in applying our model to different treatments is that some tumor cells may be incapable of regrowing a tumor, even if they carry a resistance mutation. This situation can be addressed by considering an 'effective population size' – equal to the number of cells that could seed or regrow a tumor – in place of the actual number of cells.

We note that many of our results can also be applied to the evolution of resistance to conventional chemotherapy. However, failures of chemotherapy are often due to factors other than acquired resistance, such as toxicity to the patient.

The quantitative predictions of our model are empirically testable. For instance, our model predicts a negative exponential relationship between the number of tumor cells and the probability of tumor eradication. This can be tested using data on treatment success rates for different tumor sizes. If, in addition, the tumor age [32], resistance mutation rate, and other parameters can be estimated, the formulas we present for tumor eradication probabilities can be tested directly.

CONCLUDING REMARKS AND FUTURE PERSPECTIVES

Mathematical modeling is an important tool for understanding the failure of targeted cancer therapies due to acquired resistance. Previous research on this question has focused mostly on resistance arising either during exponential tumor growth or during treatment. We show that phases of slow or no tumor growth are also clinically important in that they present an opportunity for resistance mutations to arise, thereby decreasing the chance of treatment success. Future models might incorporate more complex tumor dynamics, including different forms of density dependence [51,52] and/or alternating phases of growth and stasis.

As in the case of HIV, successful treatment of most cancers will probably require combination targeted therapy [58–61]. As more and more targeted therapies become available, the major challenge will be formulating effective combination therapies that minimize both the likelihood of resistance and toxicity to the patient. Mathematical models can help predict the success of potential combination therapies in advance of clinical trials.

APPENDIX A SUPPLEMENTARY MATERIAL

Supplementary material associated with this article can be found, in the online version, at http://www.sciencedirect.com/science/MiamiMultiMediaURL/1-s2.0-S1471491412000585/1-s2.0-S1471491412000585-mmc1.pdf/272186/FULL/S1471491412000585/9f4355fb19592f660dbf57ff6fa20bbf/mmc1.pdf.

REFERENCES

1 Sawyers, C. (2004) Targeted cancer therapy. *Nature* 432, 294–297

2 Imai, K. and Takaoka, A. (2006) Comparing antibody and small-molecule therapies for cancer. *Nat. Rev. Cancer* 6, 714–727

3 Gambacorti-Passerini, C. (2008) Part I: milestones in personalised medicine – imatinib. *Lancet Oncol.* 9, 600

4 Gerber, D.E. and Minna, J.D. (2010) ALK inhibition for non-small cell lung cancer: from discovery to therapy in record time. *Cancer Cell* 18, 548–551

5 Druker, B.J. *et al.* (2006) Five-year follow-up of patients receiving imatinib for chronic myeloid leukemia. *N. Engl. J. Med.* 355, 2408–2417

6 Gambacorti-Passerini, C. *et al.* (2011) Multicenter independent assessment of outcomes in chronic myeloid leukemia patients treated with imatinib. *J. Natl. Cancer Inst.* 103, 553–561

7 Maemondo, M. *et al.* (2010) Gefitinib or chemotherapy for non-small-cell lung cancer with mutated EGFR. *N. Engl. J. Med.* 362, 2380–2388

8 Pao, W. *et al.* (2005) Acquired resistance of lung adenocarcinomas to gefitinib or erlotinib is associated with a second mutation in the EGFR kinase domain. *PLoS Med.* 2, e73

9 Antonescu, C.R. *et al.* (2005) Acquired resistance to imatinib in gastrointestinal stromal tumor occurs through secondary gene mutation. *Clin. Cancer Res.* 11, 4182–4190

10 O'Hare, T. *et al.* (2007) Bcr-Abl kinase domain mutations, drug resistance, and the road to a cure for chronic myeloid leukemia. *Blood* 110, 2242–2249

11 Ercan, D. *et al.* (2010) Amplification of EGFR T790M causes resistance to an irreversible EGFR inhibitor. *Oncogene* 29, 2346–2356

12 Turke, A.B. *et al.* (2010) Preexistence and clonal selection of MET amplification in EGFR mutant NSCLC. *Cancer Cell* 17, 77–88

13 Nazarian, R. *et al.* (2010) Melanomas acquire resistance to B-RAF(V600E) inhibition by RTK or N-RAS upregulation. *Nature* 468, 973–977

14 Nowak, M.A. *et al.* (1991) Antigenic diversity thresholds and the development of AIDS. *Science* 254, 963–969

15 Nowak, M.A. and May, R.M. (1991) Mathematical biology of HIV infections: antigenic variation and diversity threshold. *Math. Biosci.* 106, 1–21

16 Ho, D.D. *et al.* (1995) Rapid turnover of plasma virions and CD4 lymphocytes in HIV-1 infection. *Nature* 373, 123–126

17 Wei, X. *et al.* (1995) Viral dynamics in human immunodeficiency virus type 1 infection. *Nature* 373, 117–122

18 Nowak, M.A. and May, R.M. (2000) *Virus Dynamics: Mathematical Principles of Immunology and Virology*. Oxford University Press

19 Coffin, J.M. (1995) HIV population dynamics in vivo: implications for genetic variation, pathogenesis, and therapy. *Science* 267, 483–489

20 Bonhoeffer, S. *et al.* (1997) Virus dynamics and drug therapy. *Proc. Natl. Acad. Sci. U.S.A.* 94, 6971–6976

21 Bonhoeffer, S. and Nowak, M.A. (1997) Pre-existence and emergence of drug resistance in HIV-1 infection. *Proc. R. Soc. Lond. B: Biol. Sci.* 264, 631–637

22 Nordling, C.O. (1953) A new theory on the cancer-inducing mechanism. *Br. J. Cancer* 7, 68–72

23 Armitage, P. and Doll, R. (1954) The age distribution of cancer and a multi-stage theory of carcinogenesis. *Br. J. Cancer* 8, 1–12

24 Armitage, P. and Doll, R. (1957) A two-stage theory of carcinogenesis in relation to the age distribution of human cancer. *Br. J. Cancer* 11, 161–169

25 Moolgavkar, S.H. and Knudson, A.G. (1981) Mutation and cancer: a model for human carcinogenesis. *J. Natl. Cancer Inst.* 66, 1037–1052

26 Moolgavkar, S.H. and Luebeck, E.G. (2003) Multistage carcinogenesis and the incidence of human cancer. *Genes Chrom. Cancer* 38, 302–306

27 Beerenwinkel, N. *et al*. (2007) Genetic progression and the waiting time to cancer. *PLoS Comput. Biol.* 3, e225

28 Dingli, D. *et al*. (2007) Stochastic dynamics of hematopoietic tumor stem cells. *Cell Cycle* 6, 461–466

29 Gatenby, R.A. and Gillies, R.J. (2008) A microenvironmental model of carcinogenesis. *Nat. Rev. Cancer* 8, 56–61

30 Meza, R. *et al*. (2008) Age-specific incidence of cancer: phases, transitions, and biological implications. *Proc. Natl. Acad. Sci. U.S.A.* 105, 16284–16289

31 Bozic, I. *et al*. (2010) Accumulation of driver and passenger mutations during tumor progression. *Proc. Natl. Acad. Sci. U.S.A.* 107, 18545–18550

32 Yachida, S. *et al*. (2010) Distant metastasis occurs late during the genetic evolution of pancreatic cancer. *Nature* 467, 1114–1117

33 Goldie, J.H. and Coldman, A.J. (1998) *Drug Resistance in Cancer: Mechanisms and Models*. Cambridge University Press

34 Michor, F. *et al*. (2005) Dynamics of chronic myeloid leukaemia. *Nature* 435, 1267–1270

35 Komarova, N.L. and Wodarz, D. (2005) Drug resistance in cancer: principles of emergence and prevention. *Proc. Natl. Acad. Sci. U.S.A.* 102, 9714–9719

36 Iwasa, Y. *et al*. (2006) Evolution of resistance during clonal expansion. *Genetics* 172, 2557–2566

37 Attolini, CS-O. and Michor, F. (2009) Evolutionary theory of cancer. *Ann. N.Y. Acad. Sci.* 1168, 23–51

38 Goldie, J.H. and Coldman, A.J. (1979) A mathematic model for relating the drug sensitivity of tumors to their spontaneous mutation rate. *Cancer Treat. Rep.* 63, 1727–1733

39 Goldie, J.H. *et al*. (1982) Rationale for the use of alternating non-cross-resistant chemotherapy. *Cancer Treat. Rep.* 66, 439–449

40 Coldman, A.J. and Goldie, J.H. (1983) A model for the resistance of tumor cells to cancer chemotherapeutic agents. *Math. Biosci.* 65, 291–307

41 Goldie, J.H. and Coldman, A.J. (1983) Quantitative model for multiple levels of drug resistance in clinical tumors. *Cancer Treat. Rep.* 67, 923–931

42 Coldman, A.J. *et al*. (1985) The effect of cellular differentiation on the development of permanent drug resistance. *Math. Biosci.* 74, 177–198

43 Coldman, A.J. and Goldie, J.H. (1986) A stochastic model for the origin and treatment of tumors containing drug-resistant cells. *Bull. Math. Biol.* 48, 279–292

44 Goldie, J.H. and Coldman, A.J. (1986) Application of theoretical models to chemotherapy protocol design. *Cancer Treat. Rep.* 70, 127–131

45 Haeno, H. *et al*. (2007) The evolution of two mutations during clonal expansion. *Genetics* 177, 2209–2221

46 Durrett, R. and Moseley, S. (2010) Evolution of resistance and progression to disease during clonal expansion of cancer. *Theoret. Pop. Biol.* 77, 42–48

47 Iwasa, Y. *et al*. (2003) Evolutionary dynamics of escape from biomedical intervention. *Proc. R. Soc. Lond. B: Biol. Sci.* 270, 2573–2578

48 Iwasa, Y. *et al.* (2004) Evolutionary dynamics of invasion and escape. *J. Theoret. Biol.* 226, 205–214

49 Michor, F. *et al.* (2006) Evolution of resistance to cancer therapy. *Curr. Pharm. Des.* 12, 261–271

50 Komarova, N. (2006) Stochastic modeling of drug resistance in cancer. *J. Theoret. Biol.* 239, 351–366

51 Hart, D. *et al.* (1998) The growth law of primary breast cancer as inferred from mammography screening trials data. *Br. J. Cancer* 78, 382–387

52 Spratt, J.A. *et al.* (1993) Decelerating growth and human breast cancer. *Cancer* 71, 2013–2019

53 Vaupel, P. *et al.* (1989) Blood flow, oxygen and nutrient supply, and metabolic microenvironment of human tumors: a review. *Cancer Res.* 49, 6449–6465

54 Jones, S. *et al.* (2008) Comparative lesion sequencing provides insights into tumor evolution. *Proc. Natl. Acad. Sci. U.S.A.* 105, 4283–4288

55 Jiang, Y. *et al.* (2005) A multiscale model for avascular tumor growth. *Biophys. J.* 89, 3884–3894

56 Yun, J. *et al.* (2009) Glucose deprivation contributes to the development of KRAS pathway mutations in tumor cells. *Science* 325, 1555–1559

57 Athreya, K.B. and Ney, P. (2004) *Branching Processes*. Dover Pubns

58 Azad, N.S. *et al.* (2008) Combination targeted therapy with sorafenib and bevacizumab results in enhanced toxicity and antitumor activity. *J. Clin. Oncol.* 26, 3709–3714

59 Komarova, N.L. *et al.* (2009) Combination of two but not three current targeted drugs can improve therapy of chronic myeloid leukemia. *PLoS ONE* 4, e4423

60 Sosman, J.A. *et al.* (2007) Opportunities and obstacles to combination targeted therapy in renal cell cancer. *Clin. Cancer Res.* 13, 764s–769s

61 Nimeiri, H.S. *et al.* (2008) Efficacy and safety of bevacizumab plus erlotinib for patients with recurrent ovarian, primary peritoneal, and fallopian tube cancer: a trial of the Chicago, PMH, and California Phase II consortia. *Gynecol. Oncol.* 110, 49–55

ends in Pharmacological Sciences

Targeting Protein–Protein Interactions as an Anticancer Strategy

Andrei A. Ivanov[1,3], Fadlo R. Khuri[2,3], Haian Fu[1,2,3],*

[1]*Department of Pharmacology, Emory University, Atlanta, GA 30322, USA,*
[2]*Department of Hematology and Medical Oncology, Emory University, Atlanta,
GA 30322, USA,* [3]*Emory Chemical Biology Discovery Center, Emory University,
Atlanta, GA 30322, USA*
*Correspondence: hfu@emory.edu

Trends in Pharmacological Sciences, Vol. 34, No. 7, July 2013 © 2013 Elsevier Inc.
http://dx.doi.org/10.1016/j.tips.2013.04.007

SUMMARY

The emergence and convergence of cancer genomics, targeted therapies,
and network oncology have significantly expanded the landscape of protein–
protein interaction (PPI) networks in cancer for therapeutic discovery. Extensive biological and clinical investigations have led to the identification of
protein interaction hubs and nodes that are critical for the acquisition and
maintenance of characteristics of cancer essential for cell transformation.
Such cancer-enabling PPIs have become promising therapeutic targets.
With technological advances in PPI modulator discovery and validation of
PPI-targeting agents in clinical settings, targeting of PPI interfaces as an anticancer strategy has become a reality. Future research directed at genomics-
based PPI target discovery, PPI interface characterization, PPI-focused
chemical library design, and patient-genomic subpopulation-driven clinical
studies is expected to accelerate the development of the next generation
of PPI-based anticancer agents for personalized precision medicine. Here
we briefly review prominent PPIs that mediate cancer-acquired properties,
highlight recognized challenges and promising clinical results in targeting
PPIs, and outline emerging opportunities.

RISING INTEREST IN TARGETING PPIS

PPI interfaces represent a highly promising, although challenging, class of
potential targets for therapeutic development [1]. In cancer, PPIs form signaling nodes and hubs that transmit pathophysiological cues along molecular

175

CellPress

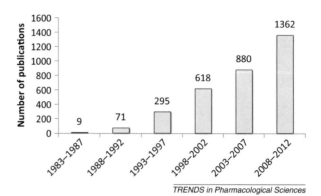

FIGURE 1 Rising number of publications in the field of cancer-related protein–protein interactions.
The PubMed database was searched using the following keywords: protein–protein interaction, tumor, cancer, and inflammation.

networks to achieve an integrated biological output, thereby promoting tumorigenesis, tumor progression, invasion, and/or metastasis. Thus, pathway perturbation, through disruption of PPIs critical for cancer, offers a novel and effective strategy for curtailing the transmission of oncogenic signals. As our understanding of cancer biology has significantly increased in recent years, interest in targeting of PPIs as anticancer strategies has increased as well (Figure 1).

PPI INTERFACES CONSTITUTE BASIC UNITS IN ONCOGENIC SIGNALING NETWORKS

A variety of environmental, genetic, and epigenetic factors induce the reprogramming of cancer-initiating cells and the acquisition of physical and molecular features that promote tumorigenesis and provide resistance to therapeutics. These characteristics, including sustained proliferative signaling and evasion of growth suppressors, permit the development and progression of cancer and have been recognized as distinctive hallmarks of cancer (Figure 2) [2]. These hallmarks provide a molecular framework for our understanding of cancer, linking molecular signaling events to pathological outcomes. The oncogenic potential of cells is determined by a combination of genetic and epigenetic alterations through the operation of well-orchestrated signaling networks. Importantly, PPIs represent the basic units within such vital networks.

On oncogenic stimulation, PPIs play essential roles in linking networks that relay oncogenic signals, allow the acquisition of hallmark features of cancer, and serve diverse roles in driving and maintaining the growth of

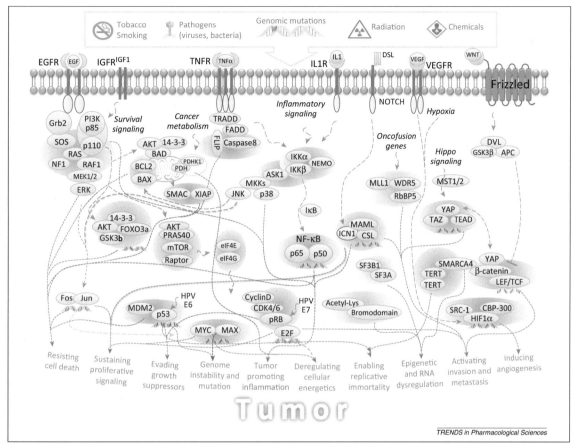

FIGURE 2 Representative PPIs in oncogenic signaling networks that drive the acquisition and development of hallmarks of cancer.

Grey broken arrows connect PPIs to corresponding cancer hallmarks. Some PPIs contribute to multiple features of cancer. It should be noted that some PPIs may impact global processes of cell growth and their precise connections to cancer remain to be established.

cancer cells (Figure 2). From the engagement of receptors with dysregulated growth factors to dimerization of receptor tyrosine kinases triggered by gene amplification or mutations, PPIs initiate a cascade of reactions to promote uncontrolled cell proliferation [3]. Activated Ras, due to perturbations such as epidermal growth factor receptor (EGFR) activation, neurofibromin 1 (NF1) deletion, or intrinsic mutation, assumes a conformation that allows it to bind to multiple regulatory proteins and results in enforced proliferation and survival. Survival signaling, activated by proteins such as insulin-like growth factor 1 (IGF1) and phosphoinositide-3-kinase (PI3K) or disabled by the negative regulator PTEN, enables tumors to resist cell death through a number of different mechanisms. For example, the Akt–FOXO3a–14-3-3 complex mediates a transcription-dependent mechanism, whereas the Akt–Bad–14-3-3 interaction mediates a transcription-independent antiapoptotic mechanism [3]. In addition to providing resistance to cell death, Akt also regulates the mTOR complex to control cap-dependent translation, through the eIF4E–eIF4G PPI, of a large number of growth-promoting genes, including c-Myc. In turn, amplified c-Myc favors binding to Max over Mad and thereby drives transcription of growth-promoting genes such as cyclin D that modulate cell cycle progression [4].

For cancer progression, cells must acquire mechanisms to evade growth suppression. Several PPI complexes, including MDM2–p53 and CDK4–pRB, play key roles in neutralizing such tumor suppressive functions [2]. These tumor suppressor mechanisms are often hijacked by viral oncoproteins, such as human papillomavirus E7 protein, which binds to pRb, and E6 protein, which binds to p53, that allow the virus to induce tumors. Such PPIs offer tumor-specific targets. In addition to the examples given above, a large number of PPIs dictate signaling networks that allow the acquisition or maintenance of other hallmarks of cancer. For instance, the VEGF–VEGFR and HIF1α–CBP PPIs mediate signals that induce angiogenesis, the catalytic activity of TERT dimers enables replicative immortality, and a variety of reprogrammed enzyme–substrate interactions, such as the onco-fusion gene-regulated PDHK1–PDHA1 PPI, play integral roles in dysregulated cellular metabolism by controlling a metabolic switch between glycolysis and oxidative phosphorylation [5]. In addition, mutated p53 and Myc also play key roles in the regulation of cancer metabolism. The IKK–NEMO–ASK1 complex integrates the proinflammatory function with stress response signaling initiated by reactive oxygen species [6]. It has recently been shown that epigenomic reprogramming is a critical part of cancer development [7], and PPIs involved in epigenomic dysregulation, such as SMARCA4 interactions, have been described [8].

As a result of oncogenic network reprogramming, some PPIs contribute to distinct features of cancer, whereas other PPIs are vital for multiple characteristics of cancer. For example, the MDM2–p53 and Myc–Max PPIs play key roles in evading growth suppression and cell death, as well as in promoting genomic instability and cancer metabolism. Thus, it is expected that interception of certain critical PPIs may disable multiple mechanisms that cancer cells rely on for survival. A large number of PPIs are involved in driving tumorigenesis through the regulation of oncogenic networks, so these PPI interfaces represent fertile ground for anticancer therapeutic discovery and development.

OVERCOMING CHALLENGES AND CURRENT STRATEGIES FOR TARGETING OF PPIS

Challenges in Discovering PPI Modulators

A number of challenges and concerns exist regarding targeting of PPIs, some of which include: (i) large PPI interface areas, (ii) a lack of deep pockets, (iii) the presence of noncontiguous binding sites, and (iv) a general lack of natural ligands. In addition, PPI surfaces differ from small-molecule binding sites in their shape and amino acid residue composition. In contrast to the well-defined and normally hydrophilic ligand-binding cavities observed in the crystal structures of enzymes and G-protein-coupled receptors, the interface surfaces of many protein–protein complexes are typically hydrophobic and relatively flat and often lack deep grooves where a small molecule can dock. Recent studies have addressed some of these concerns, as detailed in several publications [1,9,10]. Although the PPI interface generally covers an average area of 1150–10000 $Å^2$, which is larger than small-molecule binding sites of 100–600 $Å^2$, the presence of hot spots (small subsets of amino acid residues that contribute the most to the free binding energy) makes PPIs amenable to small-molecule perturbations [10,11]. In addition, PPIs are often mediated by post-translational modifications, such as the binding of 14-3-3 to phosphorylated Ser/Thr motifs and the interaction of bromodomain proteins to acetylated lysine, which simplifies definition of the targeting interfaces [12–14]. The typical lack of natural ligands for PPIs as starting points poses a significant challenge for structure-based drug design. However, promising examples of natural products that can act on PPIs do exist, such as rapamycin for mTOR and taxol for tubulin. There are also significant differences in the chemical space between PPI modulators (PPIMs) and conventional drug-like compounds [9,15,16]. In general, PPIMs have a higher molecular weight (> 400 Da) than that of typical drug-like compounds (200–500 Da) and they often violate the 'Rule of Five' [9]. Therefore, the application of commonly used high-throughput screening (HTS) methods for PPIMs has been limited because of the biased chemical composition towards classical target classes in current chemical libraries.

Current Approaches for the Discovery of PPI Modulators
Structure-based design

Structural studies allow for the identification of peptide fragments and amino acid residues that are critical for PPI. This information, combined with that from functional assays, provides a basis for the rational design of PPIMs. Not surprisingly, mimicking the structure of binding peptides is one of the approaches widely used to design novel PPIMs [17]. Application of this approach led to the identification of potent inhibitors of BCL2, XIAP, NOTCH, and MDM2 [18–22]. General structure-based approaches include computational molecular modeling [23,24], peptide engineering with display technologies such as phage display [25], design of small molecules based on α-helix and β-sheet scaffolds [26,27], and synthesis of conformationally constrained 'stapled' peptides with a stabilized α-helical structure [28–31].

Small-molecule screening methods

In contrast to structure-based design, the screening approach allows the discovery of small-molecule PPIMs even if structural information is very limited or unavailable. Various screening approaches, including HTS, have been used to identify compounds that target 'hot spots' of PPI interfaces [1]. Furthermore, HTS can be used to reveal inducible pockets in PPI interfaces, as well as allosteric modulators. In many cases, the screening approach is combined with structure-based design to further enhance the physicochemical and pharmacological properties of identified PPI modulators.

The most widely used HTS techniques for PPIs include fluorescence polarization (FP) and Föster/fluorescence resonance energy transfer (FRET). FP measures the change in emitted polarization signals in solution on association of a small fluorescent molecule (such as a peptide) with a relatively large binding partner. FRET is a nonradioactive, photophysical effect in which energy absorbed by a donor fluorophore is transferred to an acceptor fluorophore. By coupling the donor and acceptor fluorophores with appropriate spectral properties to two interacting molecules, the fluorophores can be brought into close proximity (10–100 Å) and induce a FRET signal. Both FP and FRET methods are extensively used in HTS campaigns for the discovery of PPIMs [32]. Other HTS methods include ELISA, flow cytometry, surface plasma resonance (SPR), and label-free platforms [33]. In addition to these biochemical HTS assays, intracellular PPIs can be coupled to readouts for cell-based reporter assays, which incorporate a more physiological cellular context and identify compounds that are cell-permeable. For example, the p53–MDM2 PPI has been linked to a reporter assay based on cytoplasm-nuclear redistribution [34]. Various biosensors, such as protein complementation assays, can also be used to monitor PPIs in a HTS format.

In addition to biochemical and cell-based HTS assays, fragment-based screening (FBS) is another approach commonly used for discovery of PPIMs. FBS aims to identify molecular fragments with binding activity for a target protein. Once the fragments have been identified, they, or the interactions they identify, are built into a drug-like compound [35]. The main advantage of the FBS approach is that a large chemical space can be targeted with approximately 10^3 fragments. The ligand efficiency (LE) of fragment hits is high. Moreover, FBS can be successfully utilized for many targets that were found to be challenging using traditional HTS [36]. Owing to their low molecular weight (~200 Da) and thus limited contact area with a protein, the binding affinity of the fragments is relatively low (often in the millimolar range). Therefore, to detect a binding event, sensitive biophysical methods are required, including X-ray crystallography, nuclear magnetic resonance (NMR), and SPR [35,37,38]. Key advantages of NMR include automated sampling, high sensitivity, quantitative data on binding affinity, and the ability to obtain structural information about the binding site [39]. Examples of successful application of NMR in FBS include discovery of XIAP [40], BCL2 [41], ZipA–FtsZ [42], and K-RAS–SOS [43] PPI inhibitors. A combination of FBS assays is often employed for a screening campaign. For instance, a combination of X-ray and NMR FB HTS led to identification of the Hsp90 inhibitor AT13387, which recently entered clinical trials [44,45]. Another assay commonly used for FBS is SPR, which detects changes in the refractive index near a sensor surface. SPR FBS has been utilized to identify novel inhibitors of Hsp90 interactions [46,47]. In contrast to methods for non-covalent binding, a 'tethering' approach is used to detect reversible covalent bonds formed between the cysteine of a target protein and fragment molecules containing a disulfide bond [48]. Tethering FBS has been employed to identify PPIMs such as IL2–IL2-αR inhibitors [1].

Two general approaches are used to perform computational screening of a 3D compound library: the ligand-based (also known as the pharmacophore-based) approach and structure-based virtual screening. A pharmacophore model represents the chemical features of a set of compounds critical for efficient protein binding [49]. These features or pharmacophore points (such as H-bond donors or acceptors, aromatic rings, and charges) have certain coordinates in a 3D space. The aim of this procedure is to identify compounds with a certain conformation and chemical composition that match the requirements of a pharmacophore model. The pharmacophore-based approach has been successfully utilized to identify novel inhibitors of the MDM2–p53 interaction [50,51], PPIs of BCL2 family proteins [52], and 14-3-3 inhibitors [53]. Conversely, the structure-based approach relies on structural information for the binding site on the target protein. For this type of screening, each compound in a chemical library has to be computationally docked to the binding site, and a binding affinity is estimated in terms of energy scoring functions. Application of structure-based virtual screening has led to the discovery of PPI inhibitors of Ubc13–Uev1 [54], MDM2–p53 [55], and TCF4–β-catenin [56].

CLINICAL VALIDATION OF PPI TARGETING IN CANCER

Thousands of compounds have already been tested as potential inhibitors of various PPIs and the results are promising. Titrobifan, a glycoprotein IIb/IIIa inhibitor, and Maraviroc, an inhibitor of the CCR5–gp120 interaction, are currently available on the market as cardiovascular and anti-HIV drugs,

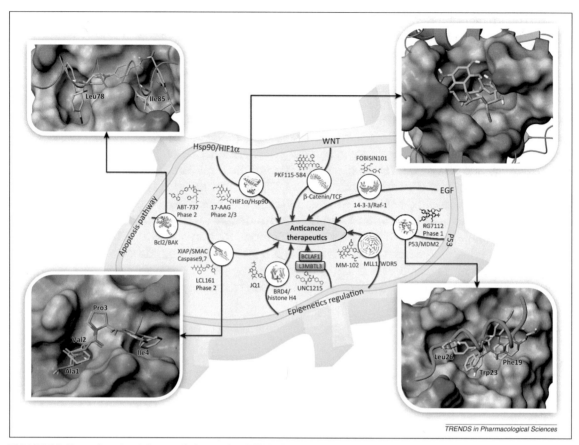

TRENDS in Pharmacological Sciences

FIGURE 3 Examples of protein–protein interaction (PPI) inhibitors that have entered clinical trials and emerging agents. Inhibitors of MDM2–p53, BCL2, XIAP, and Hsp90 PPIs are in Phase I–III trials. Examples of promising PPI targets with recently identified novel inhibitors include MLL1–WDR5, β-catenin–TCF, BCLAF1–L3MBTL3, BRD4–histone H4, and 14-3-3 interactions. Chemical structures of representative inhibitors are shown, along with available crystal structures for protein–protein complexes. Top left: superimposition of Bak peptide (orange, carbon atoms colored cyan) and ABT-737 (carbon atoms colored green) bound at BCL-xL (PDB ID: 1BXL, 2YXJ). ABT-737 occupies the same hydrophobic pocket on the BCL-xL surface as the peptide, overlapping with Leu78 and Ile85 Bak residues critical for peptide binding. Top right: crystal structure of Hsp90 in a complex with geldanamycin (PDB ID: 1YET), the first Hsp90 inhibitor to enter clinical trials. Bottom right: superimposition of p53 peptide and Nutlin-2 (one of the first identified potent MDM2–p53 inhibitors) bound to the N-terminal domain of MDM2 (PDB ID: 1YCR, 1RV1). Phe19, Trp23, and Leu26 residues of p53 occupy the hydrophobic pocket of MDM2. The ethoxy and chlorophenyl groups of Nutlin-2 match the positions of Phe19, Trp23, and Leu26, respectively. Bottom left: superimposition of SMAC AVPI peptide (carbon atoms colored orange) and GDC-0152, the first SMAC mimetic to enter clinical trials (carbon atoms colored green) (PDB ID: 3UW5, 1G73). The molecular surfaces of Hsp90, BCL-xL, MDM2, and XIAP are colored according to their electrostatic potential (blue, positive; white, neutral; red, negative).

respectively. These drugs demonstrate the feasibility of PPI targeting for the treatment of various diseases. In addition, several anticancer compounds have entered clinical trials, highlighting the potential of the PPI targeting approach in cancer.

Inhibitors of the MDM2–p53 Interaction: A Breakthrough in PPI Targeting

p53 plays a critical role in cell cycle regulation, DNA repair, angiogenesis, and apoptosis [57,58]. Activation of p53 increases the expression of human protein double minute 2 (HDM2, MDM2 in mouse), which in turn directly binds to p53 and inhibits its tumor suppressive activity (Figure 2) [59,60]. Structurally, the N-terminal domain of MDM2 binds a short 15-residue α-helical peptide of p53 [61]. Three hydrophobic residues of p53 (Phe19, Trp23, and Leu26) occupy a well-defined hydrophobic pocket of MDM2 (Figure 3). These structural features allow a strategy to target the MDM2–p53 PPI. Various approaches were used to identify drug-like inhibitors of MDM2–p53 interactions, including the design of peptidomimetics, HTS, and computational drug design. As a result, several MDM2–p53 PPI inhibitors have entered clinical trials [60,62]. For example, a series of *cis*-imidazoline analogs named Nutlins were identified by screening of compound libraries [59]. Nutlins employ the same binding mode as the p53 Phe19, Trp23, and Leu26 residues in the MDM2 binding pocket (Figure 3). Further chemical optimization of Nutlin-3 led to RG7112, the first MDM2 inhibitor to enter clinical trials in patients with advanced solid tumors in 2007 [63]. Another MDM2–p53 PPI inhibitor, RO5503781, is currently in a Phase I trial in patients with advanced malignancies (http://clinicaltrials. gov/ct2/show/NCT01462175%3Fterm%3DRO5503781%26rank%3D1). The exciting success of potent MDM2–p53 PPI inhibitors has significantly accelerated studies to target other PPIs with small chemical compounds as anticancer drugs.

Mimicking the Structure of Smac Peptides Resulted in New XIAP Antagonists in the Clinic

One strategy widely used to identify lead PPIM compounds is to mimic the structure of binding peptides [26,27]. For example, the BIR3 domain of the XIAP binds to and inhibits proapoptotic caspase-9. In turn, the anti-apoptotic activity of XIAP can be neutralized by Smac, which is released from mitochondria during apoptosis (Figure 2). The XIAP–caspase-9 interaction can be disrupted by a Smac tetrapeptide (AVPI) [64,65], which provides a novel PPI target. A combination of structure-based design and targeted compound library generation led to the identification of GDC-0152, the first Smac mimetic to enter clinical trials in patients with locally advanced or metastatic malignancies [19]. GDC-0152 binds to BIR domains with low nanomolar affinity at the same binding site on IAPs as the SMAC AVPI peptide (Figure 3). A Phase I trial with another Smac mimetic, GDC-0917 (CUDC-427), has been completed in patients with advanced solid tumors and lymphomas. The structurally related LCL161 Smac mimetic has entered Phase II trials in patients with triple-negative breast cancer (Figure 3) [20,66]. Other non-peptide XIAP–Smac inhibitors include an orally available derivative, AT-406, and two bivalent Smac mimetics, TL32711 and HGS1029, all of which are currently in clinical trials for various cancers [67,68].

FBS-Based Discovery of Mitochondrial Apoptosis Pathway Modulators

BH3-containing proapoptotic proteins, such as Bax, bind to the hydrophobic pocket of antiapoptotic BCL2 proteins through a single α-helix (the BH3 domain). Mimicking of the BH3 domain with small-molecule compounds has shown significant therapeutic potential [18]. Several BH3 mimetics have been identified using NMR-based FBS combined with structure-based optimization. For instance, ABT-737 has high binding affinity (nanomolar range) for BCL2 and BCL-xL [69]. This compound occupies the same hydrophobic pocket on BCL-xL as a Bak-derived peptide, overlapping with the Leu78 and Ile85 of Bak, which are critical residues for peptide binding (Figure 3) [69–71]. ABT-737 was further improved to generate orally bioavailable ABT-263 (Navitoclax, Phase I) and ABT-199 (Phase I) with enhanced water solubility [72]. Another oral BH3-mimetic, Obatoclax (GS-01570), discovered by screening of natural product libraries, is currently in Phase II clinical trials in patients with small-cell lung cancer [18,73,74].

Allosteric Regulation of PPIs: Hsp90 Inhibitors

The Hsp90 chaperone protein regulates the activity and stability of numerous client proteins. Inhibition of Hsp90 can simultaneously shut down multiple oncogenic pathways, which has sparked interest in targeting of Hsp90 PPIs for cancer treatment [75–77]. The natural product geldanamycin (GM) inhibits Hsp90–Src complex formation by binding to a 15-Å deep ATP-binding site in the N-terminal domain of Hsp90 (Figure 3). A 17-allylamino,17-demethoxy-substituted GM derivative, 17-AAG, was later developed as a clinical candidate, and is currently in Phase I and II trials in patients with multiple myeloma, lymphoma, stage IV pancreatic cancer, non-small-cell lung cancer, and solid tumors [78]. Other Hsp90 inhibitors in clinical testing include IPI-504 (Phase II), BIIB021 (Phase I and II), PU-H71 (Phase I), NVP-AUY922 (Phase II), AT13387 (Phase I/II), and KW-2478 (Phase I/II) [78].

EMERGING OPPORTUNITIES FOR TARGETING OF PPIS

Although validated PPIs remain active targets for therapeutic development, new concepts and promising PPIs have emerged for anticancer drug discovery (Figure 2). For example, increased knowledge of cancer genomics and PPI-mediated epigenetic mechanisms and identification of cancer-specific onco-fusion proteins have revealed a large number of new PPIs that are directly associated with pathology of cancer. Recent insight into the consequences of various cancer therapeutics and the induced therapeutic resistance offers unanticipated PPIs as potential cancer targets to enhance therapeutic efficacy.

Cancer Genomics

Large-scale genomics initiatives, such as The Cancer Genome Atlas (http://cancergenome.nih.gov) and The International Cancer Genome Consortium (http://icgc.org/icgc/cgp), have led to the discovery of a plethora of genomic alternations that drive tumorigenesis and/or progression [79,80]. It is not hard to imagine that such changes will lead to alteration of protein interaction networks that regulate cell growth. For example, Akt-activating mutations often rewire downstream phosphorelay systems via altered PPIs, such as enhanced 14-3-3 interactions with FOXO3a, Bad, and PRAS40 (Figure 2). To systematically examine PPI network changes in cancer, we have conducted large-scale experiments to establish cancer-associated PPI network maps based on genomic information from glioblastoma multiforme [81] and other tumor types. Such studies, along with predicted new PPIs [82], have revealed novel PPIs that act as major drivers of cancer and thus are potential targets for therapeutic exploration.

PPIs that Regulate Epigenetic Mechanisms

Cancer genomics studies have not only validated the importance of classical hallmarks of cancer but have also revealed new characteristics that are intricately associated with cancer, such as epigenetic dysregulation and RNA splicing [8,79,80]. Recent advances outlining the contribution of dysregulated epigenetic mechanisms to cancer offer new opportunities for PPI targeting. For instance, it has been found that dysregulated histone methylation and acetylation are associated with tumorigenesis. These changes in turn dictate the specific recognition of modified histones by methyllysine-binding proteins and by acetlylysine-binding bromodomains (Figure 2) [7,13]. A potent and selective compound, UNC1215, was recently identified that effectively disrupts the interaction of methylated histone with the L3MBTL3 methyllysine binding protein [83]. UNC1215 demonstrated significant selectivity against more than 200 other analogous methyllysine recognition domains, making it a highly promising agent for probing L3MBTL3 function in cancer. For the interaction of acetylated histone with bromodomain-containing proteins, two small molecules, JQ1 and I-BET, that are pan-bromodomain and extraterminal domain (BET) family inhibitors have been developed [84].

Antitumor activity has been observed for JQ1 in a patient-derived xeno-graft animal model. It is particularly valuable for Myc-driven tumors [85]. I-BET-151 exhibited promising efficacy against onco-fusion-driven leukemia [86]. Cancer-associated mutations in the RNA-splicing machinery indicate the importance of PPIs in the regulation of RNA processing in cancer, such as the association of frequently mutated splicing factor 3b (SF3B1) with 3a (SF3A) in theU2 small nuclear ribonucleoprotein complex [80].

Onco-Fusion PPIs Offer Cancer Selectivity

PPIs are important for the catalytic activity of many enzymes, including epigenetic-modifying enzymes, which can also be targeted. One example is the development of high-affinity peptidomimetics that antagonize the interaction of the histone methyltransferase mixed lineage leukemia (MLL1) and its activator WDR5. Dysregulated MLL1 is associated with various leukemias. Disruption of the MLL1–WDR5 PPI by peptidomimetics effectively decreased MLL1-fusion-mediated leukemogenesis [87]. Similarly, targeting of the MLL1–menin PPI led to the development of a series of lead compounds with therapeutic potential [88]. Importantly, fusion proteins such as MLL1 offer tumor-selective targets; thus, future efforts targeting onco-fusion-protein-specific PPIs are not only warranted but are much needed [89].

PPIs in Protein Complexes

As indicated for MLL1 and many other hub proteins that mediate oncogenic signaling, PPIs often involve multiprotein complexes. Selective inhibition of a particular PPI in the complex for a desired therapeutic effect is challenging. However, selective disruption of MLL1–WDR5 gave rise to promising anti-leukemogenesis activity [87]. Inhibition of MAML interaction with ICN1/CSL by a stapled peptide in NOTCH1 signaling is another example that offers a novel strategy for treating NOTCH1-dependent cancer [22]. Another challenge is experimental identification of selective agents via HTS. For example, 14-3-3 proteins interact with multiple partners, such as Raf-1, Bad, and FOXO [14]. Although these interactions engage a common binding groove, some partner-selective residues have been suggested. Technologies that can identify pan and specific modulators are expected to greatly accelerate the development of selective PPI inhibitors.

PPIs in Combination Therapies

Another emerging opportunity for PPI targeting in cancer is rewired PPIs in oncogenic signaling networks triggered by therapeutic insults. For example, inhibition of mTOR induces paradoxic activation of Akt [90]. Activated Akt triggers phosphorylation-dependent PPIs, such as 14-3-3-mediated PPIs [3,12]. Such induced PPIs may yield new cancer dependence and serve as new targets to overcome pharmacologically induced drug resistance. Interestingly, treatment of cancer cells with an MEK inhibitor renders them sensitive to the BCL2 PPI inhibitor ABT-263 [91]. PPI modulation is expected to have broad and important roles in future mechanism-based combination therapies.

CONCLUDING REMARKS

Future efforts aimed at targeting of PPIs will be greatly accelerated by a number of recent advances. Understanding the nature of PPI interfaces and successful PPIMs may provide rationale design strategies for PPI-focused libraries. PPI assay technologies that closely reflect physiological conditions and address multiprotein complex issues are likely to shorten the process of lead discovery. PPI target discovery coupled with functional validation in genetically defined model systems is vital in moving PPIMs into the pipeline for clinical evaluation. These activities are fueled by a new US national initiative, the Cancer Target Discovery and Development (CTD^2) network (http://ocg.cancer.gov/programs/ctdd.asp). CTD^2 aims to bridge the gap between the vast amount of cancer genomics information and limited therapeutics by accelerating the discovery of new promising targets, including PPIs. Emphasizing collaborative interactions among members with complementary and unique expertise, the CTD^2 network focuses on rapid identification and characterization of potential targets for the development of cancer therapeutics. These scientific efforts will significantly accelerate the expansion of the PPI target landscape, which we hope will lead to a paradigm shift in targeting the once 'undruggable' for personalized cancer therapy and precision medicine.

ACKNOWLEDGMENT

Work in our laboratory was supported in part by US National Institutes of Health grants P01CA116676 and U01 CA168449. F.R.K. and H.F. are Georgia Cancer Coalition Distinguished Scholars.

REFERENCES

1 Wells, J.A. and McClendon, C.L. (2007) Reaching for high-hanging fruit in drug discovery at protein-protein interfaces. *Nature* 450, 1001–1009

2 Hanahan, D. and Weinberg, R.A. (2011) Hallmarks of cancer: the next generation. *Cell* 144, 646–674

3 Hennessy, B.T. *et al*. (2005) Exploiting the PI3K/AKT pathway for cancer drug discovery. *Nat. Rev. Drug Discov.* 4, 988–1004

4 Prochownik, E.V. and Vogt, P.K. (2010) Therapeutic targeting of Myc. *Genes Cancer* 1, 650–659

5   Hitosugi, T. *et al.* (2011) Tyrosine phosphorylation of mitochondrial pyruvate dehydrogenase kinase 1 is important for cancer metabolism. *Mol. Cell* 44, 864–877

6   Puckett, M.C. *et al.* (2013) Integration of the apoptosis signal-regulating kinase 1-mediated stress signaling with the Akt/PKB-IkappaB kinase cascade. *Mol. Cell. Biol.* http://dx.doi.org/10.1128/MCB.00047-13

7   Kelly, T.K. *et al.* (2010) Epigenetic modifications as therapeutic targets. *Nat. Biotechnol.* 28, 1069–1078

8   Imielinski, M. *et al.* (2012) Mapping the hallmarks of lung adenocarcinoma with massively parallel sequencing. *Cell* 150, 1107–1120

9   Buchwald, P. (2010) Small-molecule protein–protein interaction inhibitors: therapeutic potential in light of molecular size, chemical space, and ligand binding efficiency considerations. *IUBMB Life* 62, 724–731

10 Moreira, I.S. *et al.* (2007) Hot spots – a review of the protein–protein interface determinant amino-acid residues. *Proteins* 68, 803–812

11 Perot, S. *et al.* (2010) Druggable pockets and binding site centric chemical space: a paradigm shift in drug discovery. *Drug Discov. Today* 15, 656–667

12 Watanabe, N. and Osada, H. (2012) Phosphorylation-dependent protein-protein interaction modules as potential molecular targets for cancer therapy. *Curr. Drug Targets* 13, 1654–1658

13 Muller, S. and Brown, P.J. (2012) Epigenetic chemical probes. *Clin. Pharmacol. Ther.* 92, 689–693

14 Fu, H. *et al.* (2000) 14-3-3 proteins: structure, function, and regulation. *Annu. Rev. Pharmacol. Toxicol.* 40, 617–647

15 Sperandio, O. *et al.* (2010) Rationalizing the chemical space of protein–protein interaction inhibitors. *Drug Discov. Today* 15, 220–229

16 Barker, A. *et al.* (2012) Expanding medicinal chemistry space. *Drug Discov. Today* 18, 298–304

17 Mason, J.M. (2010) Design and development of peptides and peptide mimetics as antagonists for therapeutic intervention. *Future Med. Chem.* 2, 1813–1822

18 Billard, C. (2012) Design of novel BH3 mimetics for the treatment of chronic lymphocytic leukemia. *Leukemia* 26, 2032–2038

19 Flygare, J.A. *et al.* (2012) Discovery of a potent small-molecule antagonist of inhibitor of apoptosis (IAP) proteins and clinical candidate for the treatment of cancer (GDC-0152). *J. Med. Chem.* 55, 4101–4113

20 Houghton, P.J. *et al.* (2012) Initial testing (stage 1) of LCL161, a SMAC mimetic, by the Pediatric Preclinical Testing Program. *Pediatr. Blood Cancer* 58, 636–639

21 Rew, Y. *et al.* (2012) Structure-based design of novel inhibitors of the MDM2-p53 interaction. *J. Med. Chem.* 55, 4936–4954

22 Moellering, R.E. *et al.* (2009) Direct inhibition of the NOTCH transcription factor complex. *Nature* 462, 182–188

23 Rubinstein, M. and Niv, M.Y. (2009) Peptidic modulators of protein–protein interactions: progress and challenges in computational design. *Biopolymers* 91, 505–513

24 Vanhee, P. *et al.* (2011) Computational design of peptide ligands. *Trends Biotechnol.* 29, 231–239

25 Rothe, A. *et al.* (2006) *In vitro* display technologies reveal novel biopharmaceutics. *FASEB J.* 20, 1599–1610

26 Cummings, C.G. and Hamilton, A.D. (2010) Disrupting protein–protein interactions with non-peptidic, small molecule alpha-helix mimetics. *Curr. Opin. Chem. Biol.* 14, 341–346

27 Whitby, L.R. and Boger, D.L. (2012) Comprehensive peptidomimetic libraries targeting protein-protein interactions. *Acc. Chem. Res.* 45, 1698–1709

28 Brown, C.J. *et al*. (2013) Stapled peptides with improved potency and specificity that activate p53. *ACS Chem. Biol.* 8, 506–512

29 Henchey, L.K. *et al*. (2008) Contemporary strategies for the stabilization of peptides in the alpha-helical conformation. *Curr. Opin. Chem. Biol.* 12, 692–697

30 Kawamoto, S.A. *et al*. (2012) Design of triazole-stapled BCL9 alpha-helical peptides to target the beta-catenin/B-cell CLL/lymphoma 9 (BCL9) protein–protein interaction. *J. Med. Chem.* 55, 1137–1146

31 Verdine, G.L. and Walensky, L.D. (2007) The challenge of drugging undruggable targets in cancer: lessons learned from targeting BCL-2 family members. *Clin. Cancer Res.* 13, 7264–7270

32 Du, Y. *et al*. (2011) A dual-readout F2 assay that combines fluorescence resonance energy transfer and fluorescence polarization for monitoring bimolecular interactions. *Assay Drug Dev. Technol.* 9, 382–393

33 Arkin, M.R. *et al*. (2012) Inhibition of protein–protein interactions: non-cellular assay formats. In *Assay Guidance Manual* (Sittampalam, G.S. *et al.* Ed.,) Eli Lilly and the National Center for Advancing Translational Sciences

34 Dudgeon, D.D. *et al*. (2010) Characterization and optimization of a novel protein–protein interaction biosensor high-content screening assay to identify disruptors of the interactions between p53 and hDM2. *Assay Drug Dev. Technol.* 8, 437–458

35 Valkov, E. *et al*. (2012) Targeting protein–protein interactions and fragment-based drug discovery. *Top. Curr. Chem.* 317, 145–179

36 Scott, D.E. *et al*. (2012) Fragment-based approaches in drug discovery and chemical biology. *Biochemistry* 51, 4990–5003

37 Meireles, L.M. and Mustata, G. (2011) Discovery of modulators of protein–protein interactions: current approaches and limitations. *Curr. Top. Med. Chem.* 11, 248–257

38 Winter, A. *et al*. (2012) Biophysical and computational fragment-based approaches to targeting protein–protein interactions: applications in structure-guided drug discovery. *Q. Rev. Biophys.* 45, 383–426

39 Maurer, T. (2011) Advancing fragment binders to lead-like compounds using ligand and protein-based NMR spectroscopy. *Methods Enzymol.* 493, 469–485

40 Wu, B. *et al*. (2013) HTS by NMR of combinatorial libraries: a fragment-based approach to ligand discovery. *Chem. Biol.* 20, 19–33

41 Lugovskoy, A.A. *et al*. (2002) A novel approach for characterizing protein ligand complexes: molecular basis for specificity of small-molecule Bcl-2 inhibitors. *J. Am. Chem. Soc.* 124, 1234–1240

42 Tsao, D.H. *et al*. (2006) Discovery of novel inhibitors of the ZipA/FtsZ complex by NMR fragment screening coupled with structure-based design. *Bioorg. Med. Chem.* 14, 7953–7961

43 Sun, Q. *et al*. (2012) Discovery of small molecules that bind to K-Ras and inhibit Sos-mediated activation. *Angew. Chem. Int. Ed. Engl.* 51, 6140–6143

44 Murray, C.W. *et al*. (2010) Fragment-based drug discovery applied to Hsp90. Discovery of two lead series with high ligand efficiency. *J. Med. Chem.* 53, 5942–5955

45 Woodhead, A.J. *et al*. (2010) Discovery of (2,4-dihydroxy-5-isopropylphenyl)-[5-(4-methylpiperazin-1-ylmethyl)-1,3-dihydrois oindol-2-yl]methanone (AT13387), a novel inhibitor of the molecular chaperone Hsp90 by fragment based drug design. *J. Med. Chem.* 53, 5956–5969

46 Miura, T. *et al*. (2011) Lead generation of heat shock protein 90 inhibitors by a combination of fragment-based approach, virtual screening, and structure-based drug design. *Bioorg. Med. Chem. Lett.* 21, 5778–5783

47 Suda, A. *et al*. (2012) Design and synthesis of novel macrocyclic 2-amino-6-arylpyrimidine Hsp90 inhibitors. *Bioorg. Med. Chem. Lett.* 22, 1136–1141

48 Erlanson, D.A. *et al*. (2004) Tethering: fragment-based drug discovery. *Annu. Rev. Biophys. Biomol. Struct.* 33, 199–223

49 Yang, S.Y. (2010) Pharmacophore modeling and applications in drug discovery: challenges and recent advances. *Drug Discov. Today* 15, 444–450

50 Lu, Y. *et al*. (2006) Discovery of a nanomolar inhibitor of the human murine double minute 2 (MDM2)-p53 interaction through an integrated, virtual database screening strategy. *J. Med. Chem.* 49, 3759–3762

51 Zhuang, C. *et al*. (2012) Discovery, synthesis, and biological evaluation of orally active pyrrolidone derivatives as novel inhibitors of p53-MDM2 protein–protein interaction. *J. Med. Chem.* 55, 9630–9642

52 Mukherjee, P. *et al*. (2010) Targeting the BH3 domain mediated protein–protein interaction of Bcl-xL through virtual screening. *J. Chem. Inf. Model.* 50, 906–923

53 Corradi, V. *et al*. (2011) Computational techniques are valuable tools for the discovery of protein–protein interaction inhibitors: the 14-3-3sigma case. *Bioorg. Med. Chem. Lett.* 21, 6867–6871

54 Scheper, J. *et al*. (2010) Protein-protein interaction antagonists as novel inhibitors of non-canonical polyubiquitylation. *PLoS ONE* 5, e11403

55 Lawrence, H.R. *et al*. (2009) Identification of a disruptor of the MDM2-p53 protein–protein interaction facilitated by high-throughput *in silico* docking. *Bioorg. Med. Chem. Lett.* 19, 3756–3759

56 Tian, W. *et al*. (2012) Structure-based discovery of a novel inhibitor targeting the beta-catenin/Tcf4 interaction. *Biochemistry* 51, 724–731

57 Wang, S. *et al*. (2012) Targeting the MDM2-p53 protein–protein interaction for new cancer therapeutics. *Top. Med. Chem.* 8, 57–80

58 Wang, X. and Jiang, X. (2012) Mdm2 and MdmX partner to regulate p53. *FEBS Lett.* 586, 1390–1396

59 Vassilev, L.T. *et al*. (2004) *In vivo* activation of the p53 pathway by small-molecule antagonists of MDM2. *Science* 303, 844–848

60 Vu, B.T. and Vassilev, L. (2011) Small-molecule inhibitors of the p53–MDM2 interaction. *Curr. Top. Microbiol. Immunol.* 348, 151–172

61 Kussie, P.H. *et al*. (1996) Structure of the MDM2 oncoprotein bound to the p53 tumor suppressor transactivation domain. *Science* 274, 948–953

62 Popowicz, G.M. *et al*. (2011) The structure-based design of Mdm2/Mdmx–p53 inhibitors gets serious. *Angew. Chem. Int. Ed. Engl.* 50, 2680–2688

63 Andreeff, M. *et al*. (2010) A multi-center, open-label, Phase I study of single agent RG7112, a first in class p53–MDM2 antagonist, in patients with relapsed/refractory acute myeloid and lymphoid leukemias (AML/ALL) and refractory chronic lymphocytic leukemia/small cell lymphocytic lymphomas (CLL/SCLL). *Blood* 116, 657

64 Franklin, M.C. *et al*. (2003) Structure and function analysis of peptide antagonists of melanoma inhibitor of apoptosis (ML-IAP). *Biochemistry* 42, 8223–8231

65 Srinivasula, S.M. *et al*. (2001) A conserved XIAP-interaction motif in caspase-9 and Smac/DIABLO regulates caspase activity and apoptosis. *Nature* 410, 112–116

66 Infante, J.R. *et al*. (2010) A phase I study of LCL161, an oral IAP inhibitor, in patients with advanced cancer. In *Abstracts of the 101st Annual AACR Meeting, Washington DC*. American Association for Cancer Research Abstract 2775

67 Cai, Q. *et al.* (2011) A potent and orally active antagonist (SM-406/AT-406) of multiple inhibitor of apoptosis proteins (IAPs) in clinical development for cancer treatment. *J. Med. Chem.* 54, 2714–2726

68 Eckhardt, S.G. *et al.* (2010) Phase I study evaluating the safety, tolerability, and pharmacokinetics (PK) of HGS1029, a small-molecule inhibitor of apoptosis protein (IAP), in patients with advanced solid tumors. In *Abstracts of the 2010 Annual ASCO Meeting, Chicago, IL*. American Society of Clinical Oncology Abstract 2580

69 Oltersdorf, T. *et al.* (2005) An inhibitor of Bcl-2 family proteins induces regression of solid tumours. *Nature* 435, 677–681

70 Lee, E.F. *et al.* (2007) Crystal structure of ABT-737 complexed with Bcl-xL: implications for selectivity of antagonists of the Bcl-2 family. *Cell Death Differ.* 14, 1711–1713

71 Sattler, M. *et al.* (1997) Structure of Bcl-xL–Bak peptide complex: recognition between regulators of apoptosis. *Science* 275, 983–986

72 Souers, A.J. *et al.* (2013) ABT-199, a potent and selective BCL-2 inhibitor, achieves antitumor activity while sparing platelets. *Nat. Med.* 19, 202–208

73 Nguyen, M. *et al.* (2007) Small molecule obatoclax (GX15-070) antagonizes MCL-1 and overcomes MCL-1-mediated resistance to apoptosis. *Proc. Natl. Acad. Sci. U.S.A.* 104, 19512–19517

74 Paik, P.K. *et al.* (2011) A phase II study of obatoclax mesylate, a Bcl-2 antagonist, plus topotecan in relapsed small cell lung cancer. *Lung Cancer* 74, 481–485

75 Maloney, A. and Workman, P. (2002) HSP90 as a new therapeutic target for cancer therapy: the story unfolds. *Expert Opin. Biol. Ther.* 2, 3–24

76 Porter, J.R. *et al.* (2010) Discovery and development of Hsp90 inhibitors: a promising pathway for cancer therapy. *Curr. Opin. Chem. Biol.* 14, 412–420

77 Neckers, L. and Workman, P. (2012) Hsp90 molecular chaperone inhibitors: are we there yet? *Clin. Cancer Res.* 18, 64–76

78 Schulte, T.W. and Neckers, L.M. (1998) The benzoquinone ansamycin 17-allylamino-17-demethoxygeldanamycin binds to HSP90 and shares important biologic activities with geldanamycin. *Cancer Chemother. Pharmacol.* 42, 273–279

79 Vogelstein, B. *et al.* (2013) Cancer genome landscapes. *Science* 339, 1546–1558

80 Garraway, L.A. and Lander, E.S. (2013) Lessons from the cancer genome. *Cell* 153, 17–37

81 Cancer Genome Atlas Research, N. (2008) Comprehensive genomic characterization defines human glioblastoma genes and core pathways. *Nature* 455, 1061–1068

82 Zhang, Q.C. *et al.* (2012) Structure-based prediction of protein–protein interactions on a genome-wide scale. *Nature* 490, 556–560

83 James, L.I. *et al.* (2013) Discovery of a chemical probe for the L3MBTL3 methyllysine reader domain. *Nat. Chem. Biol.* 9, 184–191

84 Filippakopoulos, P. *et al.* (2010) Selective inhibition of BET bromodomains. *Nature* 468, 1067–1073

85 Delmore, J.E. *et al.* (2011) BET bromodomain inhibition as a therapeutic strategy to target c-Myc. *Cell* 146, 904–917

86 Dawson, M.A. *et al.* (2011) Inhibition of BET recruitment to chromatin as an effective treatment for MLL-fusion leukaemia. *Nature* 478, 529–533

87 Karatas, H. *et al.* (2013) High-affinity, small-molecule peptidomimetic inhibitors of MLL1/WDR5 protein–protein interaction. *J. Am. Chem. Soc.* 135, 669–682

88 Zhou, H. *et al.* (2013) Structure-based design of high-affinity macrocyclic peptidomimetics to block the menin–mixed lineage leukemia 1 (MLL1) protein–protein interaction. *J. Med. Chem.* 56, 1113–1123

89 Daigle, S.R. *et al.* (2011) Selective killing of mixed lineage leukemia cells by a potent small-molecule DOT1L inhibitor. *Cancer Cell* 20, 53–65

90 Sun, S.Y. *et al.* (2005) Activation of Akt and eIF4E survival pathways by rapamycin-mediated mammalian target of rapamycin inhibition. *Cancer Res.* 65, 7052–7058

91 Sale, M.J. and Cook, S.J. (2013) The BH3 mimetic ABT-263 synergizes with the MEK1/2 inhibitor selumetinib/AZD6244 to promote BIM-dependent tumour cell death and inhibit acquired resistance. *Biochem. J.* 450, 285–294

nds in Molecular Medicine

Targeting the RAS Pathway in Melanoma

Zhenyu Ji[1,2], Keith T. Flaherty[3], Hensin Tsao[1,2,3,*]

[1]Wellman Center for Photomedicine, Massachusetts General Hospital, 55 Fruit Street, Boston, MA 02114, USA, [2]Department of Dermatology, Harvard Medical School, Edwards 211, 55 Fruit Street, Boston, MA 02114, USA, [3]MGH Cancer Center, Massachusetts General Hospital, 55 Fruit Street, Boston, MA 02114, USA
*Correspondence: htsao@partners.org

Trends in Molecular Medicine, Vol. 18, No. 1, January 2012 © 2012 Elsevier Inc.
http://dx.doi.org/10.1016/j.molmed.2011.08.001

SUMMARY

Metastatic melanoma is a highly lethal type of skin cancer and is often refractory to all traditional chemotherapeutic agents. Key insights into the genetic makeup of melanoma tumors have led to the development of promising targeted agents. An activated RAS pathway, anchored by oncogenic BRAF, appears to be the central motor driving melanoma proliferation. Although recent clinical trials have brought enormous hope to patients with melanoma, adverse effects and novel escape mechanisms of these inhibitors have already emerged. Definition of the limits of the first successful targeted therapies will provide the basis for further advances in management of disseminated melanoma. In this review, the current state of targeted therapy for melanoma is discussed, including the potent BRAF^{V600E} inhibitor vemurafenib.

CURRENT THERAPIES FOR MELANOMA

The incidence of melanoma has rapidly increased over the past several decades [1]. Approximately 10% of all patients who are diagnosed with melanoma eventually die from this cancer [2]. When it becomes metastatic, melanoma often leads to death within a year [3,4], a dismal prognosis that has resulted from a lack of highly curative therapies for advanced disease [1].

The United States Food and Drug Administration (FDA) has approved few therapies for metastatic melanoma, all of which have minimal beneficial effects on patient survival [5,6]. Many of these have been immunologic in

193

CellPress

nature, including interferon (IFN)-α2b, high-dose interleukin (IL)-2 and, as of March 2011, ipilumimab. IFN-α2b is associated with a 10–15% reduction in the risk of relapse in the adjuvant setting, whereas IL-2 produces objective response in 15% of metastatic patients [6–10]. An older FDA-approved melanoma therapy is the alkylating agent dacarbazine (DTIC), which achieves responses in less than 10% of patients [11], a profile similar to other available agents such as carmustine (BCNU), temozolomide, taxanes and platinum analogs [6,12–14]. In the face of these limited options, there has been a sea change in melanoma treatments ushered in by recent molecular advances.

Targeted agents aimed at oncogenic drivers that have been identified over the past decade provide an opportunity for novel melanoma therapeutics [15,16]. This review focuses on the central molecular network that fuels melanoma growth and recent drug development progress towards targeting these key proteins and signaling pathways.

THE CENTRAL MELANOMA AXIS AND THERAPEUTIC TARGETS

Over the past decade, much has been learned about genetic lesions that stimulate growth and signaling pathways in melanomas [17]. As shown in Figure 1, many components of the RAS pathway are either activated through oncogenic mutations or inactivated through deleterious alterations. From this composite view, activation of a KIT–NRAS–BRAF–MEK–ERK central axis (Figure 1, shaded in green) seems to be crucial in almost all forms of melanoma. Figure 1 also lists some of the drugs in the pipeline for inhibiting various components of the pathway.

Receptor Tyrosine Kinases (RTKs)

A number of growth factor RTKs such as *EGFR*, *PDGFR* and *KIT* are expressed in melanoma cells, although recurrent activating mutations are uncommon. One lineage-derived RTK is c-KIT, a receptor known to be crucial in melanocyte differentiation but whose expression appears to be lost in many melanomas [18,19]. A more direct role for c-KIT was recently recognized when genomic screens revealed that the *KIT* locus (chromosome 4q11) was amplified and/or mutated in a subset of mucosal, acral and chronically sun-damaged (CSD) melanomas (MACs) [20]. Approximately 10–20% of these melanomas harbor the same activating *KIT* mutations described in gastrointestinal stromal tumors (GISTs) [20–24].

The earlier successes of imatinib in c-KIT-mutated GISTs suggested that MAC melanomas may be particularly vulnerable to c-KIT inhibitors. The idea was initially bolstered by reports of several melanoma cases treated with imatinib [25,26]. These clinical results were subsequently confirmed in

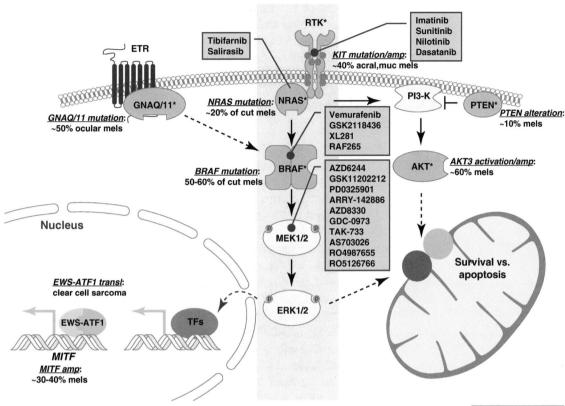

FIGURE 1 Key mutational and therapeutic targets in melanoma.
The RAS signaling network is rife with cancer-associated mutations. *BRAF* is the most commonly activated oncogene in cutaneous melanomas (cut mels), followed by *NRAS*. Upstream of RAS, the c-KIT receptor tyrosine kinase (RTK) is amplified or activated in a substantial fraction of acral and mucosal melanomas. It thus seems that the RTK–NRAS–BRAF–MEK–ERK cascade represents a central axis (highlighted in green) that is activated in nearly all melanomas. Parallel to this axis is the PI3–K pathway, which is also activated in melanomas either through loss of PTEN or activation of AKT3. In addition, it has recently been shown that GNAQ and GNA11, which are transducers of the endothelin receptor signal (ETR), are mutated in ocular melanomas. These heightened signaling events lead to both increased transcription of survival genes (such as *MITF*) and enhanced pro-survival factors in mitochondria through the regulation of BCL2 family proteins (red, pro-apoptotic; green, pro-survival). MITF, which is a master regulator of melanocyte survival, is also amplified in melanomas and a target of the EWS-ATF1 fusion protein described in clear cell sarcomas (so-called melanoma of the soft parts). Drugs known to inhibit the central axis and with a potential therapeutic impact on melanoma are listed in the purple boxes.

other melanoma cell lines sustained by an activating c-KIT mutation or an SCF–c-KIT autocrine loop [21,27]. Imatinib has minimal inhibitory effects on melanoma cell lines containing the BRAF[V600E] mutation despite evidence of c-KIT expression; furthermore, the mere presence of c-KIT receptor expression does not seem to predict response [28,29]. Thus, it appears that the potential clinical role of c-KIT inhibitors is probably restricted to those melanomas that have activating mutations and consequent c-KIT-dependent signaling. Interestingly, response seems to correlate with the site of mutation in c-KIT. For example, melanomas with mutations in the juxtamembrane region of c-KIT are associated with a better response to imatinib treatment [28]. Because imatinib is not c-KIT-specific, it is possible that a more selective agent could achieve a greater degree of inhibition and result in more profound responses.

Reports on two open-label Phase II trials of imatinib mesylate for KIT-mutated melanomas have recently been published. In the first trial, Carvajal *et al.* treated 28 patients who developed metastatic melanoma from MAC sites with 400mg of imatinib twice daily [30]. There were 2 complete responses (CRs) lasting 94 and 95 weeks, 2 durable partial responses (PRs) lasting 53 and 89 weeks, and 2 transient PRs lasting 12 and 18 weeks among 25 evaluable patients. The median progression-free survival (PFS) was 12 weeks, with a median overall survival (OS) of 46.3 weeks. At a molecular level, 23.4% of the cases harbored either *KIT* mutations or amplifications, whereas 27.8% of the tumors actually contained either *BRAF* or *NRAS* mutations. The most significant responses occurred in patients with KIT^{K642E} or KIT^{L576P} variants and those with a mutant/allele ratio >1, that is, tumors with greater activated KIT dependence. In the second trial, Guo *et al.* treated 43 metastatic melanoma patients with 400mg of imatinib per day unless intolerable toxicity or disease progression occurred [31]. Eligibility in the Guo trial required *KIT* aberrations defined as mutations in exons 9, 11, 13, 17, or 18 and/or increases in copy number. Overall, PRs, stable disease and progressive disease were observed in 10 patients (23.3%), 13 patients (30.2%) and 20 patients (46.5%), respectively. The 6-month PFS and 1-year OS rates were 36.6% and 51.0%, respectively. The median PFS time was 3.5 months (range 1.3–5.7 months) and the OS time was 14.0 months (range 10.8–17.2 months). There were no clear-cut associations between outcome and *KIT* mutation characteristics. Although the overall benefits of imatinib in these studies are encouraging, albeit modest, other RTK inhibitors (e.g. sunitinib, nilotinib and dasatanib) are emerging (Table 1) and may prove more efficacious in trials. For instance, a current Phase III trial is comparing nilotinib to dacarbazine (NCT01028222) in patients with *KIT*-mutated metastatic melanoma (exons 9, 11 or 13, or exon 17 mutations p.Tyr822Asp, p.Asp820Tyr, or p.Tyr823Asp). Masitinib is another potent and highly selective oral RTK inhibitor that has combined activity against both c-KIT and LYN. A recent small study showed some effect against another *KIT*-mediated disease, systemic mastocytosis [32], and a Phase III trial of masitinib for metastatic melanomas with juxtamembrane mutations is also currently enrolling patients (NCT01280565).

Prickett and colleagues recently scanned the tyrosine kinome and identified mutations in *ERBB4* in 19% of melanoma cases, although there were no mutational hotspots [33]. The alterations were clearly oncogenic in several *in vitro* phenotypes, such as NIH-3T3 transformation and soft-agar growth. Furthermore, inhibition of ERBB4 by lapatinib induced apoptosis, especially in *ERBB4*-mutated cells. These recent findings have led to a Phase II trial of lapatinib in stage IV melanoma for patients with *ERBB4*-mutated melanomas (NCT01264081).

RAS Inhibitors

NRAS is the second most commonly activated oncogene in melanoma after *BRAF*. Like other RAS protein members, activating changes occur on p.Gly12 or p.Gln61. The potential of NRAS as a therapeutic target has been validated in preclinical models with siRNA [34], but potent and selective pharmacologic inhibitors are not readily available. The first group of compounds used to target RAS were farnesyl transferase inhibitors (FTIs), such as R115777 (tibifarnib; Figure 1). The FTIs were designed to block the post-translational lipid modification of RAS that is required for full RAS activity [35]. FTIs inhibit tumor growth in preclinical models [36], although their performance in the clinical trial setting has been lackluster [37]. The prevailing hypothesis for the lack of efficacy is that FTIs inhibit a large family of proteins that require farnesylation as a post-translational modification. Thus, it is not possible

Table 1 Summary of RAS Pathway Inhibitors

Drug	Targets	Stage of Clinical Development
Imatinib (Gleevec, STI571)	KIT, ABL, PDGFR, NQO2 [84], V-ATPase [85]	Approved for CML and GIST [86]
Sunitinib (SU11248)	KIT, PDGFR, VEGFR [87]	Approved for RCC and GIST [87]
Nilotinib	KIT, ABL, LCK, NQO2, DDR1 [84]	Approved for CML [88]
Dasatanib (BMS-354825)	KIT, ABL, SRC [89], DDR1, BTK, TEC [84]	Approved for CML [90]
Tipifarnib (R115777)	RAS and other proteins that require farnesyl transferase [91]	Phase II/III [91]
Salirasib (FTS)	RAS, mTOR [92]	Phase II [43]
Sorafenib (BAY 43-9006)	BRAF	Approved for RCC and HCC Failed at phase II for advanced melanoma [52]
PLX4720	BRAF, CRAF, VEGF, PDGF, FLT3, KIT [46]	Precursor of PLX4032
Vemurafenib (PLX4032)	BRAF^{V600E}, BRAF^{WT}, BRK [56]	Phase III [61]
GSK2118436	BRAF^{V600E}, CRAF, BRAF^{WT}, ARAF, ACK1, SRMS and MAP4K5 [57,58]	Phase I/II [68]
PD0325901	MEK	Phase II [75]
AZD6244 (ARRY-142886)	MEK	Phase II [93]
Anthrax lethal toxin	MEK	
GSK1120212 (JTP-74057)	MEK	Phase III

to achieve sufficient downregulation of RAS without profoundly impairing the function of other farnesylated proteins, a fact that is responsible for the dose-limiting toxicity of these agents. Furthermore, KRAS and NRAS can circumvent FTIs by employing geranylgeranyltransferases in cells and thereby maintaining their function in the presence of FTIs. Nevertheless, it remains possible that FTIs might still enhance the effect of other chemotherapeutic agents in melanoma, as shown in at least one preclinical study [38].

Salirasib (S-trans,trans-farnesylthiosalicylic acid, FTS) belongs to another group of RAS antagonists and mimics the carboxy-terminal farnesylcysteine that is common to all three RAS isoforms [39]. Thus, FTS competes with the active, GTP-bound forms of RAS proteins for specific binding sites on the cytoplasmic membrane [40] and inhibits melanoma cell growth *in vitro* and in xenograft models [41,42]. Initial results from recent clinical trials with pancreatic cancer patients support the ability of FTS to suppress RAS function and possibly mediate survival benefits [43,44]. These preliminary results require validation in larger patient populations and randomized trials. RAS inhibitors still hold unrealized potential as a therapeutic approach for melanoma, especially for the ~20% of tumors in which activating mutations are found.

RAF Inhibitors

As mentioned above, the most prevalent alteration in cutaneous melanoma is activation of the serine/threonine protein kinase BRAF [45], which makes BRAF a veritable target in the therapeutic landscape. Overall, approximately 40–50% of uncultured cutaneous melanomas harbor BRAF mutations, with the p.Val600Glu missense mutation (BRAF^{V600E}), which lies in the CR3 kinase domain (Figure 2), comprising ≥90% of reported mutations.

The first BRAF inhibitor that progressed to clinical trial was sorafenib, which targets multiple protein kinases including BRAF, CRAF and the VEGF and PDGF RTKs (Figure 1) [46]. Sorafenib has minimal activity as a single agent in melanoma, although initial results were more encouraging when it was combined with carboplatin and paclitaxel or temozolomide [47–49]. However, the activity of the combination regimen did not correlate with BRAF mutational status. A Phase III trial of sorafenib in combination with carboplatin and paclitaxel in patients with advanced melanoma failed to improve overall survival [50]. The failure of sorafenib has been attributed in part to incomplete MAP kinase inhibition at the maximum tolerated dose [51] and to compensatory RAF–MEK–ERK signaling [52] or other escape mechanisms [53,54]. A recent study challenged the entire notion that sorafenib works through BRAF *in vivo*. Whittaker *et al.* engineered cells with a gatekeeper mutation (p.Thr529Asn or p.T529N) that rendered BRAF$^{T529N/V600E}$

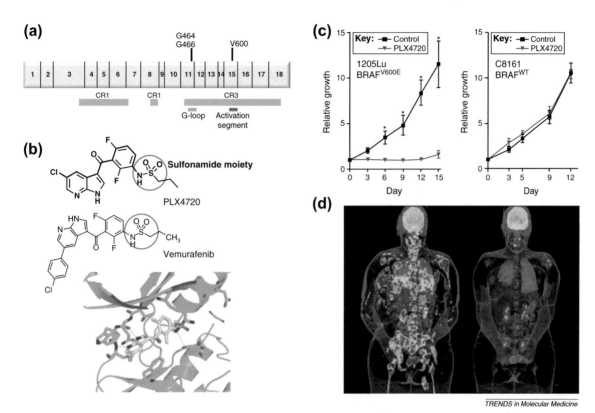

FIGURE 2 **BRAF inhibition and development of vemurafenib.**
(a) Domain structure of BRAF showing the CR domains and the location of the most common mutations in exons 11 and 15. (b) Chemical structure of PLX4720 and PLX4032 (vemurafenib). These compounds are competitive ATP protein kinase inhibitors (class I) selective for the BRAF^{V600E} mutant protein. Also shown is co-structure of PLX4720 with the BRAF kinase. It has been postulated that the key to the specificity of PLX4720 lies in the differential interaction of the sulfonamide moiety (red circle) with the Asp and Phe residues of the conserved kinase Asp-Phe-Gly (DFG) sequence. In the active conformation, the nitrogen atom of the sulfonamide moiety is deprotonated and interacts with the main-chain NH group of Asp-594, whereas the oxygen atoms of the sulfonamide form hydrogen bonds to the NH group of Phe-595 and the side chain of Lys-483. Crystallographic data support a selectivity model in which the deprotonated sulfonamide interacts more favorably with the active kinase conformation that is characteristic of BRAF^{V600E}. (c) In tumor xenografts, a BRAF^{V600E} mutated melanoma line (1205Lu) showed significantly greater response to PLX4720 (100 mg/kg by oral gavage) compared to a BRAF^{WT} line (C8161). (d) Dramatic tumor shrinkage after treatment with PLX4032 in a patient with metastatic BRAF^{V600E} melanoma. Panels b and c reproduced with permission from [56] Copyright © 2008 National Academy of Sciences, U.S.A. Panel d reproduced from [83] Copyright © 2011 Massachusetts Medical Society. Reprinted with permission from Massachusetts Medical Society.

cells resistant to sorafenib *in vitro* and yet sensitive to sorafenib *in vivo*. By contrast, truly selective BRAF inhibitors (SBIs) such as PLX4720 (see below) lost their effectiveness both *in vitro* and *in vivo* when the gatekeeper mutation was introduced. This suggests that sorafenib has BRAF-independent anti-tumor activity and may also explain why sorafenib causes side effects at a dose that does not efficiently inhibit MEK signaling [55].

More selective BRAF inhibitors have been synthesized over the past few years. The first of these, PLX4720, selectively inhibits BRAF[V600E] (IC_{50}=13 nM vs 160 nM for wild-type BRAF, BRAF[WT]) and has been thoroughly tested and validated in preclinical BRAF[V600E] models [56]. The clinical compound vemurafenib (PLX4032; Figure 2), which is an analog of PLX4720, shows higher selectivity for BRAF[V600E] (IC_{50}=31 nM) and CRAF (IC_{50}=48 nM) than wild-type BRAF (IC_{50}=100 nM) [57]. These *in vitro* parameters, however, do not necessarily predict drug activity *in vivo*, where signaling networks may produce more dynamic physiologic responses. Furthermore, vemurafenib, at least *in vitro*, also inhibits several other protein kinases, such as ACK1, SRMS and MAP4K5 [58], with similar potency (IC_{50}<50 nM) compared with BRAF[V600E]. Although speculative, suppression of these secondary targets could potentially contribute to the observed responses [58] and more investigation is needed to understand the consequences of these off-target effects.

The clinical efficacy of vemurafenib in melanoma patients with BRAF[V600E] mutations has been firmly established through three trials. The objective response exceeded 50% among those treated at the higher doses in the Phase I trial and an even higher response rate was obtained when a cohort of patients received the recommended Phase II dose [59,60]. Correlative studies also demonstrated that doses of vemurafenib that lead to >90% reductions in ERK phosphorylation are required to achieve a meaningful clinical response [58]. The pivotal Phase III study (BRIM-3) enrolled 675 patients with previously untreated metastatic melanoma that could not be surgically removed [61]. Patients with BRAF[V600E]-mutated tumors were randomly assigned to receive either vemurafenib (960mg orally twice daily) or dacarbazine. At the 6-month evaluation, OS was 84% in the vemurafenib group (n=336) and 64% in the dacarbazine group (n=336; hazard ratio 0.37, *p*<0.001). The hazard ratio for tumor progression in the vemurafenib group was 0.26 (*p*<0.001) and the estimated median PFS was 5.3 months in the vemurafenib group and 1.6 months in the dacarbazine group. The objective response rate for individuals treated with vemurafenib was 48% compared to 5% for dacarbazine-treated patients (*p*<0.001). As observed in earlier trials, keratinocytic neoplasms (cutaneous squamous cell carcinomas or a related entity, keratoacanthoma) developed in 18% of patients treated with vemurafenib. BRIM-3 represents the first prospective randomized molecular therapy trial in melanoma and the first to demonstrate a convincing survival benefit. Although its importance cannot be overstated, there are several lingering challenges worth addressing. First, patients with BRAF[WT] tumors are not eligible for vemurafenib and are in desperate need of effective agents. Second, although 48% of the patients showed an objective response, a significant fraction of BRAF-mutated tumors did not reach the RECIST criteria and therefore seem to be innately insensitive to vemurafenib; it will be important to identify these primarily resistant tumors through secondary biomarkers beyond BRAF status. Third, nearly all patients relapse with time despite ongoing treatment, and thus OS after longer follow-up will be important to establish the true rate of cure.

Secondary Resistance in RAF-mutated Tumors

One of the most pressing and exciting areas of investigation is the elucidation of primary and secondary resistance mechanisms. Biochemical and genetic studies into the development of resistance have largely focused on two specific questions: why is MEK–ERK signaling paradoxically activated by SBIs in RAS-mutated cells, and how do BRAF-mutated cells gain resistance to SBIs amid chronic suppression? There are at least two models that address the first question and both invoke RAF dimerization, although the two models differ in their molecular details (Figure 3) [57,62]. In response to growth factor receptor activation or an oncogenic *RAS* mutation, the RAS protein binds GTP, is activated (RAS*-GTP; Figure 3), localizes to the plasma membrane and induces homodimerization and heterodimerization of BRAF and CRAF. In cells driven by a BRAF^{V600E} mutation, RAS activation is bypassed and signal initiation occurs in the cytoplasm; MEK phosphorylation results from constitutive BRAF activity. Consequently, an SBI attenuates nearly all downstream MEK–ERK signal propagation. However, in BRAF^{WT} cells, this molecular assembly at the cellular membrane is a crucial step for subsequent MEK phosphorylation and downstream signaling. One unexpected observation is the paradoxical MAPK pathway stimulation by SBIs in WT BRAF cells [57,62]. How does this occur? In one model (Figure 3b), low concentrations of a RAF inhibitor inactivate only a single monomer in the RAF dimer; dimerization still transactivates the uninhibited partner RAF molecule and triggers MAPK signaling [57]. Higher RAF inhibitor levels lead to inhibition of both RAF partners and all signaling is suppressed. In a second model (Figure 3c) [62], WT BRAF remains largely inactive in the cytoplasm of RAS*-GTP cells until it binds an SBI. The SBI-bound BRAF^{WT} then translocates to the cell membrane where it dimerizes with CRAF, further stimulating CRAF signaling. A pan-RAF inhibitor (PRI) suppresses both BRAF and CRAF and thereby abrogates all MAPK signaling. Gatekeeper CRAF mutations, such as CRAF Thr421Asn, could disrupt binding of the PRI to CRAF and restore signaling. The implication for both models is that RAS-mutated cells may in fact be stimulated by an SBI, which could explain the observed squamous cell carcinomas that develop while on vemurafenib. Thus, the use of vemurafenib requires absolute genetic precision.

Several recent reports have also shed light on possible mechanisms that are responsible for resistance to vemurafenib (Figure 3). For example, reactivation of the ERK pathway by an NRAS mutation confers secondary

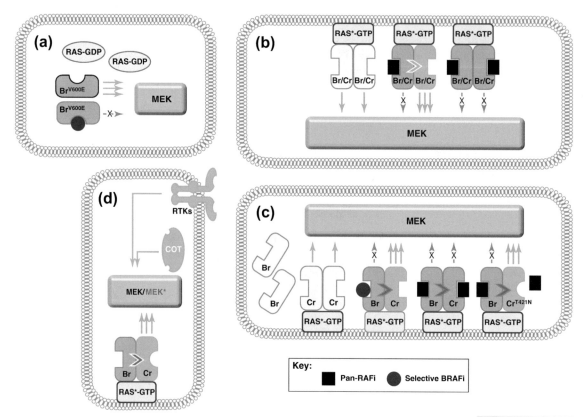

TRENDS in Molecular Medicine

FIGURE 3 Models of RAF activation and inhibition.

(a) In cells with mutated BRAF (Br^{V600E}), RAS is inactive, cytoplasmic and unable to induce RAF dimerization. MEK phosphorylation and activation occur almost exclusively via BRAF^{V600E} and thus are effectively silenced by a selective BRAF inhibitor (blue circle). This would explain the dramatic tumor shrinkage early in treatment with vemurafenib. In some cells with wild-type BRAF, RAS is activated (RAS*-GTP), presumably by mutagenesis (e.g. *NRAS* mutations) or other upstream receptor events. There are at least two models of signal transduction in these RAS-stimulated tumors. In the transactivation model **(b)**, active RAS*-GTP mobilizes to the membrane and induces homodimerization or heterodimerization of wild-type BRAF and CRAF proteins (Br/Cr), which in turn initiates signaling. At low concentrations, an ATP-competitive RAF inhibitor may occupy the active site of one of the RAF protomers. Inhibitor binding to the active site of one RAF kinase induces a conformational change that facilitates transactivation of the other RAF molecule in the dimer; therefore, there is paradoxical enhancement of MEK/ERK signaling through the uninhibited partner. At higher inhibitor concentrations, both RAF molecules are inhibited and MEK/ERK signaling is completely abolished. In the translocation model **(c)**, wild-type BRAF (Br) remains largely inactive in the cytoplasm owing to autoinhibitory signals. Activated RAS*-GTP triggers MAPK signaling through CRAF (Cr). When wild-type BRAF binds a selective ATP-competitive BRAF inhibitor, the BRAF molecule is recruited to the membrane, thereby enhancing the RAS*-GTP/CRAF interaction and signaling (as in b). With pan-RAF inhibitors, wild-type BRAF may still enhance the RAS*-GTP/CRAF interaction, but MEK/ERK signaling is suppressed because CRAF activity itself is concurrently abrogated by inhibitor binding. In situations in which CRAF harbors a gatekeeper mutation, such as p.Thr421Asn (Cr^{T421N}), a pan-RAF agent inhibits BRAF, which then transactivates CRAF^{T421N}; however, CRAF^{T421N} is not inhibited by the pan-RAF agent because the mutation prevents drug binding. Both models suggest mechanisms by which BRAF inhibition may lead to increased stimulation of CRAF in RAS-mutated cells or growth-factor-stimulated cells. **(d)** Mechanisms of BRAF inhibitor resistance include activation of receptor tyrosine kinases (RTKs) such as PDGFR and IGF1R, CRAF and other protein kinases (e.g. COT), along with mutational activation of NRAS and MEK (MEK*).

resistance to vemurafenib, as functionally shown in a relapse-derived cell line and verified in a clinically resistant melanoma sample [63]. In another study, Johannessen *et al.* undertook a kinome-wide screen for molecules that could confer resistance to PLX4720 and identified both CRAF and the kinase MAP3K8 (COT, TPL2) [64]. Naïve melanoma cell lines with elevated COT levels exhibit *de novo* resistance to PLX4720 and 2 of 3 melanoma samples taken from patients early in the course of therapy or at the time of progression also had increased COT expression. Finally, a more recent report identified a *MEK1* mutation in a single tumor that had become resistant to vemurafenib [65].

Interestingly, ERK reactivation may not to be the only means of acquiring vemurafenib resistance. Nazarian *et al.* found that PDGFRβ upregulation can be associated with vemurafenib resistance in the absence of ostensible ERK activation [63]. There is biochemical evidence to suggest that AKT activation is correlated with heightened PDGFRβ expression, although other unidentified downstream effectors may also play a role. Similarly, Villanueva *et al.* reported that increased IGF receptor signaling may also be correlated with acquired SBI resistance [66].

There are several other BRAF inhibitors currently in clinical development (Figure 1 and Table 1) [67]. GSK2118436 is a BRAF inhibitor that showed promising results in an early clinical trial [68]. Interestingly, regression of brain metastasis after treatment with GSK2118436 has been observed in several patients. A Phase II clinical trial is now ongoing to test its effect in melanoma patients with BRAF mutation (NCT01266967).

CRAF may also be an effective target for melanoma therapy [69,70], particularly in BRAF^{WT} cells, because CRAF seems to be the key mediator of MEK activation in NRAS-mutated melanomas [71]. PRIs may be more relevant for NRAS-mutated melanomas and non-V600E BRAF mutants, which tend to activate MEK through CRAF signaling [71,72].

MEK Inhibitors

MEK is the major downstream molecule of oncogenic BRAF. An early study found that melanoma cells with BRAF mutations tend to be more sensitive to MEK inhibitors than those with NRAS mutations [73]. Therefore, BRAF mutational status may also predict sensitivity to MEK inhibitors in the clinic.

MEK inhibitors reduce proliferation, colony formation and invasiveness of BRAF^{V600E} mutant human melanoma cells *in vitro* and tumor growth *in vivo* [74]. Several MEK inhibitors have been investigated in clinical trials in which patients with advanced melanoma were treated (Figure 1). PD0325901 was evaluated in a Phase I trial: 2 of 27 patients had an objective response and another 5 patients showed disease stabilization [67]. However, dose-limiting side effects such as diarrhea and rash precluded the high amount of target inhibition required to adequately suppress the MAPK pathway in tumor cells. Because MEK inhibitors inhibit MAPK pathway signaling in normal cells as well as tumor cells, it may not be possible to achieve sufficient target effects in tumors owing to normal tissue toxicity at the drug concentrations required. Phase II trials of PD0325901 in non-small-cell lung cancer were suspended because of limited activity and intolerable side effects such as visual disturbances [75]. In a recent Phase I trial, another MEK inhibitor, AZD6244, showed only moderate effects in a very small subgroup of patients with metastatic melanoma harboring BRAF^{V600E} mutations [76]. However, in the follow-up phase II trial with AZD6244, 12% of patients whose tumors harbored BRAF^{V600E} showed significant tumor regression, although the regression was not complete [67]. This limited response may be due to insufficient target inhibition or failure to induce cell death. *In vitro* studies have also demonstrated that BRAF/MEK inhibitors lead to mainly cytostatic effects in BRAF^{V600E}-mutated melanoma cells [56,73,77] and therefore AZD6244 may not be adequate as a single agent in melanoma treatment. GSK1120212 is an allosteric MEK inhibitor [78] that showed promising antitumor activity in a Phase I clinical trial [79] and is now being evaluated in a Phase III trial (NCT01245062). Finally, the lethal toxin anthrax, which selectively degrades and inactivates MEK1 and MEK2 [80], is also being tested in melanoma clinical trials [53].

CONCLUDING REMARKS

It is clear that single-agent approaches in melanoma are not capable of achieving a cure, a finding that is not surprising given the genetic complexity of melanomas and the concomitant activation of multiple signaling pathways. The experience with BRAF inhibitors has demonstrated that melanoma often resurrects itself, even after the main growth signals are abrogated. Therefore, simultaneous targeting of several pathways is likely to result in better outcomes. The redundancy within the multiple signaling pathways activated in melanoma, such as PTEN loss with consequent AKT activation, raises the possibility of combining MAPK and AKT pathway inhibitors in new formulations. Studies have shown that PI3K and MEK inhibitors synergize to reduce growth and survival of melanoma cells in 3D cell culture systems [81,82] and thus larger signaling networks may need to be considered. Moreover, melanomas with BRAF^{V600E} often have other genetic disruptions in molecules such as cyclin D1, CDK2, CDK4, MITF and AKT3 [67], which suggests that additional inhibitor combinations may enhance efficacy.

Melanomas are genetically heterogeneous, and the use of personalized cancer therapy has already been demonstrated in this cancer. To maximize success, future targeted therapy may need to be tested in patients for whom the relevant combination of genetic aberrations in the tumors have been predetermined.

ACKNOWLEDGMENTS

We want to thank the National Institutes of Health (K24 CA149202 to H.T.), the Melanoma Research Alliance, the American Skin Association and the generous donors to the Massachusetts General Hospital Millennium Melanoma Fund for their support during the writing of this review.

REFERENCES

1 Berwick, M. *et al*. (2009) Melanoma epidemiology and public health. *Dermatol. Clin.* 27, 205–214 viii

2 Siegel, R. *et al*. (2011) Cancer statistics, 2011: The impact of eliminating socioeconomic and racial disparities on premature cancer deaths. *CA Cancer J. Clin.* 61, 212–236

3 Tsao, H. *et al*. (2004) Management of cutaneous melanoma. *N. Engl. J. Med.* 351, 998–1012

4 Agarwala, S.S. (2009) Current systemic therapy for metastatic melanoma. *Expert Rev. Anticancer Ther.* 9, 587–595

5 Fecher, L.A. and Flaherty, K.T. (2009) Where are we with adjuvant therapy of stage III and IV melanoma in 2009? *J. Natl. Compr. Cancer Netw.* 7, 295–304

6 Tarhini, A.A. and Agarwala, S.S. (2006) Cutaneous melanoma: available therapy for metastatic disease. *Dermatol. Ther.* 19, 19–25

7 Kirkwood, J.M. *et al*. (2004) A pooled analysis of eastern cooperative oncology group and intergroup trials of adjuvant high-dose interferon for melanoma. *Clin. Cancer Res.* 10, 1670–1677

8 Agarwala, S.S. and Kirkwood, J.M. (1996) Interferons in melanoma. *Curr. Opin. Oncol.* 8, 167–174

9 Atkins, M.B. *et al*. (1999) High-dose recombinant interleukin 2 therapy for patients with metastatic melanoma: analysis of 270 patients treated between 1985 and 1993. *J. Clin. Oncol.* 17, 2105–2116

10 Atkins, M.B. *et al*. (2000) High-dose recombinant interleukin-2 therapy in patients with metastatic melanoma: long-term survival update. *Cancer J. Sci. Am.* 6(Suppl. 1), S11–S14

11 Ji, Z. *et al*. (2010) Molecular therapeutic approaches to melanoma. *Mol. Aspects Med.* 31, 194–204

12 Wagner, J.D. *et al*. (2000) Current therapy of cutaneous melanoma. *Plast. Reconstr. Surg.* 105, 1774–1799 quiz 1800–1771

13 Mays, S.R. and Nelson, B.R. (1999) Current therapy of cutaneous melanoma. *Cutis* 63, 293–298

14 Chapman, P.B. *et al*. (1999) Phase III multicenter randomized trial of the Dartmouth regimen versus dacarbazine in patients with metastatic melanoma. *J. Clin. Oncol.* 17, 2745–2751

15 Chin, L. *et al*. (2006) Malignant melanoma genetics and therapeutics in the genomic era. *Genes Dev.* 20, 2149–2182 (PMID:16912270)

16 Puzanov, I. *et al*. (2011) Biological challenges of BRAF inhibitor therapy. *Mol. Oncol.* 5, 116–123 (PMID:21393075)

17 Hocker, T.L. *et al*. (2008) Melanoma genetics and therapeutic approaches in the 21st century: moving from the benchside to the bedside. *J. Invest. Dermatol.* 128, 2575–2595

18 Natali, P.G. *et al*. (1992) Progression of human cutaneous melanoma is associated with loss of expression of c-kit proto-oncogene receptor. *Int. J. Cancer* 52, 197–201

19 Lassam, N. and Bickford, S. (1992) Loss of c-kit expression in cultured melanoma cells. *Oncogene* 7, 51–56

20 Curtin, J.A. *et al*. (2006) Somatic activation of KIT in distinct subtypes of melanoma. *J. Clin. Oncol*. 24, 4340–4346

21 Ashida, A. *et al*. (2009) Pathological activation of KIT in metastatic tumors of acral and mucosal melanomas. *Int. J. Cancer* 124, 862–868

22 Rivera, R.S. *et al*. (2008) C-kit protein expression correlated with activating mutations in KIT gene in oral mucosal melanoma. *Virchows Arch*. 452, 27–32

23 Antonescu, C.R. *et al*. (2007) L576P KIT mutation in anal melanomas correlates with KIT protein expression and is sensitive to specific kinase inhibition. *Int. J. Cancer* 121, 257–264

24 Smalley, K.S. *et al*. (2008) Identification of a novel subgroup of melanomas with KIT/cyclin-dependent kinase-4 overexpression. *Cancer Res*. 68, 5743–5752

25 Lutzky, J. *et al*. (2008) Dose-dependent, complete response to imatinib of a metastatic mucosal melanoma with a K642E KIT mutation. *Pigment Cell Melanoma Res*. 21, 492–493

26 Hodi, F.S. *et al*. (2008) Major response to imatinib mesylate in KIT-mutated melanoma. *J. Clin. Oncol*. 26, 2046–2051

27 Jiang, X. *et al*. (2008) Imatinib targeting of KIT-mutant oncoprotein in melanoma. *Clin. Cancer Res*. 14, 7726–7732

28 Smalley, K.S. *et al*. (2009) c-KIT signaling as the driving oncogenic event in sub-groups of melanomas. *Histol. Histopathol*. 24, 643–650

29 Hofmann, U.B. *et al*. (2009) Overexpression of the KIT/SCF in uveal melanoma does not translate into clinical efficacy of imatinib mesylate. *Clin. Cancer Res*. 15, 324–329

30 Carvajal, R.D. *et al*. (2011) KIT as a therapeutic target in metastatic melanoma. *J. Am. Med. Assoc*. 305, 2327–2334

31 Guo, J. *et al*. (2011) Phase II, open-label, single-arm trial of imatinib mesylate in patients with metastatic melanoma harboring c-Kit mutation or amplification. *J. Clin. Oncol*. 29, 2904–2909

32 Paul, C. *et al*. (2010) Masitinib for the treatment of systemic and cutaneous mastocytosis with handicap: a Phase 2a study. *Am. J. Hematol*. 85, 921–925

33 Prickett, T.D. *et al*. (2009) Analysis of the tyrosine kinome in melanoma reveals recurrent mutations in ERBB4. *Nat. Genet*. 41, 1127–1132

34 Eskandarpour, M. *et al*. (2005) Suppression of oncogenic *NRAS* by RNA interference induces apoptosis of human melanoma cells. *Int. J. Cancer* 115, 65–73

35 James, G.L. *et al*. (1993) Benzodiazepine peptidomimetics: potent inhibitors of Ras farnesylation in animal cells. *Science* 260, 1937–1942

36 Sun, J. *et al*. (1995) Ras CAAX peptidomimetic FTI 276 selectively blocks tumor growth in nude mice of a human lung carcinoma with K-Ras mutation and p53 deletion. *Cancer Res*. 55, 4243–4247

37 Caponigro, F. *et al*. (2003) Farnesyl transferase inhibitors in clinical development. *Expert Opin. Investig. Drugs* 12, 943–954

38 Smalley, K.S. and Eisen, T.G. (2003) Farnesyl transferase inhibitor SCH66336 is cytostatic, pro-apoptotic and enhances chemosensitivity to cisplatin in melanoma cells. *Int. J. Cancer* 105, 165–175

39 Weisz, B. *et al*. (1999) A new functional Ras antagonist inhibits human pancreatic tumor growth in nude mice. *Oncogene* 18, 2579–2588

40 Aharonson, Z. *et al*. (1998) Stringent structural requirements for anti-Ras activity of S-prenyl analogues. *Biochim. Biophys. Acta*. 1406, 40–50

41 Jansen, B. *et al.* (1999) Novel Ras antagonist blocks human melanoma growth. *Proc. Natl. Acad. Sci. U.S.A.* 96, 14019–14024

42 Smalley, K.S. and Eisen, T.G. (2002) Farnesyl thiosalicylic acid inhibits the growth of melanoma cells through a combination of cytostatic and pro-apoptotic effects. *Int. J. Cancer* 98, 514–522

43 Johnson, M.L. *et al.* (2009) A Phase II trial of salirasib in patients with stage IIIB/IV lung adenocarcinoma enriched for *KRAS* mutations. *J. Clin. Oncol.* 27(15 Suppl.), 8012

44 Laheru, D. *et al.* (2009) Integrated development of *S-trans,trans*-farnesylthiosalicyclic acid (FTS, salirasib) in advanced pancreatic cancer. *J. Clin. Oncol.* 27(15 Suppl.), 4529

45 Davies, H. *et al.* (2002) Mutations of the *BRAF* gene in human cancer. *Nature* 417, 949–954

46 Wilhelm, S.M. *et al.* (2004) BAY 43-9006 exhibits broad spectrum oral antitumor activity and targets the RAF/MEK/ERK pathway and receptor tyrosine kinases involved in tumor progression and angiogenesis. *Cancer Res.* 64, 7099–7109

47 Amaravadi, R. *et al.* (2007) Updated results of a randomized Phase II study comparing two schedules of temozolomide in combination with sorafenib in patients with advanced melanoma. *J. Clin. Oncol.* 25(18 Suppl.), 8527

48 Flaherty, K.T. *et al.* (2004) Phase I/II trial of BAY 43-9006, carboplatin (C) and paclitaxel (P) demonstrates preliminary antitumor activity in the expansion cohort of patients with metastatic melanoma. *J. Clin. Oncol.* 22(14 Suppl.), 7507

49 Flaherty, K.T. (2006) Chemotherapy and targeted therapy combinations in advanced melanoma. *Clin. Cancer Res.* 12, 2366s–2370s

50 Hauschild, A. *et al.* (2009) Results of a Phase III, randomized, placebo-controlled study of sorafenib in combination with carboplatin and paclitaxel as second-line treatment in patients with unresectable stage III or stage IV melanoma. *J. Clin. Oncol.* 27, 2823–2830

51 Flaherty, K.T. *et al.* (2005) Phase I/II, pharmacokinetic and pharmacodynamic trial of BAY 43-9006 alone in patients with metastatic melanoma. *J. Clin. Oncol.* 23(16 Suppl.), 3037

52 Eisen, T. *et al.* (2006) Sorafenib in advanced melanoma: a Phase II randomised discontinuation trial analysis. *Br. J. Cancer* 95, 581–586

53 Gray-Schopfer, V. *et al.* (2007) Melanoma biology and new targeted therapy. *Nature* 445, 851–857

54 Adnane, L. *et al.* (2006) Sorafenib (BAY 43-9006, Nexavar), a dual-action inhibitor that targets RAF/MEK/ERK pathway in tumor cells and tyrosine kinases VEGFR/PDGFR in tumor vasculature. *Methods Enzymol.* 407, 597–612

55 Whittaker, S. *et al.* (2010) Gatekeeper mutations mediate resistance to BRAF-targeted therapies. *Sci. Transl. Med.* 2, 35ra41

56 Tsai, J. *et al.* (2008) Discovery of a selective inhibitor of oncogenic B-Raf kinase with potent antimelanoma activity. *Proc. Natl. Acad. Sci. U.S.A.* 105, 3041–3046

57 Poulikakos, P.I. *et al.* (2010) RAF inhibitors transactivate RAF dimers and ERK signalling in cells with wild-type BRAF. *Nature* 464, 427–430

58 Bollag, G. *et al.* (2010) Clinical efficacy of a RAF inhibitor needs broad target blockade in BRAF-mutant melanoma. *Nature* 467, 596–599

59 Flaherty, K. *et al.* (2009) Phase I study of PLX4032: proof of concept for V600E BRAF mutation as a therapeutic target in human cancer. *J. Clin. Oncol.* 27(15 Suppl.), 9000

60 Flaherty, K.T. *et al.* (2010) Inhibition of mutated, activated BRAF in metastatic melanoma. *N. Engl. J. Med.* 363, 809–819

61 Chapman, P.B. *et al.* (2011) Improved survival with vemurafenib in melanoma with BRAF V600E mutation. *N. Engl. J. Med.* 364, 2507–2516

62 Heidorn, S.J. *et al.* (2010) Kinase-dead BRAF and oncogenic RAS cooperate to drive tumor progression through CRAF. *Cell* 140, 209–221

63 Nazarian, R. *et al*. (2010) Melanomas acquire resistance to B-RAF[V600E] inhibition by RTK or N-RAS upregulation. *Nature* 468, 973–977

64 Johannessen, C.M. *et al*. (2010) COT drives resistance to RAF inhibition through MAP kinase pathway reactivation. *Nature* 468, 968–972

65 Wagle, N. *et al*. (2011) Dissecting therapeutic resistance to RAF inhibition in melanoma by tumor genomic profiling. *J. Clin. Oncol.* 29, 3085–3096

66 Villanueva, J. *et al*. (2010) Acquired resistance to BRAF inhibitors mediated by a RAF kinase switch in melanoma can be overcome by cotargeting MEK and IGF-1R/PI3K. *Cancer Cell* 18, 683–695

67 Smalley, K.S. and Flaherty, K.T. (2009) Integrating BRAF/MEK inhibitors into combination therapy for melanoma. *Br. J. Cancer* 100, 431–435

68 Kefford, R.A. *et al*. (2010) Phase I/II study of GSK2118436, a selective inhibitor of oncogenic mutant BRAF kinase, in patients with metastatic melanoma and other solid tumors. *J. Clin. Oncol.* 28(15 Suppl.), 8503

69 Garnett, M.J. *et al*. (2005) Wild-type and mutant B-RAF activate C-RAF through distinct mechanisms involving heterodimerization. *Mol. Cell* 20, 963–969

70 Gray-Schopfer, V.C. *et al*. (2005) The role of B-RAF in melanoma. *Cancer Metastasis Rev.* 24, 165–183

71 Dumaz, N. *et al*. (2006) In melanoma, RAS mutations are accompanied by switching signaling from BRAF to CRAF and disrupted cyclic AMP signaling. *Cancer Res.* 66, 9483–9491

72 Smalley, K.S. *et al*. (2009) CRAF inhibition induces apoptosis in melanoma cells with non-V600E BRAF mutations. *Oncogene* 28, 85–94

73 Solit, D.B. *et al*. (2006) BRAF mutation predicts sensitivity to MEK inhibition. *Nature* 439, 358–362

74 Collisson, E.A. *et al*. (2003) Treatment of metastatic melanoma with an orally available inhibitor of the Ras–Raf–MAPK cascade. *Cancer Res.* 63, 5669–5673

75 Haura, E.B. *et al*. (2010) A phase II study of PD-0325901, an oral MEK inhibitor, in previously treated patients with advanced non-small cell lung cancer. *Clin. Cancer Res.* 16, 2450–2457

76 Adjei, A.A. *et al*. (2008) Phase I pharmacokinetic and pharmacodynamic study of the oral, small-molecule mitogen-activated protein kinase kinase 1/2 inhibitor AZD6244 (ARRY-142886) in patients with advanced cancers. *J. Clin. Oncol.* 26, 2139–2146

77 Haass, N.K. *et al*. (2008) The mitogen-activated protein/extracellular signal-regulated kinase kinase inhibitor AZD6244 (ARRY-142886) induces growth arrest in melanoma cells and tumor regression when combined with docetaxel. *Clin. Cancer Res.* 14, 230–239

78 Gilmartin, A.G. *et al*. (2011) GSK1120212 (JTP-74057) is an inhibitor of MEK activity and activation with favorable pharmacokinetic properties for sustained *in vivo* pathway inhibition. *Clin. Cancer Res.* 17, 989–1000

79 Infante, J.R. *et al*. (2010) Safety and efficacy results from the first-in-human study of the oral MEK 1/2 inhibitor GSK1120212. *J. Clin. Oncol.* 28(15 Suppl.), 2503

80 Koo, H.M. *et al*. (2002) Apoptosis and melanogenesis in human melanoma cells induced by anthrax lethal factor inactivation of mitogen-activated protein kinase kinase. *Proc. Natl. Acad. Sci. U.S.A.* 99, 3052–3057

81 Meier, F. *et al*. (2007) Combined targeting of MAPK and AKT signalling pathways is a promising strategy for melanoma treatment. *Br. J. Dermatol.* 156, 1204–1213

82 Smalley, K.S. *et al*. (2006) Multiple signaling pathways must be targeted to overcome drug resistance in cell lines derived from melanoma metastases. *Mol. Cancer Ther.* 5, 1136–1144

83 McDermott, U. *et al*. (2011) Genomics and the continuum of cancer care. *N. Engl. J. Med.* 364, 340–350

84 Rix, U. *et al*. (2007) Chemical proteomic profiles of the BCR-ABL inhibitors imatinib, nilotinib, and dasatinib reveal novel kinase and nonkinase targets. *Blood* 110, 4055–4063

85 dos Santos, S.C. and Sa-Correia, I. (2009) Genome-wide identification of genes required for yeast growth under imatinib stress: vacuolar H^+-ATPase function is an important target of this anticancer drug. *Omics* 13, 185–198

86 Capdeville, R. *et al*. (2002) Glivec (STI571, imatinib), a rationally developed, targeted anticancer drug. *Nat. Rev. Drug Discov.* 1, 493–502

87 Atkins, M. *et al*. (2006) Sunitinib maleate. *Nat. Rev. Drug Discov.* 5, 279–280

88 Jarkowski, A. and Sweeney, R.P. (2008) Nilotinib: a new tyrosine kinase inhibitor for the treatment of chronic myelogenous leukemia. *Pharmacotherapy* 28, 1374–1382

89 Weisberg, E. *et al*. (2005) Characterization of AMN107, a selective inhibitor of native and mutant Bcr-Abl. *Cancer Cell* 7, 129–141

90 Wei, G. *et al*. (2010) First-line treatment for chronic myeloid leukemia: dasatinib, nilotinib, or imatinib. *J. Hematol. Oncol.* 3, 47

91 Braun, T. and Fenaux, P. (2008) Farnesyltransferase inhibitors and their potential role in therapy for myelodysplastic syndromes and acute myeloid leukaemia. *Br. J. Haematol.* 141, 576–586

92 Charette, N. *et al*. (2010) Salirasib inhibits the growth of hepatocarcinoma cell lines *in vitro* and tumor growth *in vivo* through ras and mTOR inhibition. *Mol. Cancer* 9, 256

93 Board, R.E. *et al*. (2009) Detection of *BRAF* mutations in the tumour and serum of patients enrolled in the AZD6244 (ARRY-142886) advanced melanoma phase II study. *Br. J. Cancer* 101, 1724–1730

ends in Pharmacological Sciences

Targeting Tumor Cell Motility as a Strategy against Invasion and Metastasis

Alan Wells[1,*], Jelena Grahovac[1], Sarah Wheeler[1], Bo Ma[1],
Douglas Lauffenburger[2]

[1]Department of Pathology, University of Pittsburgh and Pittsburgh VAHS, Pittsburgh,
PA 15213, USA, [2]Department of Biological Engineering, MIT, Cambridge, MA 15213, USA
*Correspondence: wellsa@upmc.edu

Trends in Pharmacological Sciences, Vol. 34, No. 5, May 2013 © 2013 Elsevier Inc.
http://dx.doi.org/10.1016/j.tips.2013.03.001

SUMMARY

Advances in diagnosis and treatment have rendered most solid tumors largely curable if they are diagnosed and treated before dissemination. However, once they spread beyond the initial primary location, these cancers are usually highly morbid, if not fatal. Thus, current efforts focus on both limiting initial dissemination and preventing secondary spread. There are two modes of tumor dissemination – invasion and metastasis – each leading to unique therapeutic challenges and likely to be driven by distinct mechanisms. However, these two forms of dissemination utilize some common strategies to accomplish movement from the primary tumor, establishment in an ectopic site, and survival therein. The adaptive behaviors of motile cancer cells provide an opening for therapeutic approaches if we understand the molecular, cellular, and tissue biology that underlie them. Herein, we review the signaling cascades and organ reactions that lead to dissemination, as these are non-genetic in nature, focusing on cell migration as the key to tumor progression. In this context, the cellular phenotype will also be discussed because the modes of migration are dictated by quantitative and physical aspects of the cell motility machinery.

BASIC MECHANISMS OF TUMOR DISSEMINATION

Tumors of solid organs (carcinomas, sarcomas, and central nervous system tumors) kill patients mainly by dissemination from the primary site. Surgical and radiological advances have rendered localized cancers largely

211

Table 1 Properties that Distinguish Invasive from Metastatic Dissemination

Invasion	Metastasis
Local extension through tissue and into adnexa	Distant travel through conduits
Contiguous tumor	Distinct tumor
Orthotopic microenvironment	Ectopic microenvironment

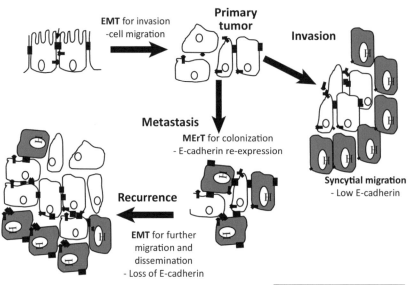

TRENDS in Pharmacological Sciences

FIGURE 1 Schematic of phenotypic changes from a normal, cell–cell connected epithelium to a disseminated carcinoma.
The cells downregulate their E-cadherin (solid bars) to allow for motility in a process denoted as 'epithelial-to-mesenchymal transition' (EMT) which allows for migration as a syncytial mass for invasion that displaces the normal parenchyma (gray cells) or as singular cells for metastasis. The survival in the distant site probably requires a reversion of the phenotype, a 'mesenchymal-to-epithelial reverting transition' (MErT) to reside among ectopic tissue epithelial cells (gray cells).

manageable, if not curable. However, once the cells migrate beyond the primary site into adjacent or distant tissue, the cells are difficult to extirpate. This dissemination may take two forms: (i) localized invasion throughout the tissue (especially for glioblastoma cerebri) and into the adnexa (most carcinomas), or (ii) metastatic dissemination (Table 1).

These two modes of dissemination require distinct sets of cellular behaviors, some of which are shared and others are distinct (Figure 1). In localized invasion, the tumor cell must acquire properties (i) enabling at least partial separation from the primary mass, (ii) recognition and (iii) reorganization of the barrier matrices, (iv) active migration through these matrices, and (v) survival in the adjacent tissues [1]. The invasive tumor then exists as a physical extension of the primary mass, with the cancer cells moving into the tissue in a syncytial manner.

The steps of the second mode of dissemination, that of metastatic seeding, have been described [2]. Initially, carcinoma cells must acquire properties that allow at least partial, if not complete, separation from the original mass to escape through boundary matrices and intravasate into a conduit [3]. Although there are copious clinical correlations and experimental animal studies showing that acquisition of mesenchymal-associated phenotypes and markers promote this escape, it is still possible that epithelioid carcinoma cells occasionally do reach the conduits and may provide for metastatic loci [4,5]. Survival during transit via vasculature is important because the shear stresses and lack of supportive signals from adhesive sites challenge these cells; still, the rapid nature of this dissemination and the mesenchymal properties probably aid in viable cells reaching target organs. At the metastatic locale the cells must recognize the endothelial cells, initiate separation from this monolayer, and then extravasate first onto and then through the basement membrane to gain access to the tissue parenchyma, a process that has been captured in experimental settings [6]. What occurs next is known only by extrapolation because metastatic nodules are not clinically evident until weeks to years later. For rapidly presenting tumor nodules, it is assumed that the carcinomas continue to proliferate as in the primary site. However, for delayed emergences, it is still open as to whether there is balanced proliferation and death/apoptosis, or individual cell dormancy, although the latter has been shown to be more statistically probable via advanced computer modeling systems [7]. Lastly, there is emerging evidence that the carcinoma cells may undergo a phenotypic switch towards a more epithelioid cell to establish metastases [5,8–11], although this may revert again to a mesenchymal phenotype as the metastasis becomes clinically evident [12].

It is assumed mainly by extrapolation that the cell aspect of invasion is syncytial, whereas metastasis occurs as escape of singular cells. The histopathological findings of invasive tumors often retaining cell–cell connections and singular metastatic tumor cells are what led originally to these assumptions, but these observations are post-hoc and may represent convergence of cells in the case of presumed syncytial invasion and death of co-disseminated cells in metastatic seeding. Experimental tracking of melanoma cell invasion into matrices suggests mainly mass movement of communicating cells even though there are singular cells extending ahead of the front ([13], J. Grahovac et al., unpublished). Others have found breast cancer cells breaking away individually prior to hematologous metastasis [14], and circulating tumor cells (CTCs) in patients are found to be singular [15,16]. Based on these and other observations, it is generally assumed that the two modes of dissemination utilize and even require quantitative, if not qualitative, differences in cell separation, but this is not likely as absolute as presented in simplified descriptions. The key for this discussion is that whether as single cells or as a group, motility is a rate-limiting process in tumor dissemination [3].

Tumors utilize both modes of dissemination, invasion and metastasis, to cause morbidity and mortality, but select tumor types show a predilection for one or the other. For example, bladder carcinoma and glioblastoma multiforme are primarily invasive, whereas breast and lung carcinomas are initially metastatic; melanoma is invasive as a prelude metastasis [9,17]. Thus, future therapies must account for both modes of cancer cell motility to be truly effective in limiting tumor progression. As such, there are two cell properties common to invasion and metastasis. The first is the epithelial-to-mesenchymal transition (EMT) that loosens the primary tumor cell mass. However, targeting this dedifferentiation is questionable because of its seemingly transient nature. The phenotypic plasticity to a more mesenchymal phenotype is reversed in the metastatic site, indicating facile adaptation by these cells [11,18,19]. This would require treatments to capture all migratory cancer cells before seeding at the metastatic site, a requirement that is unlikely to be met in most cases. Additionally, the syncytial migration of localized invasion may occur with cells expressing a partial epithelial phenotype [20]. The second shared characteristic is growth factor-induced cell motility [1]. This common requirement may hold a key to limiting tumor progression and turning cancer from a progressively lethal disease into a manageable chronic condition. This brief review should allow for the reader to discern the aspects of cell motility so as to derive approaches to targeting this process.

MOTILITY CASCADE

Tumors move under two stimuli, a basal motility from adhesion receptors and a faster rate from soluble growth factors. Growth factor receptor-mediated motility is a major driver of tumor cell dissemination via invasion or metastasis [1,14]. Thus, understanding the key molecular controls of this behavior should provide novel targets to limit initial or secondary dissemination. Solid tumors produce both autocrine and paracrine factors that, in turn, generate reciprocal paracrine signaling networks which actuate motility machinery [21]. It is possible to examine tumor cell movement even in the absence of exogenously added signals. Although far from optimal in deciphering the richness of the tumor microenvironment, these controlled contextual situations allow for parsing of key molecular switches.

BOX 1 KEY MOLECULAR SWITCHES DURING FACTOR-INDUCED MESENCHYMAL MOTILITY

Tumor cells move under the influence of autocrine and paracrine signaling. This cell migration can be characterized as a persistent random walk in the absence of attractant concentration gradients, and as a biased persistent random walk in the presence of such gradients. In either case, two key descriptors of migration behaviors are translational speed (distance traveled per time) and directional

substratum attachment intracellular force generation, and detachment of cell/substratum adhesions [66] (Figure I). Overriding the haptokinetic controls of adhesion are key targetable molecular switches (in italics) that impose a faster but less persistent motility. Inhibiting any one of these will limit cell motility and tumor invasion in experimental models [67,68]. However, coordination of these processes into cell

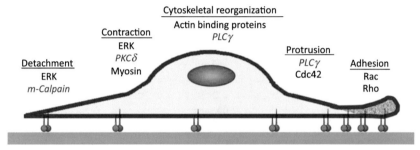

TRENDS in Pharmacological Sciences

FIGURE I

During mesenchymal migration the specific steps of cytoskeletal reorganization, lamellipodial protrusion, new forward adhesions, transcellular contractility, and rear detachment are controlled by key signaling nexi. The key integrating molecules for each process are shown, in black for those operative during both adhesion- and growth factor-driven motility, whereas those only activated during the enhanced movement promoted by growth factor receptor activation are shown in blue italics. Abbreviations: ERK, extracellular signal-regulated kinase; PKCδ, protein kinase Cδ; PLCγ, phospholipase Cγ.

persistence (average time interval between major changes in direction). These properties can be measured experimentally using individual cell video tracking methods, and have been shown to be influenced by environmental stimuli such as soluble growth factors and insoluble extracellular matrix components [64,65]. In turn, cell migration speed and directional persistence result from a highly integrated set of biophysical processes underlying locomotion. At a minimum, these biophysical processes include lamellipodial (and/or filopodial) membrane protrusion, cell/

translocation further requires a biophysical asymmetry between the effective front of the cell and its rear, so that attachment can remain strong at the front while detachment occurs at the rear. Thus, the key nodes for intervention were sought by unbiased systems biology approaches. A decision tree analysis found myosin-light chain and protein kinase C δ (PKCδ) activation status as key [22], a finding supported by principle component analyses [23]. These suggest that the ratio of adhesion to transcellular contrality governs mesenchymal motility.

The best understood mode of induced tumor cell migration is that of mesenchymal motility which is seen as cells move across stiff substrata (Box 1). However, *in vivo*, tumor cells move through matrices in three dimensions. Emerging research indicates that tumor cells vacillate between mesenchymal motility and a less-adherent amoeboid-like motility [13,21]. Still, the findings from mesenchymal motility have been validated in the more complex context of organismal dissemination in animal models. Numerous studies have shown that targeting individual intracellular effectors of motility as defined in 2D contexts can limit tumor invasion and metastasis in animal models [1,3]. More importantly, the key signaling nexus of transcellular contractility/cell-substratum adhesion [22] has been shown to govern 3D migration and integrate matrix stiffness [23,26].

To move *in vivo*, cancer cells must break away from the primary tumor mass and remodel the extracellular matrix (ECM). In a 3D environment, the force of protrusive actin polymerization drives cells into spindle-shaped mesenchymal morphology that can allow cell migration, and this process requires proteases in dense ECM [25,26]. A hallmark of tumor cell invasion is upregulation of proteolytic enzymes generated both by migrating cells and stromal cells [27]. Our appreciation of proteases in cancer progression has progressed from an early simplified conception of dissolving barrier matrices to one of not only judicious 'loosening' of matrices, primarily for syncytial invasion [28], but also of releasing or modulating pro-motility signals [29]. In addition to providing space for invasive cells to move forward, even minimal proteolytic processing changes the matrix composition, which in turn alters tumor cell migration [30]. This matrix degradation is required to a greater extent during syncytial migration or even migration of individual cells in a mesenchymal mode [31].

Depending on the 3D context, some tumor cells can alternatively invade without a requirement for proteolytic activity. They can move through the ECM without its degradation by either following natural cleavage planes or acquiring a rounded morphology and using actin contractile force to generate amoeboid bleb-like protrusions that push and squeeze cells through the ECM [25,26] (Figure 2). During amoeboid motility, little matrix proteolysis is needed as the cells 'bleb' through tight passages in the matrices; this is noted *in vitro* by imaging and suggested *in vivo* by finding rounded cells in the process of traversing barrier matrices (Figure 2). Migration is possible in the absence of proteolysis if both the porosity of the matrix and the cell body deformability can support it. Molecular and structural characteristics of both tissue microenvironment and cell behavior determine whether cells will migrate collectively or individually in mesenchymal or amoeboid mode. ECM stiffness, fiber orientation, density, and gap size provide parameters that modulate cell adhesion and cytoskeletal organization [32]. Although mesenchymal migration is dependent on alternative pushing and pulling cycles, amoeboid migration is equally mechanically complex and combines stronger pushing with less adhesive pulling of the substrate.

Tumor cell migration involves alteration between cellular states [32]. Intracellular switches that control cytoskeletal tension, and thus cell shape, regulate

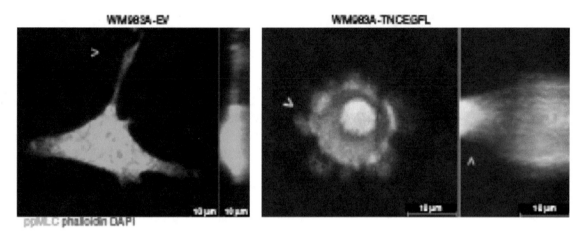

WM983A-EV   WM983A-TNCEGFL

ppMLC phalloidin DAPI

TRENDS in Pharmacological Sciences

FIGURE 2

Invading melanoma cells can passage through tight matrices either in the mesenchymal state requiring extensive matrix remodeling (left) or the amoeboid state in which the cells 'bleb' through tight spaces (right). The right panels in each pair are vertical views of cells extending through a pore in the membrane. From [13].

the conversion between mesenchymal and rounded migration; targetable molecules involve the balance between the Rac pathway that promotes cell spreading and the Rho/ROCK signaling axis that leads to cellular contraction [25]. Silencing of the ROCK pathway induces the transition from amoeboid to mesenchymal invasiveness [33], whereas silencing of Rac induces the opposite shift [34]. In addition, active Rac negatively regulates Rho/ROCK signaling [35] and inhibits cell rounding, whereas active Rho/ROCK limits Rac activity that inhibits cell extensions [33]. Therefore, anti-motility strategies must block both modes of movement.

The signaling pathways that actuate the cell motility in tumor cells do show altered levels of expression, but individual genes are not usually mutated; this is likely as activation of one step but not the others in the migration process would lead to uncoordinated and inefficient motility. As both modes of mesenchymal and amoeboid motility are interchangeable and arise in response to cues from the microenvironment [36], the signaling systems must be intact and functional. Furthermore, this same switching mechanism is used in physiological processes such as wound healing.

EXTRACELLULAR REGULATORS OF MIGRATION

Many studies have explored the extrinsic controls of tumor cell migration. At first, researchers focused on soluble signals, such as growth factors, which could drive tumor dissemination. There are extensive recent reviews of these aspects [1,21,37], thus these will not be reviewed here. Briefly, many of the classic growth factors, with those of the epidermal growth factor (EGF) and hepatocyte growth factor (HGF) families at the forefront, have been shown to promote cell migration. This raised hope that targeting such factors or their receptors might be an avenue for rationally targeting cancers. Some limited clinical advances have been made with growth factor inhibitors (e.g., of the EGF receptor), these have mainly focused on limiting cell proliferation or driving cell death when such receptors are mutated or overexpressed. The reason for the limited successes and recurrent resistance is the redundancy of the receptor pathways and the fact that often these signals occur in epigenetic or contextual situations in which the cancer cell is not intrinsically predisposed to be dependent on any one given signal. Thus, strategies to target common intracellular signals may be more effective.

Chemokines are a second group of soluble signaling proteins which were originally investigated as a secondary class of signals in immunoregulation [38]; more recently, they have been examined in the context of cancer [39]. This association occurred after increasing appreciation that these ligand–receptor systems functioned in all cells, and were crucial to terminating wound repair [40] and to inducing angiogenesis [41]. Chemokines that signal via CXCR4 [42] can promote cancer cell metastasis by increasing cell motility via both autocrine and paracrine signaling loops [43–45]. One of the chemokine networks that has come to the fore of carcinoma motility investigation is that of the CXCR3 receptor, which physiologically functions as an immune cell chemoattractant and an adherent cell inhibitor of locomotion. This receptor enhances tumor cell motility in vitro [46–48] and metastasis in vivo [49,50]. Although this might suggest that these ligands or the cognate CXCR3 receptor may be a target for limiting spread, this is a two-edged sword because this signaling system limits the movement of more differentiated cells. The switch to pro-motogenic signaling occurs via a change in the splice variant from the adherent cell CXCR3B to the immune cell CXCR3A isoform [51]. Thus, even usual 'stop' signals may become 'go' signals, requiring the targeting of the basic motility mechanisms.

Emerging evidence has pointed to active signaling by intrinsic matrix components through growth factor receptors as major drivers in tumor progression; this is in contrast to the well-described haptokinesis, deriving on physical considerations, driven through classical matrix receptors. The cancer situation resembles a wound state in which the 'immature' ECM presents numerous pro-motility domains while lacking the anti-migratory small leucine-rich proteoglycans (SLRPs) [52,53], thus representing a 'wound that won't heal'. Central to these pro-motility matrices are the matrix components tenascin-C and laminin [54], molecules that contain EGF-like repeats that bind at low affinity but high avidity to the EGF receptor to preferentially drive motility [55]. Additionally, the decrease in the SLRPs (such as decorin) [56] leads to unbalanced signaling which promotes tumor invasion [13]. Although such molecules might be considered candidates for therapies, the redundancy aspect again argues against such a straightforward strategy.

Although matrix remodeling may occur only to a limited extent in amoeboid migration, proteases can nonetheless still play an important role in tumor cell migration by altering signaling elements. Many pro-motility growth factors either are produced as pro-factors that require proteolytic processing for release (including all EGF receptor ligands) or activation (HGF is an example), or which are sequestered in the matrix requiring mobilization, such as the heparin-binding members of the FGF family or insulin-like growth factor 1 (IGF-1) bound to its sequestration molecules). In these cases, the proteases liberate the signals to act in paracrine mode, with asymmetric expression of membrane-tethered and secreted proteinases providing a signaling gradient for directional motility. Additionally, many of these same proteases can provide access to cryptic signaling elements in the matrix, such as tenascin-C. Anti-motility chemokines are also affected by proteases. These soluble peptides are inactivated by proteolysis, thus contributing to a pro-migratory environment [57]. For instance, the CXCR3-activating CXCL11/IP-9 is abrogated by a number of matrix metalloproteinases (MMPs) present during tumor cell invasion [58].

POTENTIAL INTERVENTIONAL OPPORTUNITIES

Given the central role of induced migration in tumor progression, therapies aimed at limiting this behavior should be an obvious goal. This can be attempted by targeting signals upregulated in specific cancers (such as select MMPs), molecules common to many cancers (such as tenascin-C), or the common nodal points for the cell phenotype. Targeting specific molecules overexpressed in a specific cancer offers the opportunity to tailor the therapy to individual patients and hopefully limit toxicity. The same avoidance of toxicity can justify targeting general tumor progression-associated factors, such as tenascin-C. However, both of these approaches are limited mainly by the redundancy of these pro-migratory signals, suggesting that abrogation of one would be of limited or short-lived benefit. Such a rebound phenomenon has been noted for receptor tyrosine kinase inhibitors such as Herceptin and erlotinib [59]; this issue would be especially relevant for maintenance therapies that are needed to limit progression.

A second approach is to target key intracellular signaling nodes for motility, such as those that link transcellular contractility to cell adhesion [22] or forward protrusion [24]. In this manner the redundancy of input signals is not an issue as convergent points are inhibited. Although this might risk a high level of toxicity, the fact that the motility that drives tumor progression is distinct from the routine homeostatic haptotaxis [1] suggests that there would be a therapeutic index as long as there are no ongoing wound healing processes that also use this induced cell migration. The balance of physical considerations and active biochemical signaling governed concomitantly by ECM and growth factors can be quantitatively analyzed to determine how robust the effectiveness of a signaling pathway inhibitory drug could be [60]. Thus, as it is usual for therapies to await the postoperative healing period, this should not be an issue; however, long-term maintenance therapy must account for these comorbidities.

An alternate approach would be to provide anti-motility signals rather targeting pro-motility pathways. Recent investigations have highlighted physiological 'stop' signals such as CXCR3 and decorin, holding the promise that upregulation of these signals could be triggered in a therapeutic manner to limit invasion. We have recently found that presentation of decorin in the matrix can limit melanoma invasion (J. Grahovac et al., unpublished). The unwanted side effects of such forced expression again would be during times of active wound healing. A word of caution must come to the fore in light of the findings of splice isoform switching by carcinoma cells in that the tumor cells turn this 'stop' signal into a 'go' signal by changing the expressed receptor splice variant [51]. Thus, one cannot readily extrapolate the efficacy of these 'stop' signals from non-tumor counterparts, but need to validate these signals in the relevant tumor cells and models.

A general cautionary note also must be expressed when targeting an integrated cellular behavior in a long-term manner. This relates to the issue that near-complete inhibition of a pathway may achieve the desired effect, but partial inhibition may be counterproductive. For instance, targeting the individual components of motility in a model of prostate cancer invasion and metastasis abrogated dissemination; however, when we challenged these same cells with a gradation of inhibition, the invasiveness was increased at partially diminished signaling [61]. This was predictable because the actuation of locomotion depends on a balance of signaling elements and changes to any aspect short of complete inhibition can be compensated by similar directional changes in other signaling cascades. Compensation such as this can be a major limiting factor during long-term maintenance therapies rather than the usual short-term ablative therapies currently used to kill cancer cells. A second caveat relates to the possibility that tumor cells have disseminated fairly early prior to detection and initial treatments [62]. In such a circumstance the opportunity to prevent metastasis may be missed, with many of these disseminated cells potentially entering a long period of quiescent dormancy [7]. However, as metastatic deposits can give rise to subsequent disseminations [63], even secondary prevention may be beneficial.

This leads to the major challenge facing progression-targeted therapies – how to design clinical trials to determine efficacy. There are several current trials examining tumor cell migration but these are mainly in the correlative stage and have not progressed to interventional status (clinicaltrials.gov). In addition to a few growth factor inhibition strategies already in clinical use (such as inhibitors of EGF receptor family signaling, including Herceptin and erlotinib), inhibitors of intracellular signaling nodes are beginning early stage clinical trials. However, as the target molecules are also involved in signaling proliferation and/or survival, the key mechanism of efficacy remains to be determined. Further limiting clinical testing is the open question of how to measure success of the intervention. By their nature, agents that target motility will not be useable as single agents for extant tumors and thus must be part of a regimen. Because the standard measurements for efficacy involve tumor nodule size (shrinkage or stable size being measured as response), these agents would require a different trial design. Thus, trials would need to provide for progression-free or overall survival measurements, even at the earliest stages of Phase II trials. This would be costly in terms of time and number of patients to be enrolled. The avenue of using rapidly progressive and invasive disease, such as glioblastoma multiforme is complicated by the concomitant proliferative aspects of the tumor that would be under treatment. The true target of these agents would be for limiting the spread of more indolent tumors, such as dormant breast cancer or the field effect bladder and oral carcinomas. In these cases, trials would only be manageable by allowing for surrogate endpoints such as size of field effect and time to limited localized recurrence, as determined by multiple and frequent biopsying of the bladder or oral cavity. Although these adaptations are possibly intuitive from a medical and biological perspective, the regulatory agencies would need to develop new avenues for such testing. The advantage of these trial designs would be that recruitment would potentially be easier because these treatments will probably have low toxicity and be an add-on to standard therapies.

CONCLUDING REMARKS

Tumor cell migration as induced by various soluble and matrix signals represents a novel avenue for limiting both invasion and metastasis, thus attacking the most morbid and daunting aspects of cancer. Deciphering the basic mechanisms of cell motility in 2D and 3D have highlighted intracellular cascades critical for this motility, and studies of human cancers have shown the signals in the microenvironment that drive this cellular behavior. Thus, there is no shortage of candidate targets and, importantly, there are at least lead compound inhibitors for many of these. However, rapid movement into the clinic is challenged by numerous issues. Signaling and network redundancy will require careful selection of key nodes, or even upregulation of 'stop' signals. Even if this is accomplished, the desired effect of preventing extension while not killing or shrinking the tumor *per se*, and doing this over years, will require new types of clinical trials. Still, the promise of 'stopping cancer in its tracks', literally, impels the quest to develop such treatments that limit tumor migration.

ACKNOWLEDGMENTS

We thank the Wells, Lauffenburger, Griffith, and Camacho laboratories for helpful discussions. These studies were supported by grants from the National Institutes of Health (National Institute of General Medical Sciences and National Cancer Institute) (A.W. and D.L.) and Department of Veterans Affairs (Merit Program) (A.W.).

REFERENCES

1 Wells, A. (2000) Tumor invasion: role of growth factor-induced cell motility. *Adv. Cancer Res.* 78, 31–101

2 Klein, C.A. (2008) The metastatic cascade. *Science* 321, 1785–1787

3 Wells, A. *et al*. (2011) Epithelial and mesenchymal phenotypic switchings modulate cell motility in metastasis. *Front. Biosci.* 16, 815–837

4 Tsuji, T. *et al*. (2009) Epithelial–mesenchymal transition and cell cooperativity in metastasis. *Cancer Res.* 69, 7135–7139

5 Tarin, D. (2011) Cell and tissue interactions in carcinogenesis and metastasis and their clinical significance. *Semin. Cancer Biol.* 21, 72–82

6 Kienast, Y. *et al*. (2010) Real-time imaging reveals the single steps of brain metastasis formation. *Nat. Med.* 16, 116–122

7 Taylor, D.P. *et al*. (2013) Modeling boundary conditions for balanced proliferation in metastatic latency. *Clin. Cancer Res.* 19, 1063–1070

8 Kowalski, P.J. *et al*. (2003) E-cadherin expression in primary carcinoma of the breast and its distant metastases. *Breast Cancer Res.* 5, R217–R222

9 Wells, A. *et al*. (2008) E-cadherin as an indicator of mesenchymal to epithelial reverting transitions during the metastatic seeding of disseminated carcinomas. *Clin. Exp. Metastasis* 25, 621–628

10 Chaffer, C.L. *et al*. (2006) Mesenchymal-to-epithelial transition facilitates bladder cancer metastasis: role of fibroblast growth factor receptor-2. *Cancer Res.* 66, 11271–11278

11 Chao, Y.L. *et al*. (2010) Breast carcinoma cells re-express E-cadherin during mesenchymal to epithelial reverting transition. *Mol. Cancer* 9, e179

12 Chao, Y. *et al*. (2012) Hepatocyte-induced re-expression of E-cadherin in breast and prostate cancer cells increases chemoresistance. *Clin. Exp. Metastasis* 29, 39–50

13 Grahovac, J. *et al*. (2013) Melanoma cell invasiveness is regulated at least in part by the epidermal growth factor-like repeats of tenascin-C. *J. Invest. Dermatol.* 133, 210–220

14 Wang, W. *et al*. (2005) Tumor cells caught in the act of invading: their strategy for enhanced cell motility. *Trends Cell Biol.* 15, 138–145

15 Bednarz-Knoll, N. *et al*. (2011) Clinical relevance and biology of circulating tumor cells. *Breast Cancer Res.* 13, e228

16 Yu, M. *et al*. (2011) Circulating tumor cells: approaches to isolation and characterization. *J. Cell Biol.* 192, 373–382

17 Steeg, P. (2006) Tumor metastasis: mechanistic insights and clinical challenges. *Nat. Med.* 12, 895–904

18 Tsai, J.H. *et al.* (2012) Spatiotemporal regulation of epithelial–mesenchymal transition is essential for squamous cell carcinoma metastasis. *Cancer Cell* 22, 725–736

19 Gunasinghe, N. *et al.* (2012) Mesenchymal–epithelial transition (MET) as a mechanism for metastatic colonisation in breast cancer. *Cancer Metastasis Rev.* 31, 469–478

20 Kim, H-D. *et al.* (2011) Signaling network state predicts Twist-mediated effects on breast cell migration across diverse growth factor contexts. *Mol. Cell. Proteomics* 10, eM111.008433

21 Friedl, P. and Gilmour, D. (2009) Collective cell migration in morphogenesis, regeneration and cancer. *Nat. Rev. Mol. Cell Biol.* 10, 445–457

22 Hautaniemi, S. *et al.* (2005) Modeling and prediction of signal transduction cascades using decision trees. *Bioinformatics* 21, 2027–2035

23 Kim, H.D. *et al.* (2008) Epidermal growth factor-induced enhancement of glioblastoma cell migration in 3D arises from an intrinsic increase in speed but an extrinsic matrix- and proteolysis-dependent increase in persistence. *Mol. Biol. Cell* 19, 4249–4259

24 Meyer, A.S. *et al.* (2012) 2D Protrusion but not motility predicts growth factor-induced cancer cell migration in 3D collagen. *J. Cell Biol.* 197, 721–729

25 Sahai, E. and Marshall, C.J. (2003) Differing modes of tumour cell invasion have distinct requirements for Rho/ROCK signalling and extracellular proteolysis. *Nat. Cell Biol.* 5, 711–719

26 Wolf, K. *et al.* (2003) Compensation mechanism in tumor cell migration: mesenchymal–amoeboid transition after blocking of pericellular proteolysis. *J. Cell Biol.* 160, 267–277

27 Kessenbrock, K. *et al.* (2010) Matrix metalloproteinases: regulators of the tumor microenvironment. *Cell* 141, 52–67

28 Wolf, K. *et al.* (2007) Multi-step pericellular proteolysis controls the transition from individual to collective cancer cell invasion. *Nat. Cell Biol.* 9, 893–904

29 Page-McCaw, A. *et al.* (2007) Matrix metalloproteinases and the regulation of tissue remodelling. *Nat. Rev. Mol. Cell Biol.* 8, 221–233

30 Zaman, M.H. *et al.* (2006) Migration of tumor cells in three-dimensional matrices in governed by matrix stiffness along with cell-matrix adhesion and proteolysis. *Proc. Natl. Acad. Sci. U.S.A.* 103, 10889–10894

31 Wolf, K. and Friedl, P. (2011) Extracellular matrix determinant of proteolytic and non-proteolyic cell migration. *Trends Cell Biol.* 21, 736–744

32 Friedl, P. and Wolf, K. (2010) Plasticity of cell migration: a multiscale tuning model. *J. Cell Biol.* 188, 11–19

33 Sanz-Moreno, V. *et al.* (2011) ROCK and JAK1 signaling cooperate to control actomyosin contractility in tumor cells and stroma. *Cancer Cell* 20, 229–245

34 Yamazaki, D. *et al.* (2009) Involvement of Rac and Rho signaling in cancer cell motility in 3D substrates. *Oncogene* 28, 1570–1583

35 Sanders, E.E. *et al.* (1999) Rac downregulates rho activity: reciprocal balance between both GTPases determines cellular morphology and migratory behavior. *J. Cell Biol.* 147, 1009–1021

36 Pankova, K. *et al.* (2010) The molecular mechanisms of transition between mesenchymal and amoeboid invasiveness in tumor cells. *Cell. Mol. Life Sci.* 67, 63–71

37 Birchmeier, C. *et al.* (2003) Met, metastasis, motility and more. *Nat. Rev. Mol. Cell Biol.* 4, 915–925

38 Onuffer, J.J. and Horuk, R. (2002) Chemokines, chemokine receptors and small-molecule antagonists: recent developments. *Trends Pharmacol. Sci.* 23, 459–467

39 Fulton, A.M. (2009) The chemokine receptors CXCR4 and CXCR3 in cancer. *Curr. Oncol. Rep.* 11, 125–131

40 Yates, C.C. *et al.* (2008) ELR-negative CXC chemokine CXCL11(IP-9/I-TAC) facilitates dermal and epidermal maturation during wound repair. *Am. J. Pathol.* 173, 643–652

41 Bodnar, R.J. *et al.* (2009) ELR-negative chemokine IP-10/CXCL10 induces dissociation of newly-formed vessels secondary to calpain cleavage of β3 integrin. *J. Cell Sci.* 122, 2064–2077

42 Menon, L.G. *et al.* (2007) Differential gene expression associated with migration of mesenchymal stem cells to conditioned medium from tumor cells or bone marrow cells. *Stem Cells* 25, 520–528

43 Huang, P.H. and Marais, R. (2009) Cancer: melanoma troops massed. *Nature* 459, 336–337

44 do Carmo, A. *et al.* (2010) CXCL12/CXCR4 promotes motility and proliferation of glioma cells. *Cancer Biol. Ther.* 9, 56–65

45 Dai, X. *et al.* (2013) The CXCL12/CXCR4 autocrine loop increases the metastatic potential of non-small cell lung cancer in vitro. *Oncol. Lett.* 5, 277–282

46 Martins, V.L. *et al.* (2009) Increased invasive behaviour in cutaneous squamous cell carcinoma with loss of basement-membrane type VII collagen. *J. Cell Sci.* 122, 1788–1799

47 Shin, S.Y. *et al.* (2010) TNFα-exposed bone marrow-derived mesenchymal stem cells promote locomotion of MDA-MB-231 breast cancer cells through transcriptional activation of CXCR3 ligand chemokines. *J. Biol. Chem.* 285, 30731–30740

48 Zipin-Roitman, A. *et al.* (2007) CXCL10 promotes invasion-related properties in human colorectal carcinoma cells. *Cancer Res.* 67, 3396–3405

49 Cambien, B. *et al.* (2009) Organ-specific inhibition of metastatic colon carcinoma by CXCR3 antagonism. *Br. J. Cancer* 100, 1755–1764

50 Ma, X. *et al.* (2009) CXCR3 expression is associated with poor survival in breast cancer and promotes metastasis in a murine model. *Mol. Cancer Ther.* 8, 490–498

51 Wu, Q. *et al.* (2012) Altered CXCR3 isoform expression regulates prostate cancer cell migration and invasion. *Mol. Cancer* 11, 3

52 Tran, K.T. *et al.* (2004) Extracellular matrix signaling through growth factor receptors during wound healing. *Wound Repair Regen.* 12, 262–268

53 Hood, B.L. *et al.* (2010) Proteomic analysis of laser microdissected melanoma cells from skin organ cultures. *J. Proteome Res.* 9, 3656–3663

54 Quaranta, V. (2002) Motility cues in the tumor environment. *Differentiation* 70, 590–598

55 Iyer, A.K.V. *et al.* (2008) Cell surface restriction of EGFR by a Tenascin cytotactin-encoded EGF-like repeat is preferential for motility-related signaling. *J. Cell. Physiol.* 214, 504–512

56 Neill, T. *et al.* (2012) Decorin: a guardian from the matrix. *Am. J. Pathol.* 181, 380–387

57 Rodriguez, D. *et al.* (2010) Matrix metalloproteinases: what do they not do? New substrates and biology identified by murine models and proteomics. *Biochem. Biophys. Acta* 1803, 39–54

58 Cox, J.H. *et al.* (2008) Matrix metalloproteinase processing of CXCL11/I-TAC results in loss of chemoattractant activity and altered glycosaminoglycan binding. *J. Biol. Chem.* 283, 19389–19399

59 Rosenzweig, S.A. (2012) Acquired resistance to drugs targeting receptor tyrosine kinases. *Biochem. Pharmacol.* 83, 1041–1048

60 Wu, S. *et al.* (2011) Controlling multipotential stromal cell migration by integrating "coarse-graining" materials and "fine-tuning" small molecules via decision tree signal-response modeling. *Biomaterials* 32, 7524–7531

61 Kharait, S. *et al.* (2007) Decision tree modeling predicts effects of inhibiting contractility signaling on cell motility. *BMC Syst. Biol.* 1, 9

62 Cristofanilli, M. *et al*. (2004) Circulating tumor cells, disease progression, and survival in metastatic breast cancer. *N. Engl. J. Med.* 351, 781–791

63 Fidler, I.J. and Nicolson, G.L. (1977) Fate of recirculating B16 melanoma metastatic variant cells in parabiotic syngeneic recipients. *J. Natl. Cancer Inst.* 58, 1867–1872

64 Shreiber, D. *et al*. (2001) Effects of PDGF-BB on rat dermal fibroblast behavior in mechanically stressed and unstressed collagen and fibrin gels. *Exp. Cell Res.* 266, 155–166

65 Gobin, A.S. and West, J.L. (2003) Effects of EGF on fibroblast migration through biomimetic hydrogels. *Biotechnol. Prog.* 19, 1781–1785

66 Lauffenburger, D.A. and Horwitz, A.F. (1996) Cell migration: a physically integrated molecular process. *Cell* 84, 359–369

67 Mamoune, A. *et al*. (2003) m-Calpain as a target for limiting prostate cancer invasion. *Cancer Res.* 63, 4632–4640

68 Kassis, J. *et al*. (2001) Tumor invasion as dysregulated cell motility. *Semin. Cancer Biol.* 11, 105–118

ends in Pharmacological Sciences

The Valley of Death in Anticancer Drug Development: A Reassessment

David J. Adams[*]

Department of Medicine, Duke University Health System, Duke Box # 2638, Research
Drive, Durham, NC 27710, USA
*Correspondence: adams041@mc.duke.edu

Trends in Pharmacological Sciences, Vol. 33, No. 4, April 2012 © 2012 Elsevier Inc.
http://dx.doi.org/10.1016/j.tips.2012.02.001

SUMMARY

The past decade has seen an explosion in our understanding of cancer
biology and with it many new potential disease targets. Nonetheless, our
ability to translate these advances into therapies is poor, with a failure rate
approaching 90%. Much discussion has been devoted to this so-called
'Valley of Death' in anticancer drug development, but the problem persists.
Could we have overlooked some straightforward explanations to this highly
complex problem? Important aspects of tumor physiology, drug pharmacokinetics, preclinical models, drug delivery, and clinical translation are not
often emphasized, but could be crucial. This perspective summarizes current views on the problem and suggests feasible alternatives.

THE VALLEY OF DEATH

Failure to translate our rapidly expanding knowledge of cell biology into
effective therapeutics has been a topic of lively and ongoing debate in the
scientific community [1–9] as well as popular press [10–13], which has been
particularly critical. The issue is urgent for anticancer drug development
where the late-stage attrition rate for oncology drugs is as high as 70% in
Phase II and 59% in Phase III trials [14]. Commentators have delineated
numerous underlying factors for this conundrum, often referred to as the
'Valley of Death'. Clearly, financial and other economic and nontechnical
factors play a major role (summarized in Table 1). However, there are scientific issues that, although well known in their respective fields, are not prominent in the debate – but could be equally important in bridging the divide
(summarized in Table 2). These issues fall under five main categories: (i)

227

CellPress

Table 1 Reported Factors that Contribute to the Valley of Death in Anticancer Drug Development

Factor	Cause(s)	Ref
Lack of efficacy and safety	Lack of predictive animal models and strong evidence for mechanism of action (MOA)-based efficacy; failure to eliminate compounds with MOA-based toxicity; increasing safety hurdles; poor pharmacokinetics	[3,4,14,40]
Lack of financial resources	Risk-averse mentality at the National Institutes of Healthy (NIH), pharma, venture capital; majority of NIH support funds basic research, less than 5% for translational research	[4–6]
Lack of human resources	Consolidation in pharmaceutical industry; >35 000 jobs lost in 2010 alone	[5,102,103]
Lack of required research structure	Individual investigator model versus multidisciplinary team	[6]
Lack of support expertise	Different support and management structures for basic versus translational and clinical research	[5]
Communication	Poor communication between clinical and basic scientists and between scientists and the business community	[6]
Design of clinical trials	Lack of medically and statistically meaningful endpoints; lack of and failure to incorporate validated biomarkers; lack of appropriate patient selection; lack of pharmacokinetic guidance	[7,14,104,105]
Healthcare culture	Failure to adopt results from clinical studies into clinical practice	[106]
Lack of incentives in academia	Reward/promotion structure in academia; difficulty in assessing outcomes of translational research to reward effort	[5]
Profit structure in industry	Large pharma seeks blockbuster drugs in large markets versus orphan drugs; pressure from Wall Street for short-term profits	[9,107]
Focus on high risk diseases	Trade-off high potential profits for high attrition rate; compounds with novel MOAs have higher attrition rates	[1,14,108]
Focus on technology	Human genome project has generated unlimited potential targets, but few have been validated	[102]
Choice of drug type	Focus is on small molecules whereas biologicals have higher rates of success	[14]
Lack of predictive discovery models	Current target-centric approach; many targeted agents affect essential cellular functions and behave similarly to cytotoxics	[104,109]
Lack of predictive development models	Failure to capture tumor heterogeneity and cellular complexity; inadequate understanding of pathway connectivity in tumor versus normal cells; limited support for clinically relevant model development	[2,108]
Adverse regulatory environment	Initial clinical experience is in patients with advanced, refractory disease; inadequate funding of regulatory agencies such as the FDA; lack of global regulatory harmonization	[2,9,110]
Intellectual property issues	Limited patent lifetime relative to the extended development time; overarching patents that restrict research in key fields	[111]
Aggressive pricing may create barriers to reimbursement	High attrition rates amplify costs of drug development that are passed on to patients	[2,107]
Lack of innovation	Pipelines all focus on a limited number of mechanisms and targets; lack of compelling new biology, enabling technologies, genomics-derived tumor-specific targets, and therapeutic concepts	[3,110]

Table 1 Reported Factors that Contribute to the Valley of Death in Anticancer Drug Development *Continued*

Factor	Cause(s)	Ref
Feasibility and cost of manufacture and development	Complex drug molecules or drug carriers; overall cost to bring NME to market now estimated at $1 billion	[3,7]
Commercial issues	Alignment of corporate R&D with marketing goals; awareness of competitor programs; over-management of R&D by those lacking scientific–medical expertise; resources focused on marketing	[14,107]

Table 2 Approaches to Improve Anticancer Drug Development

Traditional Approach	Proposed Approach	Refs
Employ preclinical cell culture models at pH 7.4 and ambient oxygen (21%)	Employ preclinical cell-culture models at physiological tumor pH (6.5–7.0) and oxygen (0–5%)	[24]
Rank compounds by effective concentration (IC50, ED50)	Rank compounds by effective exposure ($C^n \times T$)	[42,43]
Screen for antitumor activity against bulk tumor	Screen for antitumor activity against tumor stem cells	[81]
Screen for antitumor activity against primary tumors in murine subcutaneous xenografts	Screen for antitumor activity against metastatic disease in orthotopic models or genetically-engineered mice with cancer-specific mutations	[54]
Optimize drug candidates by standard ADMET parameters	Optimize drug candidates according to charge dynamics that exploit the tumor pH gradient	[38–40]
Rely on rodent models to predict clinical activity	Utilize 3D culture of primary human tumors and biomarker-driven Phase 0 trials to predict pharmacokinetics and activity	[47,112]
Focus on oncogenic molecular pathways as biomarkers for drug response	Incorporate tumor drug uptake and retention as a primary marker for drug response	[56]
Design drug carriers to improve formulation	Design drug carriers for tumor-specific delivery and for imaging drug delivery: theranostics	[58,113,114]
Use plasma pharmacokinetics to model drug delivery to solid tumors	Measure tumor pharmacokinetics directly for drug or drug carrier; apply miniaturized, implantable imaging technology	[56]
Dose to normal tissue toxicity	Dose to tumor drug saturation; minimize toxicity	
Treat all patients with the same dose and schedule	Treat patients based on their unique tumor physiology (e.g. level of hypoxia) and pharmacogenomics; utilize therapeutic drug monitoring for pharmacokinetically-guided dose adjustment	[115]
Create drug combinations based on discreet MOA and non-overlapping toxicities; use standard MTD dosing	Create drug combinations based on synergy in preclinical models; use ratiometric dosing with optimal sequence of administration	[116,117]
Traditional Phase I/II trial designs to establish the MTD according to Response Evaluation Criteria in Solid Tumors (RECIST)	Biomarker-driven adaptive or continual reassessment designs; randomized Phase II/III trials	[76]

tumor physiology, (ii) drug pharmacokinetics, (iii) preclinical models, (iv) drug delivery, and (iv) clinical translation. This perspective highlights underappreciated aspects of anticancer drug development that could help to reduce attrition rates and improve R&D productivity.

THE TUMOR MICROENVIRONMENT

The rise of molecular biology and the emergence of the genomics era have led to major progress in our understanding of cancer cell biology. However, this focus has also overshadowed knowledge of cancer at the tissue level. A prime example is the prevalence of hypoxic and acidic microenvironments in human solid tumors. Although hypoxia is recognized as an important target for cancer therapy [15], the concept is not routinely incorporated into preclinical models. A central issue is the definition of 'normoxia'. Normoxia is often equated with the ambient air found in tissue culture incubators. However, no tissue in the body is exposed to 20–21% oxygen, and instead tissue levels can range from zero in bone marrow to 14% in well-perfused organs (lung, liver, kidney, heart) with circulating levels of 10–12% [16,17]. Human solid tumors are typically hypoxic at 0–5% O_2 [18]. Low levels of oxygen can induce stabilization of the transcription factor hypoxia-inducible factor 1α (HIF-1α), which upregulates over sixty genes, including those controlling the glycolytic phenotype that produce lactate-mediated extracellular acidification [19,20]. Thus, gene-expression profiles are very different under hypoxia versus the standard hyperoxic conditions of traditional *in vitro* models. The impact upon drug development is illustrated by the camptothecins. The canonical mechanism of action for camptothecins is DNA damage-induced apoptosis associated with inhibition of nuclear topoisomerase I during replication, a result that was obtained under standard hyperoxic culture conditions. Moreover, topotecan and the active metabolites of irinotecan and rubitecan are also known to act via topoisomerase I-mediated inhibition of HIF-1α transcription [21–23]. In addition, the marked interaction of irinotecan and rapamycin is only observed under hypoxic conditions that induce HIF-1α [24]. Conversely, reports that camptothecins have a mechanism employing reactive oxygen species [25–28] are probably not clinically relevant in hypoxic solid tumors.

Hypoxia is also important for hematologic malignancies, given increasing evidence that leukemic stem cells are the root cause of the disease. Hematopoietic stem cells (HSCs) and leukemic stem cells both prefer a hypoxic environment. This led Eliasson and Jonsson to conclude that 'if important steps in early hematopoietic differentiation, including the regulation of primitive HSCs and progenitors, take place at hypoxia, then many of these experiments have to be redone at appropriate O_2 concentrations (1–3%)' [29].

Clearly, this conclusion could apply to much of what we know about cancer biology from traditional cell-culture models. Moreover, much of what we do know about tumor hypoxia derives from the field of radiation oncology that lacks a strong drug-development culture. The approach that currently characterizes this field – repurposing of drugs developed for other applications (e.g. bioreductive prodrugs of DNA-reactive cytotoxins) – suffers from lack of selectivity. Moreover, the identification of truly selective molecular targets that are needed to develop small-molecule inhibitors in hypoxic cells is in its infancy. Both approaches must overcome the challenge of drug penetration through poorly perfused tissue. Accordingly, experts in the field have recently referred to tumor hypoxia as 'the best validated target that has yet to be exploited in oncology' [15].

The acidic extracellular environment associated with tumor hypoxia also deserves consideration. Acidic extracellular pHe coupled with maintenance of a physiological intracellular pHi creates a pH gradient unique to tumors that can have a profound impact upon the uptake, retention and activity of anticancer drugs [30–37]. Doxorubicin, one of the most important and most utilized chemotherapy drugs in the armamentarium, is a weak base and therefore will be excluded from acidic tumors [32,34]. The pKa and associated charge dynamics of small-molecule drugs at acidic versus physiological pH should therefore be a crucial parameter in anticancer drug design [38], but this concept is not generally appreciated [39,40]. Simply screening compounds that possess ionizable groups in tumor cells adapted to growth at pH 6.8 versus standard pH 7.4 conditions can yield a completely different order of activity and hence selection of drug candidates [30].

THE IMPACT OF EXPOSURE TIME

Typically, the initial stage in drug discovery and development is target-to-hit in which a large number of compounds from combinatorial/parallel chemistry are subjected to high-throughput screening against an isolated molecular target in a cell-free assay. The endpoint is potency measured as the concentration that produces half-maximal response (IC50, ED50). Potency remains the primary endpoint when screening progresses to cell-based assays, despite the various biological barriers between drug and molecular target that could affect the interaction kinetics. Compounds are ranked by effective concentration without regard to exposure time. For example, the NCI-60 screen exposes tumor cell lines to compounds for a fixed 48h exposure time, regardless of individual tumor growth or drug action kinetics. Endpoints include concentrations that produce increasing levels of effect (growth inhibition: GI50; cytostasis: TGI; cytotoxicity: LC50). However, exposure time is crucial. For example, in the TF-1a erthyroleukemia model the

IC50 concentration of cytarabine is 373-fold lower at 72 versus 48h exposure. By contrast, the IC50 for daunorubicin is only five-fold lower. The drugs exhibit similar potency at 72h, but daunorubicin is over 80-fold more active at 48h (D.J.A., unpublished observation). Drug-response heat maps comparing such agents with others against multiple cell lines will clearly look very different at different exposure times.

As early as 1908, investigators studying antimicrobials determined that drug response for a specified effect was not simply concentration-dependent, but a complex function of both concentration and exposure time, modeled as $C^n \times T$, where n is the concentration coefficient [41]. Thus, drugs can be compared by equal effect level and exposure time (i.e. IC50 at 48h) only if their concentration coefficients are similar. A surprising number of diverse anticancer agents follow this simple, hyperbolic pharmacodynamic model and exhibit a range of n values, for example from <1 (5-FU) to >4 (etoposide) in breast [42] and bladder [43] cancer cell lines. Creation of a pharmacodynamic endpoint that accounts for variation in the concentration coefficient improves the correlation with animal [44] and clinical [45] activities. In addition, recognition of the impact of exposure time in preclinical models can alter subsequent clinical trial design. After the experimental drug crisnatol was administered by infusion instead of short-term bolus dosing, long-term responses were observed in a Phase I glioma trial [46]. Therefore, drug-development algorithms would benefit from early evaluation of the impact of exposure time on response. Although kinetic experiments are not easily amenable to high-throughput processes, the increase in data quality versus quantity is a trade-off worth considering.

CHOICE OF PRECLINICAL MODEL

3D primary tumor models have been demonstrated to provide more clinically relevant results than the typical monolayer cell-line models [47]. Because these models are necessarily low-throughput, they have been overshadowed by high-throughput technologies that can rapidly generate large databases of information for biostatistical analyses that yield the now familiar pathway or heat maps of drug response. Whether this will be a superior approach given its inherent limitations remains to be seen. In a similar vein, the use of cancer stem cell versus traditional bulk tumor models in drug development might improve our ability to hit the root cause of disease. For example, several agents that target the hedgehog pathway, which regulates cancer stem-cell survival and the tumor microenvironment, are now in clinical trials with promising early results [48]. Likewise, a novel monoclonal antibody that targets the Wnt signaling pathway has advanced to Phase I testing [49]. Models or screens based on cancer stem cells are not facile due to their small population and the difficulty of maintaining and expanding these cells in culture without losing their pluripotent nature. However, the quality of the information they yield may be more valuable than the quantity. The same may be said for *in vivo* models, in which the ability of murine xenografts to predict clinical toxicity or activity remains controversial even with the advent of genetic engineering [50–53]. A new approach is the 'co-clinical trial' concept of Pandolfi and coworkers in which preclinical trials in genetically engineered mice are conducted in parallel with human Phase I/II clinical trials [54]. This approach has proved successful in acute promyelocytic leukemia (APL), where APL mouse models recapitulated not only the biological and pathological features of human disease but also the drug-response profile.

DRUG DELIVERY TO TUMORS

The emergence of biomarker-driven drug development is an important step toward the goal of personalized medicine. However, correlation of molecular biomarker data to tumor response with parallel measurement of active drug in tumor target tissue is rare and is required for accurate interpretation. Comparatively little effort has been devoted to tumor pharmacokinetics because we continue to use plasma pharmacokinetics as a surrogate (the property of kinetic homogeneity), despite considerable evidence from the field of tumor angiogenesis that tumors develop their own unique, often compromised vasculature [55]. The few studies that have been done are instructive. For example, Wolf and colleagues assessed tumor uptake and retention of 5-fluorouracil by ^{19}F magnetic resonance spectroscopic (MRS) imaging in 60 patients with a spectrum of tumor types. Responses were observed in 60% of patients who retained the drug, but in none of the patients who did not [56].

The field of molecular imaging has expanded in response to the rapidly increasing number of molecular biomarkers. To date, imaging drug delivery has been largely restricted to radioisotope labeling of the agent itself, which is costly, can require complex radiochemistry and can alter pharmacokinetics [57]. However, emergence of the new field of theranostics, a term first coined in 2002, presents some exciting possibilities. A theranostic combines a signal emitter detectable by magnetic resonance imaging (MRI), computerized tomography (CT), positron emission tomography/single-photon emission computed tomography (PET/SPECT), ultrasound (US), fluorescence molecular tomography (FMT), photoacoustic tomography (PAT), or optical spectroscopy and a carrier (e.g. liposomes, micelles, viruses, antibodies, dendrimers, polymers, nanocomposites with magnetic iron and silica cores) with a therapeutic payload coupled to a targeting ligand to treat and image a tumor simultaneously (reviewed in [58]). Theranostics have not yet reached clinical evaluation. Barriers include difficulty in matching the dose required for therapy with that required for imaging, limited nanoparticle tissue penetration, complex manufacturing processes, storage and shelf-life issues, and biocompatibility and toxicity concerns [59]. In addition, most theranostic uptake studies *in vitro* are performed under traditional culture conditions; some studies fail to report this experimental detail altogether. Nevertheless, rapid advances are being made to overcome limitations. For example, optical imaging with near-infrared labels applied to drug carriers is a promising alternative, but suffers from poor deep-tissue penetration of external light sources. One approach to enhancing optical imaging is the development of miniature optical biosensors implanted at the tumor site. The technology for such implantable biosensors is becoming available, driven by its application to glucose monitoring in diabetes [60,61]. Such sensors can be multichannel, permitting real-time quantitation of tumor pharmacokinetics/pharmacodynamics (PK/PD). The impact could be substantial as illustrated by recent market entry of an implantable radiation dosimeter (DVS®, Sicel Technologies) that has revealed systematic variance in radiation dose administered to patients with breast and prostate cancers [62].

CHANGING CLINICAL PRACTICE

The previous categories all reflect missed connections on the 'bench' side of the bridge across the Valley of Death. There are similar issues on the 'bedside' of the bridge. For example, consider the concept of pharmacokinetically-guided dosing. Although there is consensus that such individualized dosing of cancer chemotherapeutics is of value, particularly for reducing toxicity, methotrexate is the only oncology drug in which therapeutic drug monitoring (TDM) for dose adjustment is standard practice [63]. TDM is normally associated with older cytotoxic drugs, but several newer targeted agents (e.g. imatinib) meet the traditional criteria for TDM, spurring calls for its application [64]. Slow acceptance of TDM into clinical practice has prompted some authors to question its future [65] even though it clearly addresses the current goal of personalized medicine.

Just as the effective concentration is stressed over exposure time in preclinical development, dose, and particularly the maximum tolerated dose (MTD), is the focus of clinical development. Exposure time is secondary. This practice derives from the concept of dose-intensity formulated in 1984, which is based on the assumption that dose scheduling or infusion time, does not directly determine effectiveness in killing tumor cells [66]. However, in 1986, Collins argued that drugs should be developed based on exposure –

as the AUC (area under the plasma concentration time curve) – instead of on the maximum tolerated dose, because there was a better correlation between mouse and human data. The idea was modified and supported by the European Organization for Research and Treatment of Cancer (EORTC) [67]. Carboplatin is one drug where AUC is much better than mg/m^2 dosing as a predictor of toxicity. There is also evidence that AUC is prognostic for toxicity and/or tumor response for busulfan, methotrexate, 5-fluorouracil, etoposide, irinotecan and docetaxel [68]. Although pharmacokinetic guidance has been used successfully, it is not easy to implement and not standard practice in clinical trial design.

Another challenge in translational research is the design of combination chemotherapy regimens. This area is of increased importance in the era of targeted agents, recently prompting the USA Food and Drug Administration (FDA) to issue guidance on the evaluation of drug combinations up-front, instead of requiring prior approval of one or both drugs as single agents. Preclinical evaluation of drug combinations has a long history. One of the most-utilized methods, the combination index of Chou and Talalay, dates to 1984 [69]. An important endpoint for clinical translation of this method is the dose-reduction index – the amounts that each agent can be reduced in a synergistic combination while maintaining the efficacy of the drugs when used alone. The sequence of administration and the drug ratio are additional crucial determinants. Despite extensive preclinical studies, combination clinical trials continue to be based primarily on MTD dosing with near-simultaneous administration. Drug ratio is not considered because until recently [70] it could not be controlled *in vivo*. The potential of ratiometric rather than MTD dosing is illustrated by CPX-351, a liposomal drug formulation that delivers cytarabine and daunorubicin in a synergistic 5:1 molar ratio with demonstrated activity in relapsed or refractory AML [71]. Failure to appreciate the determinants of drug interaction presents the very real possibility that patients could be exposed to antagonistic drug combinations that produce toxicity without efficacy. Moreover, once this result is observed clinically, the combination will probably never be tried again, even though it could be active under the proper exposure conditions.

Determination of patient sensitivity and resistance to chemotherapy to guide treatment is a hot topic in clinical drug development, and is driven by the advent of genomic signatures of drug response. Before the genomics era this important aspect of personalized medicine was the domain of *in vitro* chemosensitivity testing, which included the human tumor cloning (HTCA), histoculture drug response (HDRA), extreme drug resistance (EDRA), and differential staining cytotoxicity (DiSC) assays. Again, a significant amount of preclinical work was devoted to this task and the results were impressive. The HTCA, for example, was 69% accurate in predicting drug sensitivity and a remarkable 91% accurate in predicting drug resistance in the clinic. HTCA outperformed clinician choice in head-to-head trials, but was never incorporated into clinical practice [72]. Likewise, the EDRA assay (Oncotech) was not widely used even though it was FDA-approved and reimbursed by Medicare. In operation since the 1980s, Oncotech went out of business in June of last year. Perhaps companies that utilize genomic signatures will replace this service to cancer patients. Certainly, such signatures can help to determine whether a patient tumor expresses the oncogenic pathway targeted by the respective therapeutic(s). To date, two genomic signatures for cancer have been validated: the 21 gene Oncotype DX and 70 gene MammaPrint tests to predict prognosis and benefit from chemotherapy for women with breast cancer. In addition, recent work indicates that much smaller signatures may be equally effective [73]. These results are promising, but challenges remain. It remains an open question whether gene signatures can predict drug pharmacokinetics or improve upon direct exposure of primary specimens to drugs that have multiple and often unanticipated mechanisms of action.

SOME PATHS FORWARD

The Valley of Death in anticancer drug development is a highly complex problem with numerous driving forces. The need for solutions is urgent as illustrated by recent financial reports. Of the twelve pharmaceutical companies that spent the most on R&D, return on investment fell 3.4 percentage points in 2010, a 29% decline, while the cost of bringing a new molecular entity (NME) to market rose from $830 million to $1.05 billion [74].

The pharmacological audit trail (PhAT) of Workman and colleagues, first proposed in 2003 [75] and updated in 2010 [76], represents a current path forward based on the molecular profiles of patient tumors and their associated PK and PD endpoints. This approach emphasizes the application of molecular biomarkers of clinical response early in the development process. The risk of failure can be assessed at each of nine sequential stages in a hierarchy as development proceeds. First, a patient population is identified whose tumors express a cancer-selective mutation or pathway. Next, a targeted agent that exploits this biology is identified. Genetic tests for the specific mutations are validated and drug pharmacokinetics defined. Pharmacodynamic assays are then validated and the induction of tumor apoptosis is confirmed. Biomarkers of clinical response are established. Upon disease progression, molecular mechanisms of drug resistance are identified such that appropriate alternative targeted therapies can be initiated. Successful applications of this approach include the BRAF inhibitor vemurafenib, the ALK inhibitor crizotinib, and the poly(ADP) ribose polymerase (PARP) inhibitor olaparib. For the latter agent, clinical responses were observed to monotherapy in proof-of-concept Phase II trials in advanced breast and recurrent ovarian cancers that expressed BRCA1 and 2 mutations [77,78]. However, a subsequent Phase II trial in women with ovarian and triple negative breast cancer indicated response to oliparib was not entirely BRCA1/2-dependent [79]. Such results suggest that reliance on molecular biomarkers alone may not be sufficient to predict the population of patients who would benefit from therapy [80].

As summarized in Table 2, there are other paths that should be considered. A key principle is that tumor physiology is as important as molecular biology. Principles of tumor physiology should be applied to preclinical development in which drug response as a function of both concentration and exposure time is measured under clinically relevant conditions: pH 7.4 and 10% oxygen for blood-borne tumors and acidic pH and 1–5% oxygen for solid tumors. Studies performed under non-physiological conditions should be reported as such. Preclinical models should focus on primary or low-passage human tumors preferably maintained in 3D culture. Alternatively, studies should be conducted in the appropriate cancer stem-cell model if one is available. Gupta et al. recently reported a high-throughput screen that revealed selective inhibitors of breast cancer stem cells [81]. A notable hit was salinomycin, an agricultural antibiotic in use for over thirty years. Salinomycin has now been shown to inhibit Wnt signaling and induce apoptosis in stem cells from leukemia, osteosarcoma and colon, lung, gastric, prostate and pancreatic carcinomas [82–88].

In addition to assessing molecular targets and associated pathways, a clinical workup should include imaging of tumor pH (e.g. by dynamic nuclear polarization enhanced magnetic resonance [89]), hypoxia (by PET and MRI [90]), and tumor perfusion (by contrast-enhanced, microbubble ultrasound [91]). This information can help guide treatment selection including use of hypoxia-activated prodrugs or agents with acidic pKas that can exploit the tumor pH gradient. For tumors such as pancreatic cancer in which tissue perfusion is impaired by desmoplastic tumor stroma, combination of chemotherapy with hedgehog-pathway inhibitors could be considered to improve drug delivery [92]. Ideally, imaging of drug delivery to tumor tissue will be possible and permit real-time therapeutic drug monitoring. This would allow dosing to tumor-tissue saturation rather than toxicity. As stated previously, rapid advances in the field of nanotechnology-driven theranostics coupled with external fiber-optic or implantable biosensors could make this a reality and should be a priority. Progress is being made as exemplified by the work of Kim et al. with deformable, tumor-homing chitosan-based nanoparticles loaded with paclitaxel and labeled with the near-infrared dye Cy5.5 [93]. Moreover, targeting tumor physiology (specifically pH regulation) is emerging as a therapeutic strategy [94]. Phase I/II trials of a monoclonal antibody that targets carbonic anhydrase IX (Girentuximab; Wilex AG) combined with interferon-α in metastatic clear-cell renal carcinoma indicated that the therapy was well tolerated and led to both disease stabilization and a significant increase in 2-year survival [95]. A Phase III trial (ARISER) will report results later this year. Of note, carbonic anhydrase IX has both diagnostic and prognostic value in this disease [96].

Changes in clinical practice are also needed. Clinical evaluation should move from a focus on MTD dosing to exposure (AUC) that produces changes in response biomarkers. In addition to traditional outcome biomarkers (e.g. myelosuppression), these predictive biomarkers can now be assessed in circulating tumor cells rather than traditional biopsies. In the case of drug combinations, the current MTD model should be replaced with a synergy model applying optimized drug concentrations, ratio and sequence of exposure.

Rethinking clinical trial design should be another focus. Incorporation of Phase 0 clinical trials into clinical development is an important first step and one that is increasingly embraced; there are currently more than 4500 registered Phase 0 clinical cancer trials listed on clincialtrials.gov. The primary goal of Phase 0 trials is to obtain preliminary PK/PD and proof-of-mechanism data at non-therapeutic drug exposures with minimal toxicity. Such trials can compress drug development timelines as illustrated by the PARP inhibitor ABT888 [97,98]. Follow-on Phase I trials of ABT888 have demonstrated drug-response biomarker utility in both peripheral blood mononuclear and circulating tumor cells [99]. The value of genetically-informed Phase II/III trials is illustrated by crizotinib (Xalkori; Pfizer), which was approved just four years after discovery of the *EML4–ALK* gene fusion in lung cancer patients [100]. Adaptive trial designs that allow rapid dose escalation, expansion cohorts in Phase I for hypothesis testing, and introduction of randomized Phase II/III trials with early-stopping rules, represent new approaches to increase efficiency and reduce cost [76]. Lessons learned from our experience in pediatric cancer should also be considered. A key point is that most of the progress – a 30% improvement in overall cure rate since 1971 – has come from clinical research that was made possible because over 90% of pediatric cancer patients are enrolled on treatment protocols versus only 3% of adults [101]. Barriers to participation in clinical research, particularly the insurance healthcare payment system, must be reduced and better regulatory guidance provided if particular study designs are required for specific agent approvals.

CONCLUDING REMARKS

In summary, given the continuing and probably increasing restrictions on resources to support bringing new therapies to cancer patients, it is worth taking stock of our current processes and re-thinking our assumptions. The current emphasis on genetic molecular approaches has certainly paid dividends. However, advances in understanding and modeling tumor physiology and in drug delivery and monitoring also have important roles to play. Multidisciplinary approaches to both translational and clinical research must evolve along with the respective reward systems. Both the public and private sectors that support the enterprise and the cancer patients who bear the enormous burden of this disease would benefit.

ACKNOWLEDGMENTS

This work was supported in part by Small Business Innovation Research Grant CA125871 from the National Cancer Institute. The author acknowledges the invaluable scientific contributions and mentorship in anticancer drug development of Drs O. Michael Colvin, David Rizzieri, and Mark Dewhirst from Duke University, Durham, NC; Drs Mansukh Wani and Govindarajan Manikumar from Research Triangle Institute International, Research Triangle Park, NC and Dr Lee Roy Morgan from DEKK-TEC, Inc., New Orleans, LA. This paper is dedicated to the memory of Drs Robert Silber and Monroe Wall whose commitment to bringing new therapies to cancer patients inspired this author and many others.

REFERENCES

1 Pammolli, F. *et al.* (2011) The productivity crisis in pharmaceutical R&D. *Nat. Rev. Drug Discov.* 10, 428–438

2 Hait, W.N. (2010) Anticancer drug development: the grand challenges. *Nat. Rev. Drug Discov.* 9, 253–254

3 Paul, S.M. *et al.* (2010) How to improve R&D productivity: the pharmaceutical industry's grand challenge. *Nat. Rev. Drug Discov.* 9, 203–214

4 Wehling, M. (2009) Opinion. Assessing the translatability of drug projects: what needs to be scored to predict success? *Nat. Rev. Drug Discov.* 8, 541–546

5 FasterCures (2011) *Crossing Over the Valley of Death*, FasterCures/The Center for Accelerating Medical Solutions. http://www.fastercures.org/documents/file/Valley%20of%20Death%20-%20Translational%20Research(1).pdf

6 Butler, D. (2008) Translational research: crossing the valley of death. *Nature* 453, 840–842

7 Coller, B.S. and Califf, R.M. (2009) Traversing the Valley of Death: a guide to assessing prospects for translational success. *Sci. Transl. Med.* 1, 1–5

8 Friedl, K.E. (2006) Overcoming the 'valley of death': Mouse models to accelerate translational research. *Diabetes Technol. Ther.* 8, 413–414

9 Reed, J.C. (2011) NCATS could mitigate pharma Valley of Death. *Genet. Eng. Biotechnol. News* 31, 6–8

10 Leaf, C. (2004) Why we're losing the war on cancer and how to win it. *Fortune* 149, 76

11 Kolata, G. (2009) Advances elusive in the drive to cure cancer. *New York Times* 24 April

12 Begley, S. and Carmichael, M. (2010) Desperately seeking cures. *Newsweek* 155, 38–42

13 Begley, S. *et al.* (2008) We fought cancer... and cancer won. *Newsweek* 152, 42–66

14 Kola, I. and Landis, J. (2004) Can the pharmaceutical industry reduce attrition rates? *Nat. Rev. Drug Discov.* 3, 711–715

15 Wilson, W.R. and Hay, M.P. (2011) Targeting hypoxia in cancer therapy. *Nat. Rev. Cancer* 11, 393–410

16 Atkuri, K.R. and Herzenberg, L.A. (2005) Culturing at atmospheric oxygen levels impacts lymphocyte function. *Proc. Natl. Acad. Sci. U.S.A.* 102, 3756–3759

17 Ivanovic, Z. (2009) Hypoxia or *in situ* normoxia: the stem cell paradigm. *J. Cell. Physiol.* 219, 271–275

18 Vaupel, P. *et al.* (2007) Detection and characterization of tumor hypoxia using pO_2 histography. *Antioxid. Redox Signal.* 9, 1221–1235

19 Stubbs, M. *et al.* (2003) Understanding the tumor metabolic phenotype in the genomic era. *Curr. Mol. Med.* 3, 49–59

20 Stubbs, M. and Griffiths, J.R. (2010) The altered metabolism of tumors: HIF-1 and its role in the Warburg effect. *Advances in Enzyme Regulation* 50, 44–55

21 Puppo, M. *et al.* (2008) Topotecan inhibits vascular endothelial growth factor production and angiogenic activity induced by hypoxia in human neuroblastoma by targeting hypoxia-inducible factor-1 alpha and -2 alpha. *Mol. Cancer Ther.* 7, 1974–1984

22 O'Leary, J.J. *et al.* (1999) Antiangiogenic effects of camptothecin analogues 9-amino-20(S) camptothecin, topotecan, and CPT-11 studied in the mouse cornea model. *Clin. Cancer Res.* 5, 181–187

23 Kamiyama, H. *et al.* (2005) Anti-angiogenic effects of SN38 (active metabolite of irinotecan): inhibition of hypoxia-inducible factor 1 alpha (HIF-1 alpha)/vascular endothelial growth factor (VEGF) expression of glioma and growth of endothelial cells. *J. Cancer Res. Clin. Oncol.* 131, 205–213

24 Pencreach, E. *et al.* (2009) Marked activity of irinotecan and rapamycin combination toward colon cancer cells *in vivo* and *in vitro* is mediated through cooperative modulation of the mammalian target of rapamycin/hypoxia-inducible factor-1 alpha axis. *Clin. Cancer Res.* 15, 1297–1307

25 Cordero, M.D. *et al.* (2010) Acute oxidant damage promoted on cancer cells by amitriptyline in comparison with some common chemotherapeutic drugs. *Anti-Cancer Drugs* 21, 932–944

26 Li, Y. *et al.* (2009) Camptothecin and Fas receptor agonists synergistically induce medulloblastoma cell death: ROS-dependent mechanisms. *Anti-Cancer Drugs* 20, 770–778

27 Timur, M. *et al.* (2005) The effect of Topotecan on oxidative stress in MCF-7 human breast cancer cell line. *Acta Biochim. Pol.* 52, 897–902

28 Simizu, S. *et al.* (1998) Requirement of caspase-3(-like) protease-mediated hydrogen peroxide production for apoptosis induced by various anticancer drugs. *J. Biol. Chem.* 273, 26900–26907

29 Eliasson, P. and Jonsson, J.I. (2010) The hematopoietic stem cell niche: low in oxygen but a nice place to be. *J. Cell. Physiol.* 222, 17–22

30 Adams, D.J. *et al.* (2006) Camptothecin analogs with enhanced activity against human breast cancer cells. II. Impact of the tumor pH gradient. *Cancer Chemother. Pharmacol.* 57, 145–154

31 Gerweck, L.E. and Seetharaman, K. (1996) Cellular pH gradient in tumor versus normal tissue: Potential exploitation for the treatment of cancer. *Cancer Res.* 56, 1194–1198

32 Gerweck, L.E. *et al.* (2006) Tumor pH controls the *in vivo* efficacy of weak acid and base chemotherapeutics. *Mol. Cancer Ther.* 5, 1275–1279

33 McSheehy, P.M.J. *et al.* (2000) Role of pH in tumor-trapping of the anticancer drug 5-fluorouracil. *Adv. Enzyme Regul.* 40, 63–80

34 Raghunand, N. *et al.* (1999) Enhancement of chemotherapy by manipulation of tumour pH. *Br. J. Cancer* 80, 1005–1011

35 Raghunand, N. *et al.* (2003) Tumor acidity, ion trapping and chemotherapeutics I. pH-dependent partition coefficients predict importance of ion trapping on pharmacokinetics of weakly basic chemotherapeutic agents. *Biochem. Pharmacol.* 66, 1219–1229

36 Gabr, A. *et al.* (1997) Cellular pharmacokinetics and cytotoxicity of camptothecin and topotecan at normal and acidic pH. *Cancer Res.* 57, 4811–4816

37 Stubbs, M. *et al.* (1999) Causes and consequences of acidic pH in tumors: a magnetic resonance study. *Adv. Enzyme Regul.* 39, 13–30

38 Adams, D.J. and Morgan, L.R. (2011) Tumor physiology and charge dynamics of anticancer drugs: implications for camptothecin-based drug development. *Curr. Med. Chem.* 18, 1367–1372

39 Gleeson, M.P. *et al.* (2011) Probing the links between *in vitro* potency, ADMET and physicochemical parameters. *Nat. Rev. Drug Discov.* 10, 197–208

40 van de Waterbeemd, H. and Gifford, E. (2003) ADMET *in silico* modelling: towards prediction paradise? *Nat. Rev. Drug Discov.* 2, 192–204

41 Chick, H. (1908) An investigation of the laws of disinfection. *J. Hyg. (Lond.)* 8, 92–158

42 Adams, D.J. (1989) *In vitro* pharmacodynamic assay for cancer drug development – application to crisnatol, a new DNA intercalator. *Cancer Res.* 49, 6615–6620

43 Millenbaugh, N.J. *et al.* (2000) A pharmacodynamic analysis method to determine the relative importance of drug concentration and treatment time on effect. *Cancer Chemother. Pharmacol.* 45, 265–272

44 Adams, D.J. *et al.* (1990) Evaluation of arylmethylaminopropanediols by a novel *in vitro* pharmacodynamic assay – correlation with antitumor activity *in vivo*. *Cancer Res.* 50, 3663–3669

45 Adams, D.J. (1992) Antineoplastic drug screening. *J. Natl. Cancer Inst.* 84, 1288–1289

46 New, P. *et al.* (1997) Long-term response to crisnatol mesylate in patients with glioma. *Invest. New Drugs* 15, 343–352

47 Furukawa, T. *et al.* (1995) Clinical applications of the histoculture drug response assay. *Clin. Cancer Res.* 1, 305–311

48 Harris, L.G. *et al.* (2011) Hedgehog signaling: networking to nurture a promalignant tumor microenvironment. *Molecular Cancer Research* 9, 1165–1174

49 Oncomed Pharmaceuticals (2011) OncoMed pharmaceuticals announces FDA clearance to commence Phase 1 testing of anti-cancer stem cell therapeutic OMP-18R5. OncoMed Pharmaceuticals, Inc. http://www.oncomed.com/news/pr/press_release_2011_04_28.pdf

50 Bracken, M.B. (2009) Why animal studies are often poor predictors of human reactions to exposure. *J.R. Soc. Med.* 102, 120–122

51 Cheon, D.J. and Orsulic, S. (2011) Mouse models of cancer. *Annu. Rev. Pathol.: Mech. Disease* 6, 95–119

52 Olson, H. *et al.* (2000) Concordance of the toxicity of pharmaceuticals in humans and in animals. *Regul. Toxicol. Pharmacol.* 32, 56–67

53 Rangarajan, A. and Weinberg, R.A. (2003) Opinion – comparative biology of mouse versus human cells: modelling human cancer in mice. *Nat. Rev. Cancer* 3, 952–959

54 Nardella, C. *et al.* (2011) The APL paradigm and the 'Co-Clinical Trial' project. *Cancer Discov.* 1, 108–116

55 Siemann, D.W. (2011) The unique characteristics of tumor vasculature and preclinical evidence for its selective disruption by tumor-vascular disrupting agents. *Cancer Treat. Rev.* 37, 63–74

56 Wolf, W. *et al*. (2000) F-19-MRS studies of fluorinated drugs in humans. *Adv. Drug Del. Rev.* 41, 55–74

57 Ahn, B.C. (2011) Applications of molecular imaging in drug discovery and development process. *Curr. Pharm. Biotechnol.* 12, 459–468

58 Kelkar, S.S. and Reineke, T.M. (2011) Theranostics: combining imaging and therapy. *Bioconjug. Chem.* 22, 1879–1903

59 Pan, D.P.J. *et al*. (2010) Nanomedicine strategies for molecular targets with MRI and optical imaging. *Future Med. Chem.* 2, 471–490

60 O'Sullivan, T. *et al*. (2010) Implantable semiconductor biosensor for continuous *in vivo* sensing of far-red fluorescent molecules. *Optics Express* 18, 12513–12525

61 Valdastri, P. *et al*. (2011) Wireless implantable electronic platform for chronic fluorescent-based biosensors. *IEEE Trans. Biomed. Eng.* 58, 1846–1854

62 Scarantino, C.W. *et al*. (2008) The observed variance between predicted and measured radiation dose in breast and prostate patients utilizing an *in vivo* dosimeter. *Int. J. Radiat. Oncol. Biol. Phys.* 72, 597–604

63 Alnaim, L. (2007) Therapeutic drug monitoring of cancer chemotherapy. *J. Oncol. Pharm. Pract.* 13, 207–221

64 Buclin, T. *et al*. (2011) Who is in charge of assessing therapeutic drug monitoring? The case of imatinib. *Lancet Oncol.* 12, 9–11

65 Bach, D.M. *et al*. (2010) Therapeutic drug monitoring in cancer chemotherapy. *Bioanalysis* 2, 863–879

66 Hryniuk, W. and Bush, H. (1984) The importance of dose intensity in chemotherapy of metastatic breast cancer. *J. Clin. Oncol.* 2, 1281–1288

67 EORTC Pharmacokinetics, Metabolism Group. (1987) Pharmacokinetically guided dose escalation in phase I clinical trials. Commentary and proposed guidelines. *Eur. J. Cancer Clin. Oncol.* 23, 1083–1087

68 Meza-Junco, J. and Sawyer, M.B. (2009) Drug exposure: still an excellent biomarker. *Biomarkers Med.* 3, 723–731

69 Chou, T. and Talalay, P. (1984) Quantitative analysis of dose-effect relationships: the combined effects of multiple drugs or enzyme inhibitors. *Adv. Enzyme Regul.* 22, 27–55

70 Dicko, A. *et al*. (2010) Use of nanoscale delivery systems to maintain synergistic drug ratios *in vivo*. *Expert Opin. Drug Deliv.* 7, 1329–1341

71 Feldman, E.J. *et al*. (2011) First-in-man study of CPX-351: a liposomal carrier containing cytarabine and daunorubicin in a fixed 5:1 molar ratio for the treatment of relapsed and refractory acute myeloid leukemia. *J. Clin. Oncol.* 29, 979–985

72 Vonhoff, D.D. (1990) He's not going to talk about *in vitro* predictive assays again, is he? *J. Natl. Cancer Inst.* 82, 96–101

73 Cuzick, J. *et al*. (2011) Prognostic value of a combined estrogen receptor, progesterone receptor, ki-67, and human epidermal growth factor receptor 2 immunohistochemical score and comparison with the genomic health recurrence score in early breast cancer. *J. Clin. Oncol.* 29, 4273–4278

74 Armstrong, D. (2011) Drugmakers' returns on research fall as pipeline projects fail. Bloomberg. http://www.bloomberg.com/news/2011-11-21/drugmakers-returns-on-research-fall-as-pipeline-projects-fail.html

75 Workman, P. (2003) How much gets there and what does it do?: The need for better pharmacokinetic and pharmacodynamic endpoints in contemporary drug discovery and development. *Curr. Pharm. Des.* 9, 891–902

76 Yap, T.A. *et al.* (2010) Envisioning the future of early anticancer drug development. *Nat. Rev. Cancer* 10, 514–525

77 Audeh, M.W. *et al.* (2010) Oral poly(ADP-ribose) polymerase inhibitor olaparib in patients with BRCA1 or BRCA2 mutations and recurrent ovarian cancer: a proof-of-concept trial. *Lancet* 376, 245–251

78 Tutt, A. *et al.* (2010) Oral poly(ADP-ribose) polymerase inhibitor olaparib in patients with BRCA1 or BRCA2 mutations and advanced breast cancer: a proof-of-concept trial. *Lancet* 376, 235–244

79 Gelmon, K.A. *et al.* (2011) Olaparib in patients with recurrent high-grade serous or poorly differentiated ovarian carcinoma or triple-negative breast cancer: a phase 2, multicentre, open-label, non-randomised study. *Lancet Oncol.* 12, 852–861

80 Ratain, M.J. and Glassman, R.H. (2007) Biomarkers in phase I oncology trials: Signal, noise, or expensive distraction? *Clin. Cancer Res.* 13, 6545–6548

81 Gupta, P.B. *et al.* (2009) Identification of selective inhibitors of cancer stem cells by high-throughput screening. *Cell* 138, 645–659

82 Dong, T.T. *et al.* (2011) Salinomycin selectively targets 'CD133+' cell subpopulations and decreases malignant traits in colorectal cancer lines. *Ann. Surg. Oncol.* 18, 1797–1804

83 Kim, K.Y. *et al.* (2011) Salinomycin-induced apoptosis of human prostate cancer cells due to accumulated reactive oxygen species and mitochondrial membrane depolarization. *Biochem. Biophys. Res. Commun.* 413, 80–86

84 Lu, D.S. *et al.* (2011) Salinomycin inhibits Wnt signaling and selectively induces apoptosis in chronic lymphocytic leukemia cells. *Proc. Natl. Acad. Sci. U.S.A.* 108, 13253–13257

85 Tang, Q.L. *et al.* (2011) Salinomycin inhibits osteosarcoma by targeting its tumor stem cells. *Cancer Lett.* 311, 113–121

86 Wang, Y. (2011) Effects of salinomycin on cancer stem cell in human lung adenocarcinoma A549 cells. *Med. Chem.* 7, 106–111

87 Zhang, G.N. *et al.* (2011) Combination of salinomycin and gemcitabine eliminates pancreatic cancer cells. *Cancer Lett.* 313, 137–144

88 Zhi, Q.M. *et al.* (2011) Salinomycin can effectively kill ALDH(high) stem-like cells on gastric cancer. *Biomed. Pharmacother.* 65, 509–515

89 Gallagher, F.A. *et al.* (2011) Imaging pH with hyperpolarized ^{13}C. *NMR Biomed.* 24, 1006–1015

90 Chitneni, S.K. *et al.* (2011) Molecular imaging of hypoxia. *J. Nucl. Med.* 52, 165–168

91 Sboros, V. and Tang, M.X. (2010) The assessment of microvascular flow and tissue perfusion using ultrasound imaging. *Proc. Inst. Mech. Eng. H-J. Eng. Med.* 224, 273–290

92 Olive, K.P. *et al.* (2009) Inhibition of hedgehog signaling enhances delivery of chemotherapy in a mouse model of pancreatic cancer. *Science* 324, 1457–1461

93 Kim, K. *et al.* (2010) Tumor-homing multifunctional nanoparticles for cancer theragnosis: Simultaneous diagnosis, drug delivery, and therapeutic monitoring. *J. Control. Release* 146, 219–227

94 Neri, D. and Supuran, C.T. (2011) Interfering with pH regulation in tumours as a therapeutic strategy. *Nat. Rev. Drug Discov.* 10, 767–777

95 Siebels, M. *et al.* (2011) A clinical phase I/II trial with the monoclonal antibody cG250 (RENCAREX®) and interferon-alpha-2a in metastatic renal cell carcinoma patients. *World J. Urol.* 29, 121–126

96 Tostain, J. *et al.* (2010) Carbonic anhydrase 9 in clear cell renal cell carcinoma: A marker for diagnosis, prognosis and treatment. *Eur. J. Cancer* 46, 3141–3148

97 Kummar, S. *et al*. (2009) Phase 0 clinical trial of the poly (ADP-ribose) polymerase inhibitor ABT-888 in patients with advanced malignancies. *J. Clin. Oncol.* 27, 2705–2711

98 Kummar, S. *et al*. (2007) Compressing drug development timelines in oncology using phase '0' trials. *Nature Reviews Cancer* 7, 131–139

99 Kummar, S. *et al*. (2011) Phase I study of PARP inhibitor ABT-888 in combination with topotecan in adults with refractory solid tumors and lymphomas. *Cancer Res.* 71, 5626–5634

100 Editorial (2011) Upcoming battles in the war on cancer. *Cancer Discov.* 1, 544–546

101 Brennan, R. *et al*. (2010) The war on cancer: have we won the battle but lost the war? *Oncotarget* 1, 77–83

102 Williams, M. (2011) Productivity shortfalls in drug discovery: contributions from the pre-clinical sciences? *J. Pharmacol. Exp. Ther.* 336, 3–8

103 LaMattina, J.L. (2011) The impact of mergers on pharmaceutical R&D. *Nat. Rev. Drug Discov.* 10, 559–560

104 Kamb, A. *et al*. (2007) Opinion – why is cancer drug discovery so difficult? *Nat. Rev. Drug Discov.* 6, 115–120

105 Orloff, J. *et al*. (2009) A guide to drug discovery – opinion. The future of drug development: advancing clinical trial design. *Nat. Rev. Drug Discov.* 8, 949–957

106 Woolf, S.H. (2008) The meaning of translational research and why it matters. *JAMA* 299, 211–213

107 Cuatrecasas, P. (2006) Drug discovery in jeopardy. *J. Clin. Invest.* 116, 2837–2842

108 Kamb, A. (2005) What's wrong with our cancer models? *Nat. Rev. Drug Discov.* 4, 161–165

109 Swinney, D.C. and Anthony, J. (2011) How were new medicines discovered? *Nat. Rev. Drug Discov.* 10, 507–519

110 Kamb, A. (2010) At a crossroads in oncology. *Curr. Opin. Pharm.* 10, 356–361

111 Roin, B.N. (2009) Unpatentable drugs and the standards of patentability. *Texas Law Rev.* 87, 503–570

112 LoRusso, P.M. (2009) Phase 0 clinical trials: an answer to drug development stagnation? *J. Clin. Oncol.* 27, 2586–2588

113 Rosen, H. and Abribat, T. (2005) The rise and rise of drug delivery. *Nat. Rev. Drug Discov.* 4, 381–385

114 Yoo, J.W. *et al*. (2011) Bio-inspired, bioengineered and biomimetic drug delivery carriers. *Nat. Rev. Drug Discov.* 10, 521–535

115 van den Bongard, H. *et al*. (2000) Pharmacokinetically guided administration of chemotherapeutic agents. *Clin. Pharmacokinet.* 39, 345–367

116 Mayer, L.D. *et al*. (2006) Ratiometric dosing of anticancer drug combinations: controlling drug ratios after systemic administration regulates therapeutic activity in tumor-bearing mice. *Mol. Cancer Ther.* 5, 1854–1863

117 Mayer, L.D. and Janoff, A.S. (2007) Optimizing combination chemotherapy by controlling drug ratios. *Mol. Interventions* 7, 216–223

nds in Pharmacological Sciences

Metformin and Cancer: From the Old Medicine Cabinet to Pharmacological Pitfalls and Prospects

Arian Emami Riedmaier[1,*], Pascale Fisel[1,*], Anne T. Nies[1], Elke Schaeffeler[1], Matthias Schwab[1,2,†]

[1]Dr Margarete Fischer-Bosch Institute of Clinical Pharmacology, Stuttgart, Germany,
[2]Department of Clinical Pharmacology, Institute of Experimental and Clinical Pharmacology and Toxicology, University Hospital, Tübingen, Germany
[†]Correspondence: matthias.schwab@ikp-stuttgart.de

Trends in Pharmacological Sciences, Vol. 34, No. 2, February 2013 © 2013 Elsevier Inc.
http://dx.doi.org/10.1016/j.tips.2012.11.005

SUMMARY

Metformin is a biguanide derivative used in the treatment of type II diabetes (T2D) and one of the world's most widely prescribed drugs. Owing to its safety profile, it has been recently promoted for a range of other indications, particularly for its role in cancer prevention. There is evidence from studies in diabetic cohorts, as well as laboratory studies, that the action of metformin depends on a balance between the concentration and duration of exposure, which depends crucially on cell- and tissue-specific pharmacological factors. Mechanistic studies have revealed the involvement of increasingly complex pathways. Yet, there are several missing links regarding the role of drug transporters and drug–drug interactions, as well as the expression levels of transporters in normal versus tumor tissues, which may affect patient exposure and dosing when metformin is used in cancer prevention. This review highlights the current knowledge on metformin action and pharmacology, including novel insights into genomic factors, with a specific focus on cancer prevention. Furthermore, future challenges that may influence therapeutic outcome will be discussed.

*These authors contributed equally to the writing and preparation of this review.

245

THE PAST AND PRESENT OF METFORMIN

Metformin, the gold standard of antidiabetic treatment, was prescribed for a millennium by herbalists for the relief of polyuria and halitosis, in the form of herbal teas made from French lilac (*Galega officinalis*), although the active guanidine-derivative content was unknown at that time [1]. Not until the 1950s were the active ingredients extracted and developed for use in the clinic as metformin and phenformin. Although phenformin was soon withdrawn due to it causing lactic acidosis, metformin showed an exceptional therapeutic index for diabetes, making it the most widely prescribed drug in patients with T2D [2].

Chemically, metformin is a synthetic biguanide. Owing to their polar guanidine fraction, biguanides are hydrophilic bases that all exist as cationic species at physiological pH, with a minimal expected passive membrane diffusion [3]. Metformin is mainly absorbed from the small intestine and is excreted unchanged in urine, suggesting a lack of metabolism [3]. As a result, absorption, uptake, and excretion of metformin are primarily mediated by membrane transporters.

The precise molecular mechanism of metformin action is still widely discussed. The primary molecular target of metformin is believed to be the mitochondria [4]. Due to its positive charge, metformin accumulates within the matrix of mitochondria, where it acts to inhibit complex I of the mitochondrial electron transport chain, resulting in a reduction in NADH oxidation and ultimately a reduction in the synthesis of ATP [4]. It must be noted that metformin binds poorly to the mitochondrial membrane compared with phenformin, which might be one of the factors associated with the up to tenfold reduced risk of lactic acidosis compared with phenformin. These changes result in activation of AMP-activated protein kinase (AMPK), which switches the cell from an anabolic to a catabolic state by restoring energy balance through transcriptional regulation of downstream genes involved in gluconeogenesis in the liver and genes that encode glucose transporters in muscle cells (e.g., GLUT1) [5]. Hence, the mechanism of action is believed to involve inhibition of gluconeogenesis and induction of glucose uptake into muscle cells, lowering fasting blood glucose and insulin in T2D patients. Metformin also causes indirect induction of insulin receptor expression and tyrosine kinase activity, enhancing insulin sensitivity and reducing insulin resistance in patients [6].

Owing to its status as a relatively safe compound, with respect to its reduced risk of side-effects (described above), metformin has also been tested in many clinical trials for its effectiveness in other conditions [e.g., polycystic ovary syndrome (PCOS)], which have demonstrated its tremendous therapeutic potential [7,8]. Most recently, several studies have demonstrated therapeutic potential for metformin in the prevention and treatment of cancer [9]. The past few years have witnessed major discoveries in metformin's mechanism of action with regard to cancer therapy and its potential use in cancer patients with or without diabetes, starting with the observation that metformin has the potential to reduce the incidence of cancer in diabetic patients and ranging to the elucidation of multiple pathways that may be responsible for antineoplastic effects. However, pharmacological studies of metformin disposition, which could significantly affect our understanding of its action and patient dosing with regards to cancer prevention, seem to be largely lacking. Thus, this review highlights the current knowledge on metformin efficacy, mechanism of action, and pharmacology with specific focus on cancer prevention and therapy. Future challenges that may influence therapeutic outcome are discussed.

RATIONALE FOR CANCER PREVENTION WITH METFORMIN

Chronic elevated levels of plasma insulin may promote tumorigenesis owing to insulin's mitogenic and prosurvival effects and the high levels of insulin receptor expression on tumor cells [10]. A potential role of metformin in cancer prevention was initially demonstrated in several retrospective studies carried out in T2D patients with various forms of cancer. These studies have demonstrated beneficial effects of metformin at standard doses on cancer incidence and mortality [11,12]. Other studies, although demonstrating antineoplastic effects of metformin, showed no effect on the long-term, relapse-free survival rate of patients [13,14]. However, the retrospective and diabetes-specific nature of these studies makes them incomparable to cases involving the general population and, therefore, limited at best. Accordingly, prospective clinical trials in non-diabetic patients have recently begun. A summary of currently available prospective and meta-analysis studies is provided in Table 1 and demonstrates promising prospects for metformin use in cancer patients. In particular, an early study by Goodwin et al. demonstrated a significant reduction in insulin levels following a standard dose (1500mg/day) of metformin in non-diabetic women with early breast cancer [15]. As obesity-associated increase in insulin levels is believed to mediate poor breast cancer outcomes through binding to insulin receptors on breast cancer cells, this observation offers a positive prospect for improved breast cancer outcome in these patients [14]. A similar study analyzed the antiproliferative effects of metformin in non-diabetic women with operable invasive breast cancer and found reduced tumor cell proliferation, although the small sample size of this study could make it prone to false positives [16]. A larger, similarly designed randomized, placebo-controlled presurgical study found different effects on tumor proliferation of metformin depending on insulin resistance, suggesting population-dependent effects of metformin, based on patient- or

Table 1 Summary of Meta-Analyses and Prospective Clinical Trials of Metformin Use in Cancer Prevention[a]

Refs	Cancer Type	No. and Type of Studies	Summary Effect of Metformin Use	Specific Outcome	Comments
[79]	Breast, colorectal, hepatocellular, lung, pancreatic, prostate	11 (7 CCS, 4 CS)	31% decrease of cancer incidence or mortality (RR = 0.69, CI 0.61–0.79)	Significant decrease in risk of pancreatic and hepatocellular cancer, but not colon, breast, or prostate cancer	Evidence for between-study heterogeneity ($P = 0.03$, $I^2 = 64\%$), no evidence of publication bias
[80]	Mixed	7 (RCT)	No effect on cancer incidence or mortality		No evidence for between-study heterogeneity
[81]	Breast	7 (3 CCS, 4 CS)	Decreased risk of breast cancer incidence (OR = 0.83, CI 0.71–0.97)	Stronger effect for longer (> 3 year) metformin use (OR 0.75, CI 0.62–0.91)	Weak evidence for between-study heterogeneity ($P = 0.06$, $I^2 = 51\%$)
[82]	Bladder, breast, colon, gastric, hepatocellular, lung, pancreatic, prostate	24 (10 CCS, 11 CS, 3 RCT)	Decreased risk of all-cancer incidence (RR = 0.67, CI 0.53–0.85) and mortality (RR = 0.66, CI 0.49–0.88)	Significant decrease in risk of colorectal, hepatocellular and lung cancer, but not bladder, breast, gastric, pancreatic, or prostate cancer	Significant evidence for between-study heterogeneity, no evidence of publication bias
[83]	Mixed	17 (9 CCS, 8 CS)	Decreased risk of all-cancer incidence (RR = 0.61, CI 0.54–0.70)	Significant decrease in risk of colorectal and pancreatic cancer, but not breast or prostate cancer	Evidence for between-study heterogeneity ($P < 0.001$, $I^2 = 84\%$), evidence of publication bias
[84]	Colorectal	4 (1 CCS, 3 CS)	Decreased risk of colorectal cancer incidence (RR = 0.63, CI 0.47–0.84)		No evidence for between-study heterogeneity
[85]	Liver	5 (2 CCS, 3 CS)	62% decrease in liver cancer incidence (OR = 0.38, CI 0.24–0.59)	Significant decrease in cancer risk, also for hepatocellular carcinoma (OR 0.30, CI 0.17–0.52)	Evidence for between-study heterogeneity ($P = 0.001$, $I^2 = 78\%$)
[86]	Postmenopausal breast	Prospective observational study	Decreased risk of breast cancer incidence (HR = 0.75, CI 0.57–0.99)		Fewer estrogen and progesterone-receptor positive and HER2-negative breast cancers diagnosed in metformin users
[17]	Breast	Prospective RCT	No significant effect on breast cancer proliferation (Ki-67 staining)	Trend to a decreased proliferation in women with elevated insulin resistance	4-week metformin intervention study in non-diabetic women with operable breast cancer

Table 1 Summary of Meta-Analyses and Prospective Clinical Trials of Metformin Use in Cancer Prevention[a] *Continued*

Refs	Cancer Type	No. and Type of Studies	Summary Effect of Metformin Use	Specific Outcome	Comments
[87]	Prostate	Prospective RCT	Significant improvement of metabolic complications (e.g., weight, BMI, hypertension) in men with androgen-deprivation therapy when combined with lifestyle changes	Biochemical markers of insulin resistance did not differ significantly	6-month metformin intervention study
[16]	Breast	Prospective RCT	Significant down-regulation of Ki-67 staining and mRNA expression of PDE3B	First biomarker evidence for antiproliferative effects of metformin in breast cancer	2-week metformin intervention study
[18]	Colorectal aberrant crypt foci (ACFs)	Prospective RCT	Significant decrease in number of ACFs and proliferating cell nuclear antigen index	First evidence for suppression of colonic epithelial proliferation	1-month metformin intervention study in non-diabetic patients

[a]*Abbreviations: CCS, case control study; CS, cohort study; RCT, randomized controlled study; RR, relative risk; OR, odds ratio; HR, hazard ratio; CI, confidence interval; I^2: percentage of total variation across studies attributable to heterogeneity rather than to chance.*

tumor-specific metabolic characteristics [17]. With respect to colorectal cancer, Hosono *et al.* have demonstrated reduced colonic epithelial proliferation and rectal crypt foci formation in non-diabetic patients at a very low dose of metformin (250mg/day) [18]. These prospective studies provide promising evidence for antineoplastic and chemopreventive potentials of metformin and are a valuable basis for transitioning into phase II and III trials of metformin, which are currently underway (http://clinicaltrials.gov/ct2/results?term=metformin+phase+III).

ROLE OF METFORMIN IN GLUCONEOGENESIS AND ITS LINK TO CHEMOPREVENTION

Insulin-Dependent (Indirect) Effects of Metformin

One of the underlying mechanisms supporting the association between metformin use and reduced cancer risk is the same as the one believed to underlie its antidiabetic effect. The indirect activation of AMPK via metformin results in lowered plasma glucose and reduced insulin levels through downregulation of key gluconeogenesis genes and activation of glucose uptake into muscle [6,19] (Figure 1). As a result, metformin acts in an indirect way to reduce the negative effects of insulin on tumor growth and progression.

AMPK-Dependent (Direct) Effects of Metformin

First clues to an alternative mechanism of metformin action in cancer came from a publication by Shaw et al. showing that AMPK is activated by an upstream gene encoding the tumor suppressor LKB1 [20]. Deletion of LKB1

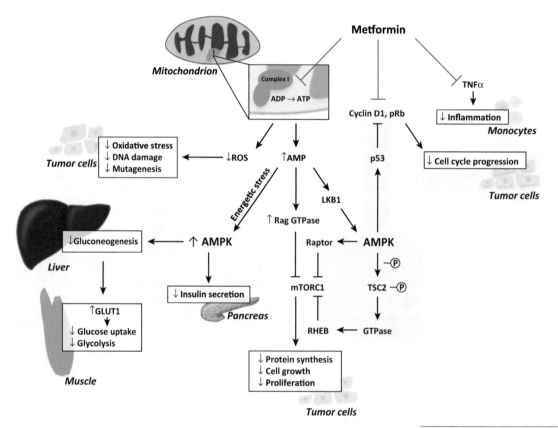

TRENDS in Pharmacological Sciences

FIGURE 1 Schematic diagram of selected proposed mechanisms of action of metformin. The background color of each pathway depicts the specific mechanism that may be involved. The blue background depicts an insulin-dependent (direct) mechanism, the red background depicts an AMP-activated protein kinase (AMPK)-dependent (indirect) mechanism, and the yellow background depicts an AMPK-independent pathway. Abbreviations: pRb, retinoblastoma protein; TSC2, tuberous sclerosis protein 2; RHEB, Ras homolog enriched in brain (GTP-binding protein); GLUT1, glucose transporter protein 1.

in mice resulted in loss of AMPK activity, in addition to hyperglycemia and increased gluconeogenic and lipogenic gene expression. These findings suggested that the primary mode of metformin action may depend on its ability to activate AMPK.

Subsequent findings linked the activation of AMPK to a reduction in mTOR signaling and protein synthesis in cancer cells. AMPK reduces mTORC1 levels through phosphorylation and activation of tuberous sclerosis complex 2 (TSC2), resulting in activation of GTPase-activating protein, which acts to inhibit a downstream small GTPase (RHEB), causing a reduction in mTOR signaling [21]. AMPK may also act directly to inhibit raptor, the regulatory associated protein of mTOR. mTOR plays a key role in controlling cell growth, proliferation, and metabolism and mediates one of the most frequently deregulated molecular networks in human cancers (i.e., the phosphoinositide 3-kinase [PI3K]/protein kinase B [PKB]/Akt signaling pathway [22]). Activation of the mTOR pathway occurs often in breast cancers, correlating with progression and poor prognosis [23]. Thus, the potential role of metformin as an anticancer agent is thought to involve metformin-mediated activation of AMPK and subsequent inhibition of downstream mTOR signaling, reduction in the expression of its major downstream effectors, inhibition of total protein synthesis, and reduction of cancer cell proliferation in various cancer cell lines (Figure 1).

AMPK is further believed to be involved in p53-mediated cell cycle arrest. In the case of cellular energy imbalance and crisis, AMPK phosphorylates p53, resulting in p53-mediated cell cycle arrest in p53-expressing cells and apoptosis in cells with mutated p53. Functional p53 has the capacity to activate various genes that negatively regulate the AKT and mTORC1 pathways, subsequently resulting in cellular senescence, quiescence, or apoptosis [24,25]. This mechanism introduces the possibility that mutated p53 might be necessary for drug response; however, given the diverse cellular pathways and genes regulated by p53, it is likely that the mechanism of metformin action would differ based on cell type and cellular stress levels [24].

AMPK-Independent Effects of Metformin

Recently, metformin was shown to inhibit hepatic gluconeogenesis in an LKB1-, TSC2-, and AMPK-independent manner [26]. It was recently proposed that induction of acute changes in gene expression is not the only means of inhibiting glucose output; rather, metformin's mechanism of action is related to its negative effects on gluconeogenic flux instead of a direct effect on gluconeogenic gene expression. Therefore, metformin-induced inhibition of glucose output may be a result of disruptions in cellular energy output (i.e., reduced cellular ATP followed by an increase in cellular AMP [26]). Kalender et al. further demonstrated that metformin can act through

AMPK-independent inhibition of mTORC1 via Rag GTPase, which, when activated, acts to sustain homeostatic mTORC1 signaling [27].

Furthermore, Arai et al. showed that metformin partially reduces chronic inflammatory responses by inhibiting the production of tumor necrosis factor alpha (TNFα) in human monocytes, probably independent of its activation of AMPK [28]. Because chronic inflammation is the basis of the progression of many cancer types, this may contribute to the role of metformin in preventing tumor development (Figure 1).

Recently, Ben Sahra et al. showed that metformin induces cell cycle arrest in human prostate cancer cells by inhibiting the expression of cyclin D1 and retinoblastoma-protein (pRb), two key regulators of the cell cycle, in a p53-dependent manner. This pathway is mediated by upregulation of p53 and p21, resulting in reduction of cyclin D1 levels and, consequently, reduction of cyclin-dependent kinases, eventually resulting in an antiproliferative effect mediated by G_1 cell cycle arrest [29,30]. Metformin also decreased E2F1 protein levels. E2F1 plays an important role in cell cycle progression and mediation of tumor-suppressor genes. Increased levels of E2F1 activate growth-receptor signaling pathways and promote an antiapoptotic environment in tumors. Hence, the inhibitory effect of metformin results in decreased cellular proliferation and metastasis in tumor tissues. These findings were confirmed by in vivo experiments in which metformin also caused a reduction in cyclin D1 protein levels in prostate tumors [30] (Figure 1).

Studies by Algire et al. demonstrated that metformin reduces production of endogenous reactive oxygen species (ROS) through inhibition of mitochondrial complex I, the cellular source of ROS production [31] (Figure 1). Because ROS also play a key role in the formation of advanced glycation end-products, which enhance oxidative stress, metformin was shown not only to reduce DNA damage and mutagenesis in normal somatic cells but also to reduce oxidative stress [31].

In 2011, Zhou et al. identified a genetic variant near the Ataxia Telangiectasia Mutated (ATM) gene locus, associated with metformin treatment success (defined as HbA1c ≤ 7%) [32]. Because ATM is involved in DNA repair and cell cycle arrest, Vanquez-Martin and colleagues looked for and found an additional link between metformin and activation of checkpoint homolog kinase 2 (CHK2), which mediates the DNA damage response (DDR) of ATM [33]. Menendez et al. further showed that metformin sensitizes cancer cells against further damage by activating ATM-mediated DDR and inhibiting premalignant cells before they gain tumor-initiating properties [34]. Because metformin has been shown to disrupt glycolysis, which protects cells from the prosenescent effects of mitochondrial respiration-mediated oxidative stress, metformin is believed to protect against cellular immortalization and inhibit the progression of premalignant cells and tumorigenesis [32].

Additional Roles for Metformin in Cancer

Interestingly, current research has implicated metformin not only in the prevention and treatment of cancer, but also in cancer treatment optimization. In endometrial cancer cells, metformin reversed resistance to progestin therapy by inhibiting the expression of glyoxalase I, an enzyme related to glucose metabolism and normally overexpressed in tumor tissues and associated with resistance to chemotherapy [35]. Metformin treatment is also implicated in improving the response of human tumor xenografts to chemotherapy by having a preferentially more cytotoxic action (at clinical doses) against cancer stem cells (CSCs), which are chemoresistant and radioresistant compared with non-cancer stem cells [36]. Furthermore, metformin was shown to improve the response of tumors to irradiation [37]. This effect is linked to metformin-induced activation of AMPK and the resulting inactivation of mTOR. Because the P13K/Akt/mTOR pathway is hyperactive in some CSCs due to its critical role in the maintenance, survival, and proliferation of these cells, CSCs are believed to be more sensitive to mTOR inhibition [37].

Although numerous pathways have been implicated in the antineoplastic effects of metformin, LKB1-dependent suppression of mTOR signaling remains the main candidate mechanism. However, because cancer and diabetes involve alterations in crucial metabolic responses, it is expected that, in the presence of metformin, the cell would manifest multiple responses simultaneously to modify or inhibit these changes. One recent area of research suggests the possibility of variable effects of metformin dependent on the metabolic and genetic characteristics of tumors, such that it may act favorably in tumors exhibiting loss of p53 or LKB1 and not in other tissues [38]. Alternatively, metformin could act at the systemic level to improve the global metabolic profile in patients, rather than having a tumor cell-specific effect.

INFLUENCE OF METFORMIN DISPOSITION ON CANCER PREVENTION

There is notable evidence for interindividual variability of metformin pharmacokinetic parameters, potentially explaining differences in patient response to metformin, defined as lowered blood glucose [3]. This variability has previously been related to disease state, body mass index (BMI), host metabolic characteristics, transporter pharmacogenetics, and many other as yet uncharacterized factors in T2D patients, which may be equally applicable in cancer prevention. In the case of cancer prevention, metformin's therapeutic effect would significantly depend on its accumulated concentration inside various affected cells, which is critically dependent on whole-organism and cellular pharmacokinetic factors. There is a major body of evidence that tissue expression of several membrane transporters implicated in the uptake and secretion of metformin, such as organic cation transporters (OCTs) 1, 2, and 3 as well as the multidrug and toxin extrusion transporters (MATE1 and MATE2/2K), differs significantly between normal and tumor tissues (Figure 2)

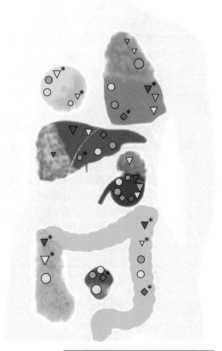

TRENDS in Pharmacological Sciences

FIGURE 2 Schematic representation of transporter mRNA and protein expression in normal and tumor tissues of selected organs.

Discolored segments of organs represent tumor tissues. Transporters of the same family are depicted with the same symbols. A reduction or an increase in the size of a transporter symbol, in a given tissue, indicates a reduction or an increase in the expression levels of the transporter, respectively. An asterisk (*) is used beside a symbol where only information on mRNA levels of the corresponding transporter is available [41,88–95].

[39,40]. Thus, variability in the expression and activity of these transport proteins may affect the pharmacological action of metformin and/or its systemic and local accumulation, thereby emphasizing the importance of elucidating the underlying mechanisms for variable expression. In particular, with regard to a complex disease such as cancer, not only genetic variations may affect transporter expression and activity, but also non-genetic, environmental, and epigenetic factors may account for a significant fraction of variability.

Genetic Variability of Metformin Disposition and Action
OCT1–SLC22A1

Human OCT1 is predominantly expressed on the basolateral membrane of hepatocytes [41] and is one of the key transporters responsible for hepatic metformin uptake, according to knockout mouse experiments and human data [42]. However, microarray data from 381 samples of 20 diverse human tissues also support highly varied expression of OCT1, with notable differences between tumor and the corresponding normal tissue [39]. Currently, at least 25 coding single nucleotide polymorphisms (cSNPs) have been described in SLC22A1, with high interethnic variability with respect to frequency distribution (for a summary, see ref [43]). Of these SNPs, 12 have been shown to affect functionality (Table 2) by in vitro and in vivo data [42,44,45]. To date, based on multivariate analysis, the only polymorphism identified in a comprehensive study of non-tumor liver tissue to significantly affect OCT1 mRNA and protein expression is OCT1-R61C (rs12208357) [41]. Because OCT1 is expressed, at low levels on the apical side of proximal and distal renal tubules [44], lower activity of OCT1 may also result in decreased tubular reabsorption and subsequent increased elimination. Low-function OCT1 polymorphisms are crucial for therapeutic response and have been shown to cause significant variations in metformin response [43,46].

OCT2–SLC22A2

OCT2 is mainly expressed on the basolateral membrane of renal tubular cells, where it mediates entry of metformin and, together with MATE1/2 (see below), mediates the secretion of metformin into urine [43]. OCT2 expression is also highly variable among normal and tumor tissues, with the highest expression in kidney and lower levels in testis and prostate [39]. Renal clearance of metformin has a strong genetic component and, because OCT2 has been shown to account for 80% of total metformin clearance, functionally relevant SLC22A2 polymorphisms result in significant variations in the pharmacokinetic and/or pharmacodynamic properties of metformin [47,48]. At least 10 cSNPs have been reported in SLC22A2 [43] (Table 2). Several functional studies of these polymorphisms have been conducted and display in vivo and in vitro changes in metformin disposition (Table 2). Interestingly, a recent metabolomics study has shown significant changes in endogenous

Table 2 Summary of Transporter Coding SNPs Associated with Metformin Disposition[a]

dbSNP ID	Amino Acid	Allele Frequencies	Functional Effect		Refs
			In vitro	*In vivo*	
SLC22A1 (OCT1)					
rs34447885	Ser14Phe	AA: 3.1% (*n* = 200)[A]	Reduced V_{max}[B]		[A][96] [B][42]
rs12208357	Arg61Cys	EA: 9.1% (*n* = 243)[A] EA: 9.7% (*n* = 150)[B]	Reduced uptake[C]	Increased AUC and C_{max} Decreased V_d Cl_{renal} similar to reference[D]	[A][97] [B][41] [C][42] [D][98]
rs34104736	Ser189Leu	EA: 0.5% (*n* = 200)[A]	Reduced V_{max}[B]		[A][96] [B][42]
rs36103319	Gly220Val	AA: 0.5% (*n* = 200)[A]	Reduced uptake[B]		[A][96] [B][42]
rs2282143	Pro341Leu	Korean: 16.7% (*n* = 150)[A] Chinese: 11% (*n* = 100)[A]	Similar functionality[B]	No impact on metformin response[C]	[A][99] [B][42] [C][100]
rs34130495	Gly401Ser	EA: 3.2% (*n* = 232)[A]	Decreased V_{max} Elevated K_m[B]	Increased glucose half-life following metformin treatment[B]	[A][97] [B][42]
rs628031	Met408Val	AA: 73.5% (*n* = 200)[A] EA: 59.7% (*n* = 232)[B] Korean: 74% (*n* = 150)[C]	Normal uptake[D]	Positive predictor for efficacy of metformin[E]	[A][96] [B][97] [C][99] [D][42] [E][100]
rs202220802	Met420del	EA: 15.7% (*n* = 232)[A]	Decreased V_{max} Elevated K_m[B]	Increase in glucose half-life following metformin treatment[B]	[A][97] [B][42]
rs34059508	Gly465Arg	EA: 1.5% (*n* = 232)[A]	Reduced uptake[C]	Increase in glucose half-life following metformin treatment[D]	[A][96] [C][42]
SLC22A2 (OCT2)					
rs201919874	Thr199Ile	Korean: 0.7% (*n* = 150)[A]	Decrease in V_{max} is greater than K_m changes[B]	Higher C_{max} and AUC Lower CL/F, V_d/F, Cl_{renal}[B]	[A][99] [B][88]
rs145450955	Thr201Met	Korean: 0.7% (*n* = 150)[A]	Decrease in V_{max} is greater than K_m changes[B]	Higher C_{max} and AUC[B] No association with metformin response[C]	[A][99] [B][88] [C][100]
rs316019	Ala270Ser	EA: 15.7% (*n* = 200)[A] AA: 11% (*n* = 200)[A]	Decreased V_{max} No K_m changes[B]	Higher C_{max} and AUC Lower CL/F, V_d/F, Cl_{renal}[B] No association with metformin response[C]	[A][89] [B][88] [C][100]
SLC22A3 (OCT3)					
rs8187715	Thr44Met	AA: 0.6% (*n* = 200)[A] EA: 0.6% (*n* = 200)[A]	IncreasedV_{max}		[A][51]

Table 2 Summary of transporter coding SNPs associated with metformin disposition[a] *Continued*

dbSNP ID	Amino Acid	Allele Frequencies	Functional Effect		Refs
			In vitro	*In vivo*	
rs8187717	Ala116Ser	AA: 1.7% ($n = 200$)[A]	Similar uptake compared to reference[A]		[A][51]
rs8187725	Thr400Ile	EA: 0.5% ($n = 200$)[A]	Increased K_m Slightly decreased V_{max}[A]		[A][51]
	Val423Phe		Increased K_m Slightly decreased V_{max}[A]		[A][51]
SLC47A1 (MATE 1)					
rs77630697	Gly64Asp	Japanese: 0.6% ($n = 89$)[A]	Complete loss of transport activity[B]	No influence on metformin disposition[C]	[A][56] [B][90]
rs77474263	Leu125Phe	AS: 0.7% ($n = 68$)[A] ME: 5.1% ($n = 68$)[A]	Significantly reduced transport[A]	No influence on metformin disposition[B]	[A][52] [B][90]
rs35646404	Thr159Met	Japanese: 1% ($n = 95$)[A]	Decreased V_{max} and K_m[A]		[A][59]
rs111060526	Ala310Val	Japanese: 2.2% ($n = 89$)[A]	Significant decreased transport activity Increased K_m Decreased V_{max}[A]		[A][56]
rs149774861	Asp328Ala	Japanese: 0.6% ($n = 89$)[A]	Significant decrease in transport activity Decreased V_{max}[A]	No influence on disposition of metformin[B]	[A][56] [B][90]
rs35790011	Val338Ile	EA: 0.4 ($n = 253$)[A]	Loss of transport activity Decreased V_{max} and K_m[A]		[A][59]
rs111060528	Asn474Ser	Japanese: 0.6% ($n = 89$)[A]	Significant decrease in transport activity Increased K_m[A]		[A][56]
rs76645859	Val480Met	AS: 0.8% ($n = 68$)[A]	Complete loss of function[A]		[A][52]
rs35395280	Cys497Phe/ Ser	AA: 2.4% (C) ($n = 68$)[A]	Significantly reduced transport[A]		[A][52]
rs78700676	Gln519His	AA: 0.8% ($n = 68$)[A]	Similar uptake[A]		[A][52]
SLC47A2 (MATE 2/2K)					
rs111060529	Lys64Asn	Japanese: 0.6% ($n = 89$)[A]	Decreased transport activity Decreased V_{max}[A]		[A][56]
rs111060532	Gly211Val	Japanese: 1.7% ($n = 89$)[A]	Complete loss of transport activity[A]		[A][56]

[a]Abbreviations: EA, European-American; AA, African-American; AS, Asian-American; ME, Mexican; PA, Pacific Islander.

tryptophan levels in the presence of the OCT2-A270S variant and thus tryptophan may serve as a promising biomarker for determining variability in the transport activity of OCT2 [49].

OCT3–SLC22A3

In contrast to OCT1/2, OCT3 is expressed at the mRNA level in multiple tissues as well as in corresponding tumors [39]. OCT3 is also believed to play an important role in metformin transport [41]. At least five cSNPs in the *SLC22A3* gene have been described [43]. Moreover, OCT3 mRNA expression in liver tissue from Asians and Caucasians containing the promoter variant rs555754 was significantly higher, consistent with reporter assays [50]. However, unlike OCT1 and OCT2, the evidence for functional consequences on metformin disposition and patient response is limited (Table 2) [51]. Of note, the *SLC22A3* gene appears to be an important risk factor for prostate cancer as well as coronary artery disease, which is not seen for OCT1 or OCT2 [43].

MATE1/2K–SLC47A1/2

Recently, polymorphisms have been detected in human MATE1 (hMATE1, *SLC47A1*), MATE2, and a human MATE2 (SLC47A2) transcript variant, MATE2-K. hMATE2-K lacks part of exon 7, resulting in a 566-amino acid protein that is shorter than the originally identified hMATE2 and is abundantly expressed in kidney. The finding of these polymorphisms suggests a more significant contribution of these transporters to the genetic variability of metformin efflux than previously believed [52,53], because both MATE1 and MATE2/2K have been shown to mediate the transcellular transport of metformin *in vitro* [40]. Additionally, loss of MATE1 function in a mouse model was shown to cause significantly increased metformin concentration in the liver and increased lactate levels in the bloodstream [54]. MATE1 is located at the bile canaliculi of the liver and mainly at the brush border membrane of renal proximal tubule cells, whereas MATE2 and its MATE2K splice variant are widely expressed in humans, with the highest mRNA expression in the kidney [40,55]. MATE1 expression at the mRNA level is highly variable between normal and tumor tissues, with predominant expression in adrenal gland, kidney, liver, and testis (M. Schwab, unpublished). To date, at least 11 cSNPs have been identified in *SLC47A1* and two in *SLC47A2* [52,56]. Of these, several variants have been reported to cause loss of function and altered localization in *in vitro* uptake studies (Table 2). Becker *et al.* described an intronic *MATE1* polymorphism (rs2289669) that was significantly associated with an increased glucose-lowering effect of metformin [57]. Additionally, promoter polymorphisms in *MATE1* (rs2252281) and *MATE2K* (rs12943590) were associated with altered luciferase activity *in vitro* [58] and in part with reduced metformin response in T2D patients [53]. Given that the MATE transporters work together with OCT2 at the renal site and with hepatic OCT1 at the major site of metformin action, *MATE* polymorphisms may have a greater effect on metformin transport than previously considered [43,59].

Additional Transporters

Recently, a novel proton-activated organic cation transporter, known as plasma membrane monoamine transporter (PMAT) encoded by *SLC29A4*, has been found to transport metformin [60]. PMAT is believed to play a role in renal reabsorption and intestinal absorption of metformin, although not much is known about its expression in other tissues (including tumor tissue) or its interindividual variability based on genetic variation [60,61].

Furthermore, metformin has been shown to sensitize glycolytic tumor cells to growth inhibition through monocarboxylate transporter (MCT) inhibitors [62]. Recently, MCTs were found to be overexpressed in several tumor types, including lung and prostate tumors [63]. Because metformin inhibits oxidative phosphorylation and stimulates glycolysis for energy production, the harmful effect of MCT inhibitors on lactic acid export-dependent tumor cells may be enhanced through concomitant administration of metformin [62]. Polymorphisms in the MCT sequence have been identified, but have not yet been functionally evaluated, particularly with respect to metformin action [62].

Downstream Targets of Metformin

As mentioned above, a variant (rs11212617) of the *ATM* gene, which plays a key role in the activation of AMPK, may contribute to variability in patient response to metformin [32]. Additionally, in 2010, the first large-scale prospective analysis of SNP associations with metformin treatment identified five variants in the gene encoding AMPK (*LKB/STK11*) and the AMPK subunit genes (*PRKAA1*, *PRKAA2*, and *PRKAB2*) [64]. Interestingly, a variant in *STK11* (rs8111699) was previously shown to be significantly associated with ovulatory response to metformin in women with PCOS [65].

Taken together, knowledge of genetic variants affecting metformin disposition and downstream targets may ultimately prove helpful in predicting metformin response. Because Phase II and III trials of the use of metformin in cancer in non-diabetic patients are currently under assessment or already under way, more effort should be put into validating the above-mentioned potential predictive genetic markers of metformin response.

Impact of Non-Genetic Factors on Metformin Disposition

One main reason behind the chemoresistance of cancer cells is upregulation of the expression of efflux transporters, thus reducing cellular accumulation. In recent years, major interest in the effects of chemotherapy on the expression of uptake transporters, independent of genetic variation, has been increasing. Particularly, OCT1 expression has been reported in colon cancer and polyps [66]. Because OCT1 is an uptake transporter, its increased expression in these cancer cells introduces the prospect of targeted therapy by OCT1 anticancer agents, such as platinum drugs. Similarly, enhanced expression of OCT2 has been demonstrated in human colorectal tumor cells [67]. Recently, it has been demonstrated that liver- and renal-specific expression of OCT1 and OCT2, respectively, are inversely regulated by methylation at CpG sites of the promoter region in *SLC22A1* and *SLC22A2* [68]. Schaeffeler *et al.* showed that decreased expression of *SLC22A1* in hepatocellular carcinoma compared with adjacent and normal liver tissues is the consequence of different methylation patterns with potential therapeutic consequences [39]. Moreover, aberrant methylation of the OCT3 promoter contributes to reduced expression in prostate cancer compared with matched normal samples [50]. Thus, knowledge of variations in methylation status and their effects on expression prompt investigation of their impact on cancer prevention with metformin.

Finally, comedication is another contributing factor in the expression or function of drug transporters. In patients being treated concomitantly with metformin and tyrosine kinase inhibitors, such as imatinib and erlotinib, a selectively potent inhibitory effect was observed with respect to OCT1, MATE1, and MATE2K transport [69]. This inhibitory effect was enhanced in patients carrying the OCT1-M420del polymorphism [69], corroborating the importance of absorption, distribution, metabolism, and excretion (ADME) pharmacogenetics and its impact on drug–drug interactions (DDIs) [70]. Several other frequently used clinical agents, such as proton pump inhibitors and antibiotics, inhibit both OCTs and MATEs [40,71,72]. Because these transporters are involved in the renal clearance of metformin, DDIs would result in higher circulating plasma levels of metformin, which would potentially affect cancer prevention with this drug. The importance of DDIs is supported by a recent observation that an ATM inhibitor, KU-55933, also inhibited OCT1-mediated uptake of metformin [73]. This may illustrate that KU-55933 inhibits metformin uptake through inhibition of OCT1 rather than ATM, emphasizing the importance of investigating all of the cellular factors involved in metformin disposition.

CONCLUDING REMARKS

Recent preclinical and clinical data have made it apparent that metformin exerts its chemopreventive and antineoplastic effects in a multitude of tissues. Aside from solid tumors, its antineoplastic effects have also been reported in hematologic tumors, such as induction of autophagy in lymphoma cells and chemosensitization of acute lymphoblastic leukemia cells [74,75]. Hence, the mechanisms of action of metformin may also be influenced by metabolic factors, which may be affected by patient obesity factors such as insulin or inflammation levels. This is in agreement with recent clinical studies, which suggest patient-specific beneficial effects of metformin [15–17,76]. Furthermore, the action of metformin may be limited by various pharmacologic parameters, especially transporter expression in tumor cells and poor uptake into target cells. Thus, a metabolomics approach may be a novel, promising tool to identify the specific targets of metformin action, which may then be utilized, in conjunction with genetic predictors, to select patient populations where metformin may be most beneficial [77]. In patients in whom metformin disposition may be reduced, especially due to variations in transporter expression, similar drugs with less dependence on transport may be used, such as phenformin, which acts through the same mechanism as metformin but is more lipophilic, reducing its dependence on active transport [78]. Alternatively, metformin may be chemically modified to create a novel agent that would be less dependent on these limiting parameters. Although many prospective studies are under way to investigate the potential of metformin as an antineoplastic agent, they should be combined with parallel pharmacological evaluations to determine variable patient response and metformin disposition. This would allow for a more comprehensive view of the *in vivo* actions of metformin, making it possible to analyze all of the parameters implicated in its anticancer effects and to more confidently determine the profile of patients who may benefit from its use in the clinic.

ACKNOWLEDGMENTS

This work was supported by the Bundesminsterium für Bildung und Forschung (Grants 03 IS 2016C and 0315755), the Deutsche Forschungsgemeinschaft (Grant SCHW 858/1-1), the FP7 EU Initial Training Network program *FightingDrugFailure* (Grant PITN-GA-2009-238132), and the Robert Bosch Stiftung, Germany. The authors wish to acknowledge their very helpful and stimulating discussions with Dr Michael N. Pollak, Québec, Canada.

REFERENCES

1 Witters, L.A. (2001) The blooming of the French lilac. *J. Clin. Invest.* 108, 1105–1107

2 UK Prospective Diabetes Study (UKPDS) Group. (1998) Effect of intensive blood-glucose control with metformin on complications in overweight patients with type 2 diabetes (UKPDS 34). *Lancet* 352, 854–865

3 Graham, G.G. *et al.* (2011) Clinical pharmacokinetics of metformin. *Clin. Pharmacokinet.* 50, 81–98

4 El-Mir, M.Y. *et al*. (2000) Dimethylbiguanide inhibits cell respiration via an indirect effect targeted on the respiratory chain complex I. *J. Biol. Chem.* 275, 223–228

5 Stephenne, X. *et al*. (2011) Metformin activates AMP-activated protein kinase in primary human hepatocytes by decreasing cellular energy status. *Diabetologia* 54, 3101–3110

6 Gunton, J.E. *et al*. (2003) Metformin rapidly increases insulin receptor activation in human liver and signals preferentially through insulin-receptor substrate-2. *J. Clin. Endocrinol. Metab.* 88, 1323–1332

7 Wang, J. *et al*. (2012) Metformin activates an atypical PKC-CBP pathway to promote neurogenesis and enhance spatial memory formation. *Cell Stem Cell* 11, 23–35

8 Pollak, M. (2010) Metformin and other biguanides in oncology: advancing the research agenda. *Cancer Prev. Res.* 3, 1060–1065

9 Giovannucci, E. *et al*. (2010) Diabetes and cancer: a consensus report. *Diabetes Care* 33, 1674–1685

10 Pollak, M. (2008) Insulin and insulin-like growth factor signalling in neoplasia. *Nat. Rev. Cancer* 8, 915–928

11 Libby, G. *et al*. (2009) New users of metformin are at low risk of incident cancer: a cohort study among people with type 2 diabetes. *Diabetes Care* 32, 1620–1625

12 Currie, C.J. *et al*. (2009) The influence of glucose-lowering therapies on cancer risk in type 2 diabetes. *Diabetologia* 52, 1766–1777

13 Patel, T. *et al*. (2010) Clinical outcomes after radical prostatectomy in diabetic patients treated with metformin. *Urology* 76, 1240–1244

14 Jiralerspong, S. *et al*. (2009) Metformin and pathologic complete responses to neoadjuvant chemotherapy in diabetic patients with breast cancer. *J. Clin. Oncol.* 27, 3297–3302

15 Goodwin, P.J. *et al*. (2008) Insulin-lowering effects of metformin in women with early breast cancer. *Clin. Breast Cancer* 8, 501–505

16 Hadad, S. *et al*. (2011) Evidence for biological effects of metformin in operable breast cancer: a pre-operative, window-of-opportunity, randomized trial. *Breast Cancer Res. Treat.* 128, 783–794

17 Bonanni, B. *et al*. (2012) Dual effect of metformin on breast cancer proliferation in a randomized presurgical trial. *J. Clin. Oncol.* 30, 2593–2600

18 Hosono, K. *et al*. (2010) Metformin suppresses colorectal aberrant crypt foci in a short-term clinical trial. *Cancer Prev. Res.* 3, 1077–1083

19 Towler, M.C. and Hardie, D.G. (2007) AMP-activated protein kinase in metabolic control and insulin signaling. *Circ. Res.* 100, 328–341

20 Shaw, R.J. *et al*. (2005) The kinase LKB1 mediates glucose homeostasis in liver and therapeutic effects of metformin. *Science* 310, 1642–1646

21 Gwinn, D.M. *et al*. (2008) AMPK phosphorylation of raptor mediates a metabolic checkpoint. *Mol. Cell* 30, 214–226

22 Inoki, K. *et al*. (2002) TSC2 is phosphorylated and inhibited by Akt and suppresses mTOR signalling. *Nat. Cell Biol.* 4, 648–657

23 Bachman, K.E. *et al*. (2004) The PIK3CA gene is mutated with high frequency in human breast cancers. *Cancer Biol. Ther.* 3, 772–775

24 Liang, J. *et al*. (2007) The energy sensing LKB1-AMPK pathway regulates p27(kip1) phosphorylation mediating the decision to enter autophagy or apoptosis. *Nat. Cell Biol.* 9, 218–224

25 Galluzzi, L. *et al*. (2010) TP53 and MTOR crosstalk to regulate cellular senescence. *Aging (Albany N.Y.)* 2, 535–537

26 Foretz, M. *et al*. (2010) Metformin inhibits hepatic gluconeogenesis in mice independently of the LKB1/AMPK pathway via a decrease in hepatic energy state. *J. Clin. Invest.* 120, 2355–2369

27 Kalender, A. *et al*. (2010) Metformin, independent of AMPK, inhibits mTORC1 in a Rag GTPase-dependent manner. *Cell Metab.* 11, 390–401

28 Arai, M. *et al*. (2010) Metformin, an antidiabetic agent, suppresses the production of tumor necrosis factor and tissue factor by inhibiting early growth response factor-1 expression in human monocytes in vitro. *J. Pharmacol. Exp. Ther.* 334, 206–213

29 Hill, R. and Wu, H. (2009) PTEN, stem cells, and cancer stem cells. *J. Biol. Chem.* 284, 11755–11759

30 Ben Sahra, I. *et al*. (2008) The antidiabetic drug metformin exerts an antitumoral effect in vitro and in vivo through a decrease of cyclin D1 level. *Oncogene* 27, 3576–3586

31 Algire, C. *et al*. (2012) Metformin reduces endogenous reactive oxygen species and associated DNA damage. *Cancer Prev. Res.* 5, 536–543

32 Zhou, K. *et al*. (2011) Common variants near ATM are associated with glycemic response to metformin in type 2 diabetes. *Nat. Genet.* 43, 117–120

33 Vazquez-Martin, A. *et al*. (2011) Metformin activates an ataxia telangiectasia mutated (ATM)/Chk2-regulated DNA damage-like response. *Cell Cycle* 10, 1499–1501

34 Menendez, J.A. *et al*. (2011) Metformin and the ATM DNA damage response (DDR): accelerating the onset of stress-induced senescence to boost protection against cancer. *Aging (Albany N.Y.)* 3, 1063–1077

35 Dong, L. *et al*. (2012) Metformin sensitizes endometrial cancer cells to chemotherapy by repressing glyoxalase I expression. *J. Obstet. Gynaecol. Res.* http://dx.doi.org/10.1111/j.1447-0756.2011.01839.x

36 Hirsch, H.A. *et al*. (2009) Metformin selectively targets cancer stem cells, and acts together with chemotherapy to block tumor growth and prolong remission. *Cancer Res.* 69, 7507–7511

37 Song, C.W. *et al*. (2012) Metformin kills and radiosensitizes cancer cells and preferentially kills cancer stem cells. *Sci. Rep.* 2, 362

38 Algire, C. *et al*. (2011) Diet and tumor LKB1 expression interact to determine sensitivity to anti-neoplastic effects of metformin in vivo. *Oncogene* 30, 1174–1182

39 Schaeffeler, E. *et al*. (2011) DNA methylation is associated with downregulation of the organic cation transporter OCT1 (SLC22A1) in human hepatocellular carcinoma. *Genome Med.* 3, 82

40 Damme, K. *et al*. (2011) Mammalian MATE (SLC47A) transport proteins: impact on efflux of endogenous substrates and xenobiotics. *Drug Metab. Rev.* 43, 499–523

41 Nies, A.T. *et al*. (2009) Expression of organic cation transporters OCT1 (SLC22A1) and OCT3 (SLC22A3) is affected by genetic factors and cholestasis in human liver. *Hepatology* 50, 1227–1240

42 Shu, Y. *et al*. (2007) Effect of genetic variation in the organic cation transporter 1 (OCT1) on metformin action. *J. Clin. Invest.* 117, 1422–1431

43 Nies, A.T. *et al*. (2011) Organic cation transporters (OCTs, MATEs), in vitro and in vivo evidence for the importance in drug therapy. *Handb. Exp. Pharmacol.* 201, 105–167

44 Tzvetkov, M.V. *et al*. (2009) The effects of genetic polymorphisms in the organic cation transporters OCT1, OCT2, and OCT3 on the renal clearance of metformin. *Clin. Pharmacol. Ther.* 86, 299–306

45 Becker, M.L. *et al*. (2009) Genetic variation in the organic cation transporter 1 is associated with metformin response in patients with diabetes mellitus. *Pharmacogenomics J.* 9, 242–247

46 Gambineri, A. *et al.* (2010) Organic cation transporter 1 polymorphisms predict the metabolic response to metformin in women with the polycystic ovary syndrome. *J. Clin. Endocrinol. Metab.* 95, E204–E208

47 Kimura, N. *et al.* (2005) Metformin is a superior substrate for renal organic cation transporter OCT2 rather than hepatic OCT1. *Drug Metab. Pharmacokinet.* 20, 379–386

48 Leabman, M.K. and Giacomini, K.M. (2003) Estimating the contribution of genes and environment to variation in renal drug clearance. *Pharmacogenetics* 13, 581–584

49 Song, I-S. *et al.* (2012) Pharmacogenetics meets metabolomics: discovery of tryptophan as a new endogenous OCT2 substrate related to metformin disposition. *PLoS ONE* 7, e36637

50 Chen, L. *et al.* (2012) Genetic and epigenetic regulation of the organic cation transporter 3, SLC22A3. *Pharmacogenomics J.* http://dx.doi.org/10.1038/tpj.2011.60

51 Chen, L. *et al.* (2010) Role of organic cation transporter 3 (SLC22A3) and its missense variants in the pharmacologic action of metformin. *Pharmacogenet. Genomics* 20, 687–699

52 Chen, Y. *et al.* (2009) Genetic variants in multidrug and toxic compound extrusion-1, hMATE1, alter transport function. *Pharmacogenomics J.* 9, 127–136

53 Choi, J.H. *et al.* (2011) A common 5′-UTR variant in MATE2-K is associated with poor response to metformin. *Clin. Pharmacol. Ther.* 90, 674–684

54 Toyama, K. *et al.* (2012) Loss of multidrug and toxin extrusion 1 (MATE1) is associated with metformin-induced lactic acidosis. *Br. J. Pharmacol.* 166, 1183–1191

55 Komatsu, T. *et al.* (2011) Characterization of the human MATE2 proton-coupled polyspecific organic cation exporter. *Int. J. Biochem. Cell Biol.* 43, 913–918

56 Kajiwara, M. *et al.* (2009) Identification of multidrug and toxin extrusion (MATE1 and MATE2-K) variants with complete loss of transport activity. *J. Hum. Genet.* 54, 40–46

57 Becker, M.L. *et al.* (2009) Genetic variation in the multidrug and toxin extrusion 1 transporter protein influences the glucose-lowering effect of metformin in patients with diabetes: a preliminary study. *Diabetes* 58, 745–749

58 Ha Choi, J. *et al.* (2009) Identification and characterization of novel polymorphisms in the basal promoter of the human transporter, MATE1. *Pharmacogenet. Genomics* 19, 770–780

59 Meyer zu Schwabedissen, H.E. *et al.* (2010) Human multidrug and toxin extrusion 1 (MATE1/SLC47A1) transporter: functional characterization, interaction with OCT2 (SLC22A2), and single nucleotide polymorphisms. *Am. J. Physiol. Renal Physiol.* 298, F997–F1005

60 Zhou, M. *et al.* (2007) Metformin transport by a newly cloned proton-stimulated organic cation transporter (plasma membrane monoamine transporter) expressed in human intestine. *Drug Metab. Dispos.* 35, 1956–1962

61 Xia, L. *et al.* (2007) Membrane localization and pH-dependent transport of a newly cloned organic cation transporter (PMAT) in kidney cells. *Am. J. Physiol. Renal Physiol.* 292, F682–F690

62 Le Floch, R. *et al.* (2012) Growth inhibition of glycolytic tumors by targeting basigin/lactate-H+ symporters (MCTs): metformin sensitizes MCT inhibition. In *Proceedings of the 103rd Annual Meeting of the American Association for Cancer Research*, pp. 72, American Association for Cancer Research

63 Pinheiro, C. *et al.* (2012) Role of monocarboxylate transporters in human cancers: state of the art. *J. Bioenerg. Biomembr.* 44, 127–139

64 Jablonski, K.A. *et al.* (2010) Common variants in 40 genes assessed for diabetes incidence and response to metformin and lifestyle intervention in the diabetes prevention program. *Diabetes* 59, 2672–2681

65 Legro, R.S. *et al.* (2008) Ovulatory response to treatment of polycystic ovary syndrome is associated with a polymorphism in the STK11 gene. *J. Clin. Endocrinol. Metab.* 93, 792–800

66 Ballestero, M.R. *et al.* (2006) Expression of transporters potentially involved in the targeting of cytostatic bile acid derivatives to colon cancer and polyps. *Biochem. Pharmacol.* 72, 729–738

67 Kitada, N. *et al.* (2008) Factors affecting sensitivity to antitumor platinum derivatives of human colorectal tumor cell lines. *Cancer Chemother. Pharmacol.* 62, 577–584

68 Aoki, M. *et al.* (2008) Kidney-specific expression of human organic cation transporter 2 (OCT2/SLC22A2) is regulated by DNA methylation. *Am. J. Physiol. Renal Physiol.* 295, F165–F170

69 Minematsu, T. and Giacomini, K.M. (2011) Interactions of tyrosine kinase inhibitors with organic cation transporters and multidrug and toxic compound extrusion proteins. *Mol. Cancer Ther.* 10, 531–539

70 Kerb, R. and Schwab, M. (2010) *Impact of pharmacogenetics on drug-drug interactions. Enzyme- and Transporter-Based Drug-Drug Interactions: Progress and Future Challenges*, Springer. pp. 51–74

71 Nies, A.T. *et al.* (2012) Multidrug and toxin extrusion (MATE) proteins as transporters of antimicrobial drugs. *Expert Opin. Drug Metab. Toxicol.* 8, 1565–1577

72 Nies, A.T. *et al.* (2011) Proton pump inhibitors inhibit metformin uptake by organic cation transporters (OCTs). *PLoS ONE* 6, e22163

73 Yee, S.W. *et al.* (2012) The role of ATM in response to metformin treatment and activation of AMPK. *Nat. Genet.* 44, 359–360

74 Pan, J. *et al.* (2012) Differential impact of structurally different anti-diabetic drugs on proliferation and chemosensitivity of acute lymphoblastic leukemia cells. *Cell Cycle* 11, 2314–2326

75 Shi, W-Y. *et al.* (2012) Therapeutic metformin/AMPK activation blocked lymphoma cell growth via inhibition of mTOR pathway and induction of autophagy. *Cell Death Dis.* 3, e275

76 Pollak, M.N. (2012) Investigating metformin for cancer prevention and treatment: the end of the beginning. *Cancer Discov.* 2, 778–790

77 Meyer, U.A. *et al.* (2012) Omics and drug response. *Annu. Rev. Pharmacol. Toxicol.* http://dx.doi.org/10.1146/annurev-pharmtox-010510-100502

78 Segal, E.D. *et al.* (2011) Relevance of the OCT1 transporter to the antineoplastic effect of biguanides. *Biochem. Biophys. Res. Commun.* 414, 694–699

79 Decensi, A. *et al.* (2010) Metformin and cancer risk in diabetic patients: a systematic review and meta-analysis. *Cancer Prev. Res.* 3, 1451–1461

80 Johnson, J. and Bowker, S.L. (2011) Intensive glycaemic control and cancer risk in type 2 diabetes: a meta-analysis of major trials. *Diabetologia* 54, 25–31

81 Col, N.F. *et al.* (2012) Metformin and breast cancer risk: a meta-analysis and critical literature review. *Breast Cancer Res. Treat.* http://dx.doi.org/10.1007/s10549-012-2170-x

82 Noto, H. *et al.* (2012) Cancer Risk in diabetic patients treated with metformin: a systematic review and meta-analysis. *PLoS ONE* 7, e33411

83 Soranna, D. *et al.* (2012) Cancer risk associated with use of metformin and sulfonylurea in type 2 diabetes: a meta-analysis. *Oncologist* 17, 813–822

84 Zhang, Z-J. *et al.* (2011) Reduced risk of colorectal cancer with metformin therapy in patients with type 2 diabetes: a meta-analysis. *Diabetes Care* 34, 2323–2328

85 Zhang, Z-J. *et al*. (2012) Metformin for liver cancer prevention in patients with type 2 diabetes: a systematic review and meta-analysis. *J. Clin. Endocrinol. Metab.* 97, 1–7

86 Chlebowski, R.T. *et al*. (2012) Diabetes, metformin, and breast cancer in postmenopausal women. *J. Clin. Oncol.* 30, 2844–2852

87 Nobes, J.P. *et al*. (2012) A prospective, randomized pilot study evaluating the effects of metformin and lifestyle intervention on patients with prostate cancer receiving androgen deprivation therapy. *BJU Int.* 109, 1495–1502

88 Song, I.S. *et al*. (2008) Genetic variants of organic cation transporter 2 (OCT2) significantly reduce metformin uptake in oocytes. *Xenobiotica* 38, 1252–1262

89 Leabman, M.K. *et al*. (2002) Polymorphisms in a human kidney xenobiotic transporter, OCT2, exhibit altered function. *Pharmacogenetics* 12, 395–405

90 Toyama, K. *et al*. (2010) Heterozygous variants of multidrug and toxin extrusions (MATE1 and MATE2-K) have little influence on the disposition of metformin in diabetic patients. *Pharmacogenet. Genomics* 20, 135–138

91 Cutler, M.J. and Choo, E.F. (2011) Overview of SLC22A and SLCO families of drug uptake transporters in the context of cancer treatments. *Curr. Drug Metab.* 12, 793–807

92 Heise, M. *et al*. (2012) Downregulation of organic cation transporters OCT1 (SLC22A1) and OCT3 (SLC22A3) in human hepatocellular carcinoma and their prognostic significance. *BMC Cancer* 12, 109–119

93 Merezhinskaya, N. and Fishbein, W.N. (2009) Monocarboxylate transporters: past, present, and future. *Histol. Histopathol.* 24, 243–264

94 Engel, K. *et al*. (2004) Identification and characterization of a novel monoamine transporter in the human brain. *J. Biol. Chem.* 279, 50042–50049

95 Duan, H. and Wang, J. (2010) Selective transport of monoamine neurotransmitters by human plasma membrane monoamine transporter and organic cation transporter 3. *J. Pharmacol. Exp. Ther.* 335, 743–753

96 Shu, Y. *et al*. (2003) Evolutionary conservation predicts function of variants of the human organic cation transporter, OCT1. *Proc. Natl. Acad. Sci. U.S.A.* 100, 5902–5907

97 Kerb, R. *et al*. (2002) Identification of genetic variations of the human organic cation transporter hOCT1 and their functional consequences. *Pharmacogenetics* 12, 591–595

98 Shu, Y. *et al*. (2008) Effect of genetic variation in the organic cation transporter 1, OCT1, on metformin pharmacokinetics. *Clin. Pharmacol. Ther.* 83, 273–280

99 Kang, H-J. *et al*. (2007) Identification and functional characterization of genetic variants of human organic cation transporters in a Korean population. *Drug Metab. Dispos.* 35, 667–675

100 Shikata, E. *et al*. (2007) Human organic cation transporter (OCT1 and OCT2) gene polymorphisms and therapeutic effects of metformin. *J. Hum. Genet.* 52, 117–122

ends in Molecular Medicine

Curing APL through PML/RARA Degradation by As$_2$O$_3$

Valerie Lallemand-Breitenbach[1,2,3], Jun Zhu[3,4], Zhu Chen[4,5], Hugues de Thé[1,2,3,4,6,*]

[1]University Paris Diderot, Sorbonne Paris Cité, Hôpital St Louis 1, Avenue Claude Vellefaux, 75475 Paris, Cedex 10, France, [2]INSERM UMR 944, Equipe labellisée par la Ligue Nationale contre le Cancer, Institut Universitaire d'Hématologie, Hôpital St Louis 1, Avenue Claude Vellefaux, 75475 Paris, Cedex 10, France, [3]CNRS UMR 7212, Hôpital St Louis 1, Avenue Claude Vellefaux, 75475 Paris, Cedex 10, France, [4]Pole Sino-Francais des Sciences du Vivant et de Génomique de l'Hôpital Rui Jin, Rui-Jin Hospital affiliated with Jiao Tong University, 197 Rui Jin Road, Shanghai 200025, China, [5]State Key Laboratory for Medical Genomics, Shanghai Institute of Hematology, Rui-Jin Hospital affiliated with Jiao Tong University, 197 Rui Jin Road, Shanghai 200025, China, [6]AP-HP, Service de Biochimie, Hôpital St Louis 1, Avenue Claude Vellefaux, 75475 Paris, Cedex 10, France
*Correspondence: hugues.dethe@inserm.fr

Trends in Molecular Medicine, Vol. 18, No. 1, January 2012 © 2012 Elsevier Inc.
http://dx.doi.org/10.1016/j.molmed.2011.10.001

SUMMARY

Acute promyelocytic leukemia (APL) is a hematological malignancy driven by the PML/RARA oncogene. The prognosis for patients with APL was revolutionized by two treatments: retinoic acid (RA) and As$_2$O$_3$ (arsenic trioxide). These were both shown *a posteriori* to target PML/RARA, explaining their exquisite specificity for APL. Arsenic, as a single agent, cures up to 70% of patients, whereas APL patients treated with the combination of RA and As$_2$O$_3$ reach a stunning 90% cure rate. Recent physiopathological models highlight the key role of RA- and As$_2$O$_3$-triggered PML/RARA degradation, and the molecular mechanisms underlying As$_2$O$_3$-induced PML/RARA degradation have been recently clarified. As discussed below, arsenic binding, oxidation, sumoylation on PML nuclear bodies, and RNF4-mediated ubiquitination all contribute to the As$_2$O$_3$-triggered catabolism of PML/RARA.

CellPress

BOTH POISON AND CURE

If one thinks arsenic, poison from a Capra movie quickly comes to mind. However, it is one of the oldest drugs known to man, continuously used from antiquity until the 19th century to treat a wide variety of diseases [1]. The chemical properties of arsenic are summarized in Box 1. The saga of arsenic trioxide (As_2O_3) as a cure for acute promyelocytic leukemia (APL) began in the 1970s with Ailing-1, a mix of As_2O_3 and trace amounts of mercury, which had a dramatic clinical efficacy. Pure arsenic was then shown to trigger complete remissions and to prolong survival in otherwise treatment-resistant APLs (reviewed in [2]). Intense research has been aimed ever since at understanding the molecular basis of arsenic therapy for APL. Recent breakthroughs have provided significant insights on how this is achieved and why arsenic is so efficient and so specific to this disease [3–5].

In the past, APL was an acute myeloid leukemia with a very severe prognosis [6]. Historically, introduction of retinoic acid (RA), its association with chemotherapy, the use of arsenic in relapse patients and finally the frontline use of the arsenic/RA combination has radically changed this poor outcome. Currently, up to 90% of APL patients are off-treatment and

BOX 1 ARSENIC CHEMISTRY, USE AND TOXICITY

Arsenic is a metalloid with three oxidation states: V (pentoxide As_2O_5, H_3AsO_4), III (trioxide As_2O_3, H_3AsO_3, $HAsO_2$), and 0 (arsine H_3As). As_2O_3 is a white, water-soluble powder. It can be generated by oxidation of other natural arsenic compounds such as As_2S_3 (orpiment), a mineral used as a yellow pigment. In humans, acute ingestions from 8 g to 20 g were reported to be lethal. As_2O_3 is efficiently absorbed (80%), but rapidly eliminated, with 60% to 70% excreted in urine after methylation or further oxidation. The toxic effects of arsenic are due to interactions with protein thiols, inhibiting cellular oxidative processes and leading to ROS production. It therefore accumulates in keratin-rich tissues. Gustave Flaubert perfectly described the acute toxic lethal effects in "Madame Bovary" (1857).

Currently, arsenic trioxide is used as pesticides (wood treatments), herbicides, and for glass or electronic manufacturing. Arsenic may be naturally found in soils, in particular in volcanic rocks, and thus in water. Chronic exposures lead to renal failure, skin, liver and cardiovascular disorders. In 1996, the World Health Organization set the upper acceptable concentration of arsenic in drinking water at 10 µg/L, although levels over 500 µg/L may be found in some wells, notably in Asia where water pollution represents a major health challenge [65]. Arsenic contamination of water yields a higher-than-normal incidence of cancers and vascular problems, notably micro-arterial obstruction that leads to "blackfoot disease".

Upon As_2O_3 treatment of APL patients (intravenously, maximal 10 mg/day), only a few side effects have been described. These include liver dysfunction and some cardiovascular dysrhythmia, possibly in predisposed individuals [66]. These can be overcome by symptomatic treatments, notably the administration of the precursor of glutathione, N-acetylcysteine. Hyperleukocytosis, an abnormally high number of leukocytes, due to the differentiation syndrome, has also been observed [41]. Importantly, recent analyses have demonstrated that As_2O_3-treated APL patients do not show evidence of long-term toxicity and that arsenic levels return with the normal range after treatment [31,32,47].

disease-free at 5 years, generally synonymous with a definitive cure [6–8]. In low-risk patients, APL eradication may even be achieved without any DNA-damaging chemotherapy [9]. This miracle was explained by the ability of these two agents to directly target PML/RARA, the oncoprotein that drives APL transformation [10,11]. Amusingly, the targeted nature of these two agents was discovered *a posteriori*, explaining both their very high efficacy and specificity.

Several recent reviews have extensively covered the historical timeline or mechanistic aspect of transformation [2,6,10]. Briefly, PML/RARA is a multifaceted protein resulting from the fusion of the retinoic acid receptor alpha gene (RARA) a nuclear receptor regulating transcription, and PML, a redox-sensitive protein that organizes PML nuclear bodies (NBs) (see Glossary). A translocation between chromosomes 15 and 17, the t(15;17) translocation, drives the expression of PML/RARA and is the only constant genomic abnormality, although some other lesions, shared with many other leukemias or malignancies, have been implicated in tumor progression [12–14]. Hence, APL can be considered a monogenic cancer whose growth is largely, if not exclusively, PML/RARA-driven.

As a nuclear receptor, RARA inhibits transcription of target genes in the absence of its ligand RA, but dramatically activates transcription when RA-bound. PML/RARA globally interferes with transcriptional control by RARA, further repressing target gene expression even with physiological RA levels (Figure 1). RARA was implicated in the tuning of myeloid differentiation, so PML/RARA-mediated transcriptional repression of RARA signaling could explain the differentiation block. PML/RARA also represses several non-RARA target genes through its ability to form homodimers, which display highly degenerated binding site specificity [15,16]. Finally, through its ability to heterodimerize with PML, PML/RARA disrupts PML NBs [17]. Among others, NBs were associated with tuning of the P53 response through concentration and/or modification of partners such as MDM2 or CBP [18,19]. Enhanced self-renewal was observed upon PML loss or NB disruption [20,21], possibly reflecting deregulation of P53 signaling because P53 was implicated in self-renewal of normal or cancer stem cells [22,23]. These pathways probably cooperate to enforce the APL-specific differentiation block and acquisition of self-renewal, thus transforming a committed hematopoietic progenitor into an immortal, fully transformed cell.

APL RESPONSE TO THERAPIES

The first targeted therapy of APL was RA, which at pharmacological concentrations induces APL differentiation *in vivo*, yielding transient clinical

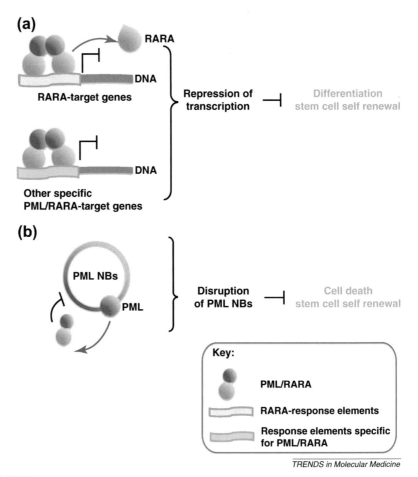

FIGURE 1

Schematic model of APL pathogenesis.

(a) PML/RARA dimers displace RARA from its binding sites or bind *de novo* sites, resulting in transcriptional repression, the differentiation block and probably enhancing self-renewal. (b) PML/RARA also delocalize PML off PML bodies, with a proposed role in cell death and self-renewal.

remissions. With very high RA levels, PML/RARA may behave as a *bona fide* transcriptional activator [24], although it is generally much less effective than RARA at activating transcription. RA also induces proteasome-mediated degradation of PML/RARA [25]. Recent studies have suggested that RA may clear PML/RARA from promoters and allow RARA to perform most of the transcriptional activation on its own particular targets [26]. The respective contributions of PML/RARA-mediated transcriptional activation and degradation to APL differentiation and clearance remains debated. PML/RARA degradation secondarily allows reformation of PML NBs, with a possible modulation of APL clonogenic activity. Indeed, evidence has suggested that loss of clonogenic activity through self-renewal of leukemia-initiating cells may be linked to PML/RARA degradation, whereas transcriptional activation accounted for differentiation [27–29]. Yet, it remains to be determined if a positive signal of transcriptional reactivation by RA is absolutely required or if the mere loss of PML/RARA and subsequent derepression may suffice for differentiation (see below).

Although very few patients with APL have been cured by RA without any DNA-damaging chemotherapy [30], different trials have demonstrated 70% cure rates with the single agent As_2O_3 [31,32]. Thus, arsenic is considerably more potent that RA for APL eradication, although its ability to differentiate APL cells is only modest. Multiple biochemical mechanisms were proposed to account for arsenic-induced apoptosis, including activation of stress signal transduction, mitochondria toxicity, and oxidative stress [33]. At high concentrations, arsenic triggers apoptosis in many transformed cells, raising hopes that it might be clinically relevant in other cancers [34]. Unfortunately, most clinical trials have been disappointing, as the extraordinary potency of arsenic currently appears limited to APL (with the notable exception of adult T cell leukemia discussed below).

This exquisite sensitivity for APL argued against pleiotropic toxic cellular effects of arsenic being directly responsible for its therapeutic effects. In striking parallel with RA, arsenic triggers PML/RARA proteolysis at therapeutic concentrations by targeting the PML moiety [35,36]. A major confirmation for the specific targeting of PML was the demonstration that rare variant APLs driven by the PLZF/RARA fusion are completely insensitive to arsenic in mouse models of APL [37,38].

A KEY ROLE OF PML/RARA DEGRADATION IN THERAPY RESPONSE

The historical, simple model whereby transcriptional reactivation of target genes clears the disease through differentiation fails to satisfactorily explain why RA must be combined with chemotherapy to cure patients [39], nor why single-agent As_2O_3 therapy cures most patients without direct transcriptional activation. Some intriguing observations have recently suggested that loss of repression may actually suffice for differentiation in vivo. Indeed, arsenic, which does not activate RARA or PML/RARA-mediated transcriptional regulation [40], induces delayed terminal differentiation in both patients and animal models [10,37,41], raising the possibility that differentiation also may result from the loss of PML/RARA. Ex vivo, arsenic-induced terminal differentiation is also observed in the presence of survival signals, such as cyclic AMP (cAMP) or granulocyte macrophage-colony stimulating factor (GM-CSF) [42]. Moreover, unpublished data have demonstrated that excision of the obligatory RXRA cofactor of both RARA and PML/RARA [16,43] paradoxically induce rapid and terminal differentiation in the absence of any external RA (J. Haftermeyer, unpublished). This probably reflects detachment of the PML/RARA–RXRA complex from DNA, resulting in derepression of targets. These observations suggest that clearance of PML/RARA from its target promoters may actually suffice to trigger differentiation.

Since our initial proposal, the idea that PML/RARA degradation may underlie therapy response to both RA and arsenic has progressively gained significant support from several studies in which mouse models played an important role [10,11,28,29]. Distinct pathways converge to enforce PML/RARA degradation (Figure 2). As_2O_3 targets the PML portion of the fusion protein and specifically induces a SUMO-dependent, ubiquitin-mediated degradation (Figure 2a). RA transactivation is directly coupled to RARA degradation in a feedback mechanism conserved for all nuclear receptors (Figure 2b) [25], whereas PML/RARA may also be cleaved by the differentiation-activated neutrophil elastase protease (Figure 2c). Finally, a cytoplasmic PML/RARA catabolic pathway that relies on autophagy is activated upon exposure to both RA and arsenic (Figure 2d) [44]. Loss of PML/RARA as the sole property

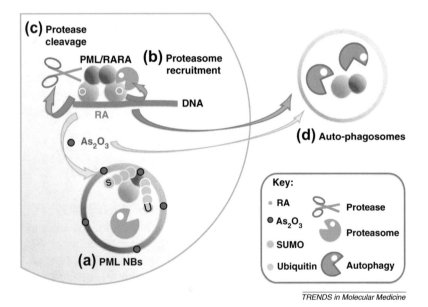

TRENDS in Molecular Medicine

FIGURE 2

Four distinct pathways enforce PML/RARA degradation.

As_2O_3 triggers PML/RARA degradation by the proteasome in NBs, via the PML portion and the SUMO-mediated/ubiquitin-dependent RNF4 pathway **(a)**. RA activates PML/RARA degradation by nuclear receptor negative feedback through proteasome interaction with the RARA portion **(b)**. RA also induces cleavage by the protease neutrophile elastase **(c)**. Both RA and As_2O_3 target PML/RARA to autophagosomes **(d)**, where the fusion oncoprotein may be degraded.

shared by RA and arsenic, together with a key role for PML/RARA in pathogenesis, supports a degradation-based mechanism for this APL cure.

Models based on degradation rather than direct transcriptional activation carried an important prediction: synergy of the two drugs for disease eradication. Models based on transcriptional reactivation predicted antagonism, which was indeed observed for differentiation [45]. Yet, a dramatic synergy between RA and As_2O_3 in APL therapy was demonstrated in animal models [37,38,46]. Because RA and As_2O_3 induce PML/RARA degradation by different mechanisms, they should not present with cross-resistance. Accordingly, treatments combining RA and As_2O_3 have shown synergistic effects, with more than 90% of the patients definitively cured [6,8–10,47].

HOW DOES ARSENIC INDUCE DEGRADATION OF PML AND PML/RARA?

Whereas RA primarily targets PML/RARA for degradation through the RARA moiety, As_2O_3 induces PML/RARA degradation by targeting the PML moiety. Accordingly, arsenic also degrades the normal PML protein(s). The biochemical pathways involved are largely the same for PML and PML/RARA. Yet, it took over 10 years to dissect these pathways, largely because they were unconventional [48]. Transfer of the nuclear diffuse fraction of PML onto nuclear matrix-associated PML NBs precedes degradation [35]. After transfer, PML is conjugated by SUMO and a single arsenic-sensitive sumoylation site initiates arsenic-induced proteolysis [49]. However, the fact that PML sumoylation could trigger the degradation process faced considerable resistance, at least in part because in other systems SUMO and ubiquitin may compete for the same conjugation site. Two independent studies then showed that polysumoylated PML was first recognized by the SUMO-dependent ubiquitin ligase RNF4 and then polyubiquitinated and degraded by the proteasome [50,51], identifying PML as the first protein degraded by this novel catabolic pathway [48] (Box 2). Moreover, dominant-negative RNF4 impedes arsenic-induced differentiation *ex vivo*, supporting the importance of RNF4-mediated degradation in this biological response [50]. PML sumoylation, ubiquitination and degradation occur on NBs, suggesting that they may be privileged sites for proteolysis [50].

HOW DOES ARSENIC TRIGGER PML SUMOYLATION?

Although RNF4 SUMO-dependent ubiquitin ligase activity explained the molecular basis for degradation of some SUMO-conjugated proteins, the

BOX 2 SUMO-TARGETED UBIQUITIN LIGASES: HOW YEAST GENETICS EXPLAINED AS$_2$O$_3$-INDUCED PML DEGRADATION

RNF4 is the central enzyme for As$_2$O$_3$-induced PML or PML/RARA degradation. It was first identified as a RING finger protein regulating transcription, later revealed to have ubiquitin E3 ligase activity [67,68]. While the existence of crosstalk between SUMO and ubiquitin was being questioned, yeast genetics came to the rescue as two reports found evidence for SUMO-mediated protein degradation [69,70]. Rfp1 and Rfp2 in *Schizosaccharomyces pombe* and Slx5 in *Saccharomyces cerevisiae* are RNF4 orthologs. These proteins lack direct E3 ligase activity but recruit Slx8, an active RING finger ubiquitin ligase, to form a functional complex with ubiquitin-conjugating activity, which specifically ubiquitinylates artificial SUMO-containing substrates *in vitro*. Rfp/Slx8-deficient cells accumulate sumoylated targets. However, yeast studies neither identified any direct substrates for these complexes nor unambiguously demonstrated proteasomal degradation of SUMO conjugates. PML was thus the first identified target for this new catabolism pathway. Yeast and *Drosophila* SUMO-dependent ubiquitin ligases are essential for genome stability, a biological function also proposed for NBs and SUMO in mammalian cells.

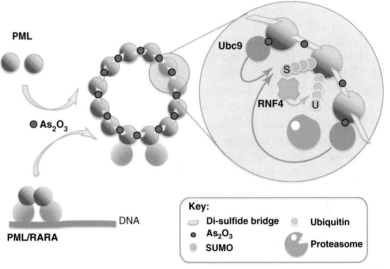

TRENDS in Molecular Medicine

FIGURE 3

As$_2$O$_3$ targets PML and PML/RARA to induce their degradation.

As$_2$O$_3$ induces PML and PML/RARA covalent dimerization by disulfide bridges and direct crosslinking. PML or PML/RARA proteins then assemble in NBs. Right scheme is a zoom on a portion of NBs describing PML sumoylation *in trans* by arsenic-enhanced UBC9 interaction, RNF4-mediated ubiquitination and proteasome degradation within the body.

initiating role of arsenic to trigger PML sumoylation remained mysterious. Two parallel studies recently clarified the molecular basis for arsenic-induced PML sumoylation (Figure 3) and led to two mechanistic insights. The first mechanism is based on the observation that arsenic directly binds

BOX 3 AS₂O₃, A TOOL FOR UNDERSTANDING PML NB BIOGENESIS

PML NBs are multiprotein complexes organized in spheres with a shell of PML and an inner core that accumulates a wide variety of partner proteins [17,64]. Different models were proposed for the formation of NBs, with sumoylation having an important role, either for PML assembling or PML partner recruitments [49,71]. PML NBs belong to the nuclear matrix, a biochemically defined nuclear skeleton resistant to combined high salt and nuclease extractions.

Arsenic was a very powerful tool to dissect NB biogenesis. PML sumoylation was first proposed to underlie NB formation [71,72], yet other studies demonstrated that NB formation does not require PML conjugation to SUMO [49]. Analysis of wild-type or mutant PML subcellular localization, matrix attachment and sumoylation in response to As₂O₃ supports a model where PML undergoes covalent, disulfide-mediated,

crosslink upon oxidative stress, generating the matrix feature of the PML mesh [3]. Thus, oxidation facilitates NB nucleation, which subsequently triggers PML sumoylation and degradation, but direct arsenic binding onto PML could also favor its sumoylation *in vivo* [4]. Experiments conducted *in vivo* with non-arsenical oxidants have similarly shown that NBs assemble in response to oxidant stress [3]. This discovery carries important consequences for the physiological role of PML NBs [17] and explains the tight association of PML NBs *in vivo* with sites of inflammation or transformation. The general function of PML may be to sense the cellular redox balance and sequester partner proteins in response. Such redox-regulated protein sequestration [17] may explain the pleiotropic functions attributed to PML and NBs because many PML-modulated pathways are also sensitive to redox stress [17,64].

PML [3,4]. At least one of the binding sites is located in the RING domain of PML, where arsenic may substitute for zinc ions that coordinate the cysteines. This results in topological changes that allow a tighter binding of the SUMO-conjugating enzyme UBC9, explaining why arsenic enhances PML sumoylation. The second proposed mechanism derives from an attempt to understand the basis for NB organization [3] (Box 3). Through its non-specific oxidative properties [52], As₂O₃ induces intermolecular disulfide bridges that result in multimerization of PML proteins into a mesh that forms NBs. As₂O₃ also binds directly to PML box B2, which harbors two adjacent cysteine residues that are the canonical binding site for arsenic. Following this As₂O₃-induced NB assembly, the mesh formed by PML multimers efficiently recruits the SUMO ligase UBC9, probably resulting in hypersumoylation of PML *in trans* (Figure 3).

In strong support of an important role for this second mechanism in the clinical APL response to arsenic are the following facts: (i) paraquat, a strong inducer of oxidative stress *in vivo* induces not only NB formation but also APL regression [3]; (ii) a point mutant in the box B2 dicysteine motif abrogates the arsenic response *ex vivo* [3]; (iii) mutations surrounding the dicysteine motif in box B2 have been observed in two arsenic-resistant patients [53]; and (iv) the vitamin E derivative α-TOS promotes degradation of PML/RARA both *ex vivo* and *in vivo* [54] and dramatically prolongs survival of APL mice [55]. Without dismissing an important role for direct arsenic binding, these observations imply that PML oxidation significantly contributes to the biological response triggered by arsenic on cells expressing PML/RARA.

WHAT IS THE ROLE FOR PML NBS IN APL THERAPY?

Because arsenic targets PML and PML/RARA, both of which are expressed in APL cells, the normal PML allele may facilitate PML/RARA degradation and/or loss of self-renewal of APL cells after PML/RARA degradation. Arsenic may degrade PML/RARA in *pml*−/− cells, although the kinetics of degradation are significantly delayed [3]. Should loss of self-renewal reflect activation of a signal downstream of PML bodies, one would expect the normal PML allele to be important. Indeed, a study has reported some alterations of RA response in PML/RARA transgenics on a *pml*−/− background [56]. Despite a report of rare PML mutations in APL patients [57], whether the loss or mutation of PML impacts on the therapy response remains to be explored.

ARE THERE OTHER ARSENIC-SENSITIVE DISEASES?

The sensitivity of APL to arsenic therapy probably reflects the fact that PML is a sensor for oxidative stress and is thus exquisitely reactive to any disturbance of the redox balance. Yet, there are some hints that arsenic may have other clinical utilities.

Several studies have demonstrated a general role for PML in the self-renewal of cancer cells. BCR/ABL-transformed cells on a *pml*−/− background cannot be eternally passaged from mouse to mouse [20]. Arsenic treatment somehow recapitulates this phenotype, raising the possibility that arsenic, by opposing PML-driven self-renewal of cancer cells, may be a general adjuvant to many anticancer drugs. Arsenicals were also shown to downregulate BCR/ABL protein levels, resulting in a greater sensitivity to kinase inhibitors [58], which may have important clinical implications. This was mechanistically explained by the stabilization of the c-CBL ubiquitin ligase, which degrades BCR/ABL [59].

A second example is adult T cell leukemia, a disease tightly associated with infection by the HTLV-I oncoretrovirus [60]. The virus expresses a potent transactivator, Tax, which behaves as an oncoprotein, activating not only the viral promoters but also many cellular genes involved in proliferation. Arsenic, in the presence of interferon α, induces Tax degradation, resulting in the selective death of ATL cells [61]. Interestingly, this interferon/arsenic combination may cure Tax-driven murine ATL, through immediate loss of self-renewal [62], as previously shown for the RA/arsenic combination in APL [27]. The molecular mechanisms may implicate PML NB-associated proteolysis [11]. Importantly, this Tax-targeted therapeutic approach to ATL has shown some success in the clinic [63]. Thus, arsenic may contribute to the degradation of other oncogenes, with resulting clinical benefit.

CONCLUDING REMARKS

The arsenic saga is the only model of an oncogene-targeted, definitive cure of leukemia [6,10]. It is amusing that such a primitive drug, known and used for over 3000 years [1], can end up being targeted to the driving PML/RARA oncoprotein. One should also remember that the development of the drug and determination of the appropriate treatment protocols was primarily, if not entirely, academic-driven. This bears important lessons to leave some room in clinical research for trials based on biological hypotheses and preclinical models, notably those using existing drugs approved for other purposes.

From a cell biology point of view, arsenic allowed dissection of the biogenesis of PML NBs, as well as some insight into their functions [17] (Box 3). These intriguing domains, so exquisitely sensitive to arsenic, have been implicated in a plethora of biological regulations [64]. In the emerging field of biological regulation by redox balance, arsenic has pinpointed PML as an unsuspected player, providing a striking illustration of chemical biology enlightening very basic biological issues.

ACKNOWLEDGMENTS

We apologize to many friends and colleagues whose work could not be cited owing to a lack of space. Work in the Paris laboratory is supported by Ligue Nationale contre le Cancer, INSERM, CNRS, University Paris Diderot, Institut Universitaire de France, Institut National du Cancer, Association pour la Recherche contre le Cancer (ARC), European Research Council (Senior grant 268729 – STEMAPL to H.d.T.) and Canceropole programs. Work in Shanghai is supported by grants from National Natural Science Foundation of China (NSFC), the Key Discipline Program of Shanghai Municipality Shanghai Municipal Education Committee (SMEC) and Shanghai Municipal Committee for Science and Technology (SMCST) programs. We thank all lab members for helpful discussions and continuous support.

REFERENCES

1 Zhu, J. et al. (2002) How acute promyelocytic leukemia revived arsenic. Nat. Rev. Cancer 2, 705–713

2 Chen, S.J. et al. (2011) From an old remedy to a magic bullet: molecular mechanisms underlying the therapeutic effects of arsenic in fighting leukemia. Blood 117, 6425–6437

3 Jeanne, M. et al. (2010) PML/RARA oxidation and arsenic binding initiate the antileukemia response of As_2O_3. Cancer Cell 18, 88–98

4 Zhang, X.W. et al. (2010) Arsenic trioxide controls the fate of the PML-RARalpha oncoprotein by directly binding PML. Science 328, 240–243

5 Kogan, S.C. (2010) Medicine. Poisonous contacts. Science 328, 184–185

6 Wang, Z.Y. and Chen, Z. (2008) Acute promyelocytic leukemia: from highly fatal to highly curable. Blood 111, 2505–2515

7 Shen, Z.X. et al. (2004) All-trans retinoic acid/As_2O_3 combination yields a high quality remission and survival in newly diagnosed acute promyelocytic leukemia. Proc. Natl. Acad. Sci. U.S.A. 101, 5328–5335

8   Tallman, M.S. and Altman, J.K. (2009) How I treat acute promyelocytic leukemia. *Blood* 114, 5126–5135

9   Estey, E. *et al*. (2006) Use of all-trans retinoic acid plus arsenic trioxide as an alternative to chemotherapy in untreated acute promyelocytic leukemia. *Blood* 107, 3469–3473

10 de The, H. and Chen, Z. (2010) Acute promyelocytic leukaemia: novel insights into the mechanisms of cure. *Nat. Rev. Cancer* 10, 775–783

11 Ablain, J. *et al*. (2011) Oncoprotein proteolysis, an unexpected Achille's heel of cancer cells? *Cancer Discov.* 1, 117–127

12 Akagi, T. *et al*. (2009) Hidden abnormalities and novel classification of t(15;17) acute promyelocytic leukemia (APL) based on genomic alterations. *Blood* 113, 1741–1748

13 Wartman, L.D. *et al*. (2011) Sequencing a mouse acute promyelocytic leukemia genome reveals genetic events relevant for disease progression. *J. Clin. Invest.* 121, 1445–1455

14 Jones, L. *et al*. (2010) Gain of MYC underlies recurrent trisomy of the MYC chromosome in acute promyelocytic leukemia. *J. Exp. Med.* 207, 2581–2594

15 Kamashev, D.E. *et al*. (2004) PML/RARA-RXR oligomers mediate retinoid- and rexinoid/cAMP in APL cell differentiation. *J. Exp. Med.* 199, 1–13

16 Martens, J.H. *et al*. (2010) PML-RARalpha/RXR alters the epigenetic landscape in acute promyelocytic leukemia. *Cancer Cell* 17, 173–185

17 Lallemand-Breitenbach, V. and de The, H. (2010) PML nuclear bodies. *Cold Spring Harb. Perspect. Biol.* 2, a000661

18 Gottifredi, V. and Prives, C. (2001) P53 and PML: new partners in tumor suppression. *Trends Cell Biol.* 11, 184–187

19 Pearson, M. *et al*. (2000) PML regulates p53 acetylation and premature senescence induced by oncogenic Ras. *Nature* 406, 207–210

20 Ito, K. *et al*. (2008) PML targeting eradicates quiescent leukaemia-initiating cells. *Nature* 453, 1072–1078

21 Regad, T. *et al*. (2009) The tumor suppressor Pml regulates cell fate in the developing neocortex. *Nat. Neurosci.* 12, 132–140

22 Zhao, Z. *et al*. (2010) p53 loss promotes acute myeloid leukemia by enabling aberrant self-renewal. *Genes Dev.* 24, 1389–1402

23 Cicalese, A. *et al*. (2009) The tumor suppressor p53 regulates polarity of self-renewing divisions in mammary stem cells. *Cell* 138, 1083–1095

24 Zhou, J. *et al*. (2006) Dimerization-induced corepressor binding and relaxed DNA-binding specificity are critical for PML/RARA-induced immortalization. *Proc. Natl. Acad. Sci. U.S.A.* 103, 9238–9243

25 Zhu, J. *et al*. (1999) Retinoic acid induces proteasome-dependent degradation of retinoic acid receptor alpha (RAR alpha) and oncogenic RAR alpha fusion proteins. *Proc. Natl. Acad. Sci. U.S.A.* 96, 14807–14812

26 Cassinat, B. *et al*. (2011) New role for granulocyte colony-stimulating factor-induced extracellular signal-regulated kinase 1/2 in histone modification and retinoic acid receptor alpha recruitment to gene promoters: relevance to acute promyelocytic leukemia cell differentiation. *Mol. Cell. Biol.* 31, 1409–1418

27 Nasr, R. *et al*. (2008) Eradication of acute promyelocytic leukemia-initiating cells through PML-RARA degradation. *Nat. Med.* 14, 1333–1342

28 Kogan, S.C. (2009) Curing APL: differentiation or destruction? *Cancer Cell* 15, 7–8

29 Ablain, J. and de The, H. (2011) Revisiting the differentiation paradigm in acute promyelocytic leukemia. *Blood* 117, 5795–5802

30 Tsimberidou, A.M. *et al*. (2006) Single-agent liposomal all-trans retinoic acid can cure some patients with untreated acute promyelocytic leukemia: an update of The University of Texas M.D. Anderson Cancer Center Series. *Leuk. Lymphoma* 47, 1062–1068

31 Mathews, V. *et al*. (2010) Single-agent arsenic trioxide in the treatment of newly diagnosed acute promyelocytic leukemia: long-term follow-up data. *J. Clin. Oncol.* 28, 3866–3871

32 Ghavamzadeh, A. *et al*. (2011) Phase II study of single-agent arsenic trioxide for the front-line therapy of acute promyelocytic leukemia. *J. Clin. Oncol.* 29, 2753–2757

33 Miller, W.H., Jr *et al*. (2002) Mechanisms of action of arsenic trioxide. *Cancer Res.* 62, 3893–3903

34 Zhu, X.H. *et al*. (1999) Apoptosis and growth inhibition in malignant lymphocytes after treatment with arsenic trioxide at clinically achievable concentrations. *J. Natl. Cancer Inst.* 91, 772–778

35 Zhu, J. *et al*. (1997) Arsenic-induced PML targeting onto nuclear bodies: implications for the treatment of acute promyelocytic leukemia. *Proc. Natl. Acad. Sci. U.S.A.* 94, 3978–3983

36 Chen, G.Q. *et al*. (1997) Use of arsenic trioxide (As_2O_3) in the treatment of acute promyelocytic leukemia (APL): I. As_2O_3 exerts dose-dependent dual effects on APL cells. *Blood* 89, 3345–3353

37 Lallemand-Breitenbach, V. *et al*. (2005) Opinion: how patients have benefited from mouse models of acute promyelocytic leukaemia. *Nat. Rev. Cancer* 5, 821–827

38 Rego, E.M. *et al*. (2000) Retinoic acid (RA) and As_2O_3 treatment in transgenic models of acute promyelocytic leukemia (APL) unravel the distinct nature of the leukemogenic process induced by the PML-RARalpha and PLZF-RARalpha oncoproteins. *Proc. Natl. Acad. Sci. U.S.A.* 97, 10173–10178

39 Sanz, M.A. *et al*. (2009) Management of acute promyelocytic leukemia: recommendations from an expert panel on behalf of the European LeukemiaNet. *Blood* 113, 1875–1891

40 Zheng, P.Z. *et al*. (2005) Systems analysis of transcriptome and proteome in retinoic acid/arsenic trioxide-induced cell differentiation/apoptosis of promyelocytic leukemia. *Proc. Natl. Acad. Sci. U.S.A.* 102, 7653–7658

41 Camacho, L.H. *et al*. (2000) Leukocytosis and the retinoic acid syndrome in patients with acute promyelocytic leukemia treated with arsenic trioxide. *J. Clin. Oncol.* 18, 2620–2625

42 Guillemin, M.C. *et al*. (2002) In vivo activation of cAMP signaling induces growth arrest and differentiation in acute promyelocytic leukemia. *J. Exp. Med.* 196, 1373–1380

43 Zhu, J. *et al*. (2007) RXR is an essential component of the oncogenic PML/RARA complex in vivo. *Cancer Cell* 12, 23–35

44 Isakson, P. *et al*. (2010) Autophagy contributes to therapy-induced degradation of the PML/RARA oncoprotein. *Blood* 116, 2324–2331

45 Shao, W. *et al*. (1998) Arsenic trioxide as an inducer of apoptosis and loss of PML/RARalpha protein in acute promyelocytic leukemia cells. *J. Natl. Cancer Inst.* 90, 124–133

46 Jing, Y. *et al*. (2001) Combined effect of all-trans retinoic acid and arsenic trioxide in acute promyelocytic leukemia cells in vitro and in vivo. *Blood* 97, 264–269

47 Hu, J. *et al*. (2009) Long-term efficacy and safety of all-trans retinoic acid/arsenic trioxide-based therapy in newly diagnosed acute promyelocytic leukemia. *Proc. Natl. Acad. Sci. U.S.A.* 106, 3342–3347

48 Geoffroy, M.C. and Hay, R.T. (2009) An additional role for SUMO in ubiquitin-mediated proteolysis. *Nat. Rev. Mol. Cell Biol.* 10, 564–568

49 Lallemand-Breitenbach, V. *et al*. (2001) Role of promyelocytic leukemia (PML) sumolation in nuclear body formation, 11S proteasome recruitment, and As(2)O(3)-induced PML or PML/retinoic acid receptor alpha degradation. *J. Exp. Med.* 193, 1361–1372

50 Lallemand-Breitenbach, V. *et al*. (2008) Arsenic degrades PML or PML-RARalpha through a SUMO-triggered RNF4/ubiquitin-mediated pathway. *Nat. Cell Biol*. 10, 547–555

51 Tatham, M.H. *et al*. (2008) RNF4 is a poly-SUMO-specific E3 ubiquitin ligase required for arsenic-induced PML degradation. *Nat. Cell Biol*. 10, 538–546

52 Kawata, K. *et al*. (2007) Classification of heavy-metal toxicity by human DNA microarray analysis. *Environ. Sci. Technol*. 41, 3769–3774

53 Goto, E. *et al*. (2011) Missense mutations in PML-RARA critical for the lack of responsiveness to arsenic trioxide treatment. *Blood* 118, 1600–1609

54 Freitas, R.A. *et al*. (2009) Apoptosis induction by (+)alpha-tocopheryl succinate in the absence or presence of all-trans retinoic acid and arsenic trioxide in NB4, NB4-R2 and primary APL cells. *Leuk. Res*. 33, 958–963

55 Dos Santos, G.A. *et al*. (2011) (+)alpha-Tocopheryl succinate inhibits the mitochondrial respiratory chain complex I and is as effective as arsenic trioxide or ATRA against acute promyelocytic leukemia in vivo. *Leukemia*. http://dx.doi.org/10.1038/leu.2011.216

56 Rego, E.M. *et al*. (2001) Role of promyelocytic leukemia (PML) protein in tumor suppression. *J. Exp. Med*. 193, 521–530

57 Gurrieri, C. *et al*. (2004) Mutations of the PML tumor suppressor gene in acute promyelocytic leukemia. *Blood* 103, 2358–2362

58 Yin, T. *et al*. (2004) Combined effects of As4S4 and imatinib on chronic myeloid leukemia cells and BCR-ABL oncoprotein. *Blood* 104, 4219–4225

59 Mao, J.H. *et al*. (2010) As4S4 targets RING-type E3 ligase c-CBL to induce degradation of BCR-ABL in chronic myelogenous leukemia. *Proc. Natl. Acad. Sci. U.S.A*. 107, 21683–21688

60 Bazarbachi, A. *et al*. (2011) How I treat adult T-cell leukemia/lymphoma. *Blood* 118, 1736–1745

61 El-Sabban, M.E. *et al*. (2000) Arsenic-interferon-alpha-triggered apoptosis in HTLV-I transformed cells is associated with tax down-regulation and reversal of NF-kappaB activation. *Blood* 96, 2849–2855

62 El Hajj, H. *et al*. (2010) Therapy-induced selective loss of leukemia-initiating activity in murine adult T cell leukemia. *J. Exp. Med*. 207, 2785–2792

63 Kchour, G. *et al*. (2009) Phase 2 study of the efficacy and safety of the combination of arsenic trioxide, interferon alpha, and zidovudine in newly diagnosed chronic adult T-cell leukemia/lymphoma (ATL). *Blood* 113, 6528–6532

64 Bernardi, R. and Pandolfi, P.P. (2007) Structure, dynamics and functions of promyelocytic leukaemia nuclear bodies. *Nat. Rev. Mol. Cell Biol*. 8, 1006–1016

65 Winkel, L.H. *et al*. (2011) Arsenic pollution of groundwater in Vietnam exacerbated by deep aquifer exploitation for more than a century. *Proc. Natl. Acad. Sci. U.S.A*. 108, 1246–1251

66 Soignet, S.L. *et al*. (2001) United States multicenter study of arsenic trioxide in relapsed acute promyelocytic leukemia. *J. Clin. Oncol*. 19, 3852–3860

67 Moilanen, A.M. *et al*. (1998) Identification of a novel RING finger protein as a coregulator in steroid receptor-mediated gene transcription. *Mol. Cell. Biol*. 18, 5128–5139

68 Hakli, M. *et al*. (2004) Transcriptional coregulator SNURF (RNF4) possesses ubiquitin E3 ligase activity. *FEBS Lett*. 560, 56–62

69 Sun, H. *et al*. (2007) Conserved function of RNF4 family proteins in eukaryotes: targeting a ubiquitin ligase to SUMOylated proteins. *EMBO J*. 26, 4102–4112

70 Prudden, J. *et al*. (2007) SUMO-targeted ubiquitin ligases in genome stability. *EMBO J*. 26, 4089–4101

71 Shen, T.H. *et al*. (2006) The mechanisms of PML-nuclear body formation. *Mol. Cell* 24, 331–339

72 Muller, S. *et al*. (1998) Conjugation with the ubiquitin-related modifier SUMO-1 regulates the partitioning of PML within the nucleus. *EMBO J.* 17, 61–70

GLOSSARY

Clonogenic activity refers to the ability of cells to self-renew and expand into a set of clones.

Nuclear bodies are proteinaceous nuclear structures organized by PML and recruiting a large number of proteins, many of which are sumoylated. The repertoire of functions influenced or performed by nuclear bodies is as yet unclear.

Nuclear receptors are a group of highly conserved transcription factors activated by hormones, such as estrogens or retinoic acid.

RING domains are a type of Zn finger domains that are commonly associated with proteins involved in ubiquitination. In general, the RING domain is found as part of E3 ligases and facilitates ubiquitination via interactions with the E2 enzyme, mediating transfer of the ubiquitin moiety to the lysine residue. RING domains are also found in proteins not involved in ubiquitination.

ends in Pharmacological Sciences

A Snapshot of Chemoresistance to PARP Inhibitors

Alberto Chiarugi*

Department of Preclinical and Clinical Pharmacology, University of Florence, Viale Pieraccini 6, 50139 Firenze, Italy
Correspondence: alberto.chiarugi@unifi.it

Trends in Pharmacological Sciences, Vol. 33, No. 1, January 2012 © 2012 Elsevier Inc.
http://dx.doi.org/10.1016/j.tips.2011.10.001

SUMMARY

The exploitation of synthetic lethality in BRCA-deficient tumor carriers using potent inhibitors of the enzyme poly(ADP-ribose) polymerase (PARP)-1 has led to an enthusiastic response among basic scientists, oncologists and pharmaceutical companies. However, accumulating evidence demonstrates that resistance to these drugs develops in tumors in both preclinical and clinical settings. Here, I focus on literature dealing with resistance to these drugs and discuss the molecular mechanisms involved, such as restoration of BRCA function, upregulation of nonhomologous end-joining-dependent DNA repair, induction of P-glycoprotein expression and epigenetic deregulation. Clinical implications of resistance to PARP1 inhibitors are also discussed.

EXPLOITING SYNTHETIC LETHALITY FOR TARGETED THERAPY

Selective toxicity against neoplastic cells is the goal of researchers and clinicians involved in developing innovative anticancer therapies. Standard chemotherapy is typically based on targeting of accelerated cell cycling, the key feature of cancer cells. Inevitably, rapidly dividing cells of normal tissues also stop growing when challenged with classic chemotherapeutics, with obvious severe side effects. Even recently developed anticancer drugs, such as those targeting tyrosine kinase receptors, have problems in terms of general efficacy and safety [1,2]. Thus, the major goal of cancer research is still to identify specific alterations in tumor cells that can be selectively targeted

283

through rational drug design. A recent anticancer strategy that is the focus of much research is so-called synthetic lethality. Two mutations are considered synthetically lethal if cells with either of the single mutations are viable, but cells with both mutations are not viable. Thus, genes or proteins that induce cell death when concomitantly inactivated are, by definition, in a synthetic lethal relationship [3–5]. The rationale for targeting synthetic lethality in cancer is that certain cancer cells lack one pathway to repair their DNA (homologous recombination, HR) but have alternative pathways (base excision or single-strand repair) that allow them to survive. Inhibition of these alternative pathways would then impair DNA repair and induce cell death [3,4].

Synthetic lethality (also known as conditional genetics) predicts that genotoxic agents leading to a particular type of DNA damage will kill cancer cells with genetic deficits in repair of that type of damage. Recently, this specific anticancer strategy has been the focus of intense investigations because of positive results from clinical trials of chemical inhibitors of the nuclear enzyme poly(ADP-ribose) polymerase (PARP)-1 in patients with breast or ovarian cancer with genetic inactivation of *BRCA1* or *BRCA2* genes [6,7]. PARP1 is the oldest member of a family of enzymes comprising both poly- and mono(ADP-ribose) transferases [8,9]. PARP1 converts βNAD^+ to long polymers of ADP-ribose [poly(ADP-ribose)] which are targeted to various chromatin-interacting proteins and to PARP1 itself. Poly(ADP-ribosyl)ation significantly affects protein function, and thus contributes to maintenance of several nuclear processes including DNA repair [10]. *BRCA1* and *BRCA2* code for factors essential in DNA repair by HR, a key process for double-strand break repair [11], which is therefore inactive in BRCA1 or -2 deficient neoplasms. The rationale for treating tumors with genetic inactivation of BRCA1 or -2 with PARP1 inhibitors (hereafter abbreviated as PARPi given their ability to inhibit other PARPs in addition to PARP1) is based on the notion that this enzyme assists single-strand break repair [12]. Under PARP1 inhibition, therefore, single-strand breaks cannot be repaired and are converted to double-strand breaks during DNA replication. Thus, in BRCA1 or -2-null cancer cells exposed to PARPi, persistence of double-strand breaks due to HR deficiency leads to replication fork collapse and death. By contrast, these double-strand breaks are readily repaired by HR in cells with wild-type BRCA or a single mutation, so cell survival is possible. Thus, PARPi selectively kill transformed BRCA-deficient cells (Figure 1) [13]. According to recent studies, however, trapping of PARP1 onto DNA lesions or impairment of PARP1-dependent reactivation of stalled replication forks can also underlie synthetic lethality of PARPi in BRCA-deficient cancer cells [14]. Further studies are needed to clearly establish the mechanism(s) of cell death induced by these drugs in cancer patients.

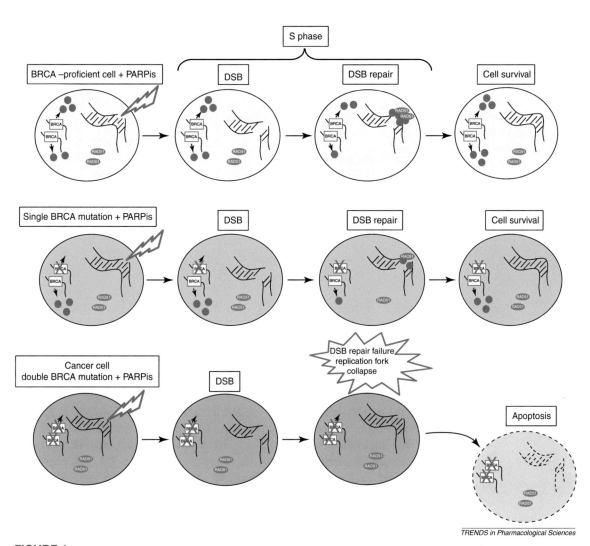

FIGURE 1

The different fate of cells exposed to PARPi according to their BRCA status.

PARP1 is a key activator of single-strand break repair. Hence, in cells exposed to PARPi, single-strand breaks due to DNA damage accumulate and are transformed into double-strand breaks (DSBs) during the S phase of the cell cycle. In a cell with functional BRCA proteins (red dots), DSBs are repaired by homologous recombination owing to BRCA-dependent recruitment of RAD51 to sites of damage, which allows cell survival. The same occurs in cells bearing a single BRCA mutation (such as those of healthy tissues in BRCA patients) because the normal allele is sufficient to carry on homologous recombination. Conversely, cells with a double BRCA mutation (such as those of tumor cells in BRCA patients) are unable to repair DSBs by homologous recombination. Thus, exposure of cells bearing a double BRCA mutation to PARPi causes synthetic lethality (because of concomitant impairment of single-strand break repair and HR), which leads to replication fork collapse and cell death by apoptosis.

Potential efficacy of PARPi as anticancer agents has been demonstrated both *in vitro* and in rodent models of *BRCA1-* or *BRACA2*-deficient neoplasms [13]. As mentioned above, these chemicals were also effective in humans carrying breast or ovarian tumors deficient in *BRCA1* or *BRCA2*. Importantly, evidence that PARPi are chemo- and radiosensitizers in different models of cancer unrelated to *BRCA1* or *-2* deficiency widens the antineoplastic potential of these chemicals [12,13]. The reader is referred to recent comprehensive reviews of PARPi in cancer for basic principles and clinical applicability [6,12,13,15,16]. The present review focuses on the emerging concept of chemoresistance to PARPi. The aim is to provide an appraisal of literature on the occurrence of PARPi resistance (PIR) and to highlight the underlying mechanisms and possible strategies to overcome this resistance.

PIR: EXPERIMENTAL AND CLINICAL EVIDENCE

Three years after the demonstration that PARPi actively kill tumor cells deficient in *BRCA1* or *-2* [17,18], three studies reporting PIR development in cancer cells were published. The Ashworth research team showed that continuous exposure of BRCA2-deficient CAPAN1 pancreatic cancer cells to the potent PARPi KU0058948 induces development of PIR clones, which also display resistance to carboplatin [19], a chemotherapy agent used for treatment of several types of cancer. These findings were confirmed in a paper by Taniguchi and his group showing that CAPAN1 cells with acquired resistance to cisplatin are also resistant to the PARPi AG14361 [20]. Further evidence of PIR stems from *in vivo* experiments to test the antitumor effects of the PARPi AZD2281 (olaparib) in mice in which both *BRCA1* and the tumor-suppressor gene *p53* have been inactivated [21]. This genetic background causes genomic instability and spontaneous occurrence of mammary adenocarcinomas. In these animals, orthotopically transplanted BRCA$^{-/-}$ p53$^{-/-}$ tumors respond to the PARPi with a dramatic reduction in tumor volume. However, tumor growth eventually relapses either during drug treatment or after its discontinuation. Interestingly, the timing of PIR development is variable, occurring as early as 40 days after treatment (i.e. on first relapse) or in later treatment cycles (~400 days from initiation of therapy) [21]. Similar results were obtained in a recent study using the same *in vivo* tumor-transplantation model to evaluate killing of BRCA2-deficient tumors by olaparib or the antimetabolite 6-thioguanine [22].

Data collected in several clinical trials also provide evidence of PIR in cancer patients. In particular, breast cancer eventually progresses (median progression-free survival of 5.7 months) in all patients with BRCA1 or BRCA2 mutations receiving the PARPi olaparib (400mg twice daily, the higher dose tested). Disease progression indicates occurrence of PIR, which apparently does not overlap with resistance to other chemotherapeutic agents [23]. However, evidence that patients progressing after platinum therapy do not respond to olaparib [23] hints that cross-resistance between platinum agents and PARPi may exist. PIR was also observed in patients with recurrent BRCA1- or -2-deficient ovarian cancer and receiving olaparib [24]. In this study, however, a response to olaparib occurred in platinum-resistant patient groups, which argues against absolute cross-resistance between the two chemotherapeutic classes. Additional evidence of PIR comes from a recent study showing that patients with triple-negative breast cancer (i.e. a neoplasm negative for estrogen and progesterone receptors and lacking overexpression of epidermal growth factor receptor type 2) treated with the PARPi iniparib plus chemotherapy eventually experienced disease progression [25].

Considering the potential clinical significance of PARPi in cancer therapy, an understanding of the molecular mechanisms of PIR clearly represents a challenge and a major goal for tumor treatment.

RESTORATION OF BRCA FUNCTION: AN INEVITABLE DRAWBACK OF PARPI?

The two studies that originally reported PIR development in BRCA-deficient cells showed that formation of RAD51 foci (a BRCA-dependent event) is resumed in nuclei of resistant cells exposed to ionizing radiations [19,20]. Similarly, chromosomal aberrations induced by mitomycin C are highly reduced in cells displaying PIR, which is indicative of improved DNA repair [19]. This evidence suggests that the HR machinery is correctly operating in resistant clones. In keeping with this, PIR clones are proficient in HR, as determined in a HR assay that measures recombination between two nonfunctional GFP genes that, when recombined, restore a functional GFP protein [19,20]. Based on the key role of BRCA1 and -2 in HR, these results taken together suggest that BRCA1 or -2 deficiency is somehow overcome in PIR cells. Indeed, careful genetic sequencing demonstrates that secondary mutations of BRCA1 or -2 occur in these cells. In particular, in CAPAN1 cells, intragenic deletion of the protein-truncating c.6174delT frameshift mutation and restoration of the open reading frame (ORF) allows expression of new BRCA2 isoforms. Importantly, when these smaller *BRCA2* genes are transfected into *BRCA2*-deficient PARPi-sensitive cells, HR is restored and PIR readily develops, which thus establishes a causative role of secondary BRCA mutations in chemoresistance to PARPi [19].

Platinum salts used for chemotherapy (such as carboplatin) cause DNA damage requiring efficient HR machinery for repair. Therefore, restoration of BRCA function might also be expected to cause resistance to platinum agents. Consistent with this, restoration of the *BRCA2* ORF is also detected in patients with ovarian carcinoma resistant to carboplatin [19]. Similarly, secondary mutations of *BRCA2* are associated with resistance to cisplatin in the BRCA2-deficient breast cancer cell line HCC1428 [20]. These findings suggest that restoration of BRCA function underlies the development of resistance to both platinum salts and PARPi. Accordingly, all the cisplatin-resistant clones of CAPAN1 cells obtained by selection pressure also display PIR [20]. Recurrent ovarian tumors in cisplatin-treated BRCA2-deficient patients show secondary mutations of BRCA1 [26] or -2 [20]. Interestingly, however, resistance to cisplatin in BRCA-deficient tumors is not always related to BRCA restoration [27], which suggests that various mechanisms may induce platinum resistance in these neoplasms. As discussed below, the same holds true for PARPi.

Theoretically, several mechanisms leading to reversion of mutations could contribute to restoration of BRCA function in humans [28]. Among these are single-strand annealing and nonhomologous end-joining (NHEJ), two error-prone DNA repair mechanisms necessitating homology regions to operate. Interestingly, Ashworth and colleagues found homology in the regions surrounding the deleted sequences of BRCA2 [19]. Theoretically, therefore, single-strand annealing and NHEJ could replace HR to repair double-strand breaks in the absence of functional BRCA. This hypothesis, along with evidence that BRCA mutations induce error-prone, homology-directed repair of DNA double-strand breaks occurring between repeat sequences [29], suggests that specific mutation mechanisms are unleashed in HR-deficient cells exposed to PARPi or platinum salts. This is consistent with the notion that both PARPi and dominant-negative mutation of PARP1 increase recombinant frequency and genomic instability [30]. The anti-recombinogenic role of PARP1 is also exemplified by the finding that PARP1 genetic suppression induces a higher rate of spontaneous sister chromatid exchange [31] and facilitates antigen receptor V(D)J recombination in defective lymphocytes [32].

Although a systematic study of the genetic basis of BRCA reversion in a large cohort of patients exposed to platinum salts or PARPi is still required, identification of the exact molecular mechanisms responsible for BRCA reversion and PIR development is of remarkable clinical importance. Nevertheless, preclinical and clinical evidence indicates that genomic instability promoted by PARPi or platinum compounds in HR-deficient cells favors BRCA reversion. Hence, the same mechanisms responsible for the anticancer effects of PARPi also promote chemoresistance to these drugs. This concept is further strengthened by recent work showing that PARPi stimulate DNA repair by the error-prone NHEJ mechanism in HR-deficient cells (Figure 2) [33]. This

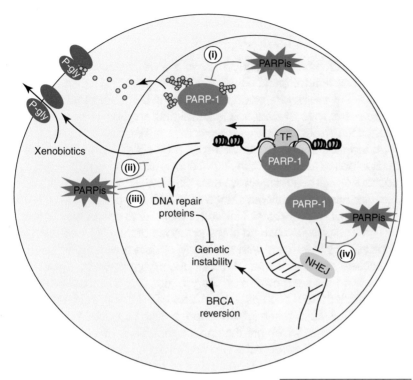

TRENDS in Pharmacological Sciences

FIGURE 2

Schematic representation of the potential mechanisms that contribute to PIR.

Under resting conditions, PARP1 plays key homeostatic roles. One of these is synthesis of poly(ADP-ribose) (chains of pink dots), which is then degraded to ADP-ribose monomers (individual pink dots). ADP-ribose is then able to inhibit the P-glycoprotein carrier (P-gly) at the plasma membrane, and thus impairs xenobiotic extrusion. PARP1 also finely tunes gene transcription by regulating assembly of supramolecular complexes containing transcription factors (TFs). By so doing, PARP1 promotes the expression of various DNA repair factors (including RAD51) and inhibits that of P-glycoprotein. In addition, PARP1 activity inhibits error-prone DNA repair mechanisms such as nonhomologous end-joining (NHEJ). On this basis, PARPi set into motion a cascade of biochemical events leading to PIR. Indeed, by inhibiting PARP1, these drugs abrogate the inhibitory effect of ADP-ribose on P-glycoprotein (1) and promote P-glycoprotein expression (2), two events that increase extracellular extrusion. PARPi molecules also impair PARP1-dependent promotion of the expression of DNA repair factors (3), and thus increase genomic instability and favor BRCA reversion in BRCA-deficient cells. Finally, by relieving the inhibitory action of PARP1 on NHEJ (4), PARPi molecules promote this error-prone DNA repair pathway, which further contributes to genomic instability and increases the probability of BRCA reversion.

event contributes to PARPi-dependent genomic instability and tumor cell death on the one hand, and on the other hand might also underlie PIR due to restoration of expression of the proteins (i.e. BRCA1 and -2) whose deficiency is responsible for HR failure.

P-GLYCOPROTEIN UPREGULATION: AUTO-DEFEATING EFFECTS OF PARPI

Increased cell extrusion of anticancer drugs is a well-recognized event underlying tumor chemoresistance. P-Glycoproteins (also known as multidrug-resistant proteins) are key transporters involved in pumping xenobiotics out of cells [34]. Several lines of evidence indicate that poly(ADP-ribosyl)ation negatively regulates P-glycoprotein expression. The Wesierska-Gadek research team reported that, in contrast to wild-type (WT) cells, PARP1 knockout (KO) fibroblasts exposed to doxorubicin do not show activation of p53. Accordingly, the drug induces cell death only in WT cells [35]. The authors demonstrated that the molecular basis of chemoresistance of PARP1 KO cells involves dramatic overexpression of P-glycoprotein. Indeed, sensitivity to doxorubicin is rescued in PARP1-null fibroblasts cultured in the presence of the P-glycoprotein inhibitor verapamil [35]. In keeping with this study, a more recent report by Rottenberg and colleagues shows that PARPi induce upregulation of P-glycoprotein expression in an *in vivo* mouse mammary tumor model without altering levels of other transporters such as ABCC1 and HPRT1 [21]. Increased expression levels of P-glycoprotein might therefore suffice to reduce the intracellular PARPi concentrations, which are then no longer able to efficiently inhibit their targets. Consistently, P-glycoprotein upregulation correlates with PIR development in mice subjected to long-term treatment with the potent PARPi olaparib, whereas tariquidar (a third-generation inhibitor of P-glycoprotein) rescues the PARPi sensitivity of recurrent tumors in mice [21]. Overall, these findings suggest that PARP1 tonically downregulates P-glycoprotein promoter activity and *trans*-activation, and are in keeping with numerous lines of evidence indicating that PARP1 activity epigenetically regulates gene expression by targeting various DNA-interacting proteins [36,37]. Conversely, PARPi unleash this tonic suppression and therefore promote their extrusion from cancer cells [21]. However, it is currently not known whether PARP1 and/or additional PARPs bind upstream regulating regions of genes coding for P-glycoprotein isoforms or whether this binding is modulated by PARPi.

Further evidence of a functional relationship between PARP1 and P-glycoprotein stems from work by Dumitriu and colleagues [38]. The authors showed that ABC transporters (the carrier family to which P-glycoprotein belongs) are inhibited by ADP-ribose in human and mouse lymphocytes. Cellular ADP-ribose mainly arises from poly(ADP-ribose) formed by PARP1 and degraded by poly(ADP-ribose) glycohydrolase (PARG) [39]. In keeping with this, both PARP1 and PARG inhibitors ameliorate ABC transporter efficiency [38]. Whether ADP-ribose also inhibits extrusion of anticancer drugs from tumor cells remains to be investigated. Interestingly, we recently reported that nudix (nucleoside diphosphate linked to X) hydrolases rapidly convert ADP-ribose to AMP in tumor cells [40]. Considering that these ancestral enzymes of bacterial origin are involved in detoxification of nucle-otides that interfere with DNA metabolism [41], it is possible that they might also be involved in detoxification of xenobiotics by reducing ADP-ribose levels and therefore increasing ABC transporter activity. A scenario could be envisaged in which the concerted actions of PARPs, PARG and nudix hydrolases regulate ADP-ribose accumulation and ABC transporter activity in neoplastic cells. Interestingly, this mechanism implies that basal degra-dation of poly(ADP-ribose) could contribute to the sensitivity threshold of a given cancer cell to antineoplastic agents. Overall, although we still need clinical evidence of ABC transporter involvement in PIR, literature data sug-gest that suppression of poly(ADP-ribosyl)ation by PARPi indirectly poten-tiates cellular ability to extrude drugs and thereby favors chemoresistance, including PIR (Figure 2). Ironically, even in this case PARPi might activate complex molecular mechanisms that counteract their antitumor effects.

FURTHER CONSIDERATIONS

There is no doubt that PIR might dampen the enthusiasm of researchers and clinicians for the antineoplastic potential of PARPi. In this regard, however, it is important to consider the following reflections. Resistance to chemother-apy is a widespread phenomenon for drugs that nonetheless retain remark-able therapeutic value. Moreover, exploitation of synthetic lethality in tumor therapy is a rather new approach [3] and at present it is not known how eas-ily transformed cells become resistant to this anticancer strategy. Additional preclinical and clinical studies are therefore needed to establish whether PIR is a real drawback that seriously biases clinical use of PARPi. Another issue of key relevance for PIR is that it occurs in patients already exposed to chemotherapy, an event that might have favored drug resistance. It will be important, therefore, to understand whether and how prior treatment with anticancer agents predisposes to PIR.

It is also worth noting that PIR seems to be due to peculiar molecular mech-anisms. Indeed, it is unusual that a chemotherapeutic drug loses its efficacy just because it sets in motion a cascade of events that reactivates the same gene whose mutation conferred the original sensitivity to the drug. Likewise, it is curious that a chemotherapeutic reduces its cytotoxic effects because, by inhibiting its target enzyme, it relieves blockage of the transporters respon-sible for its cellular extrusion (Figure 2). It is currently not known whether the peculiarity of the molecular mechanisms responsible for PIR favor the development of strategies able to overcome it. On a purely theoretical basis, however, PIR might be self-limiting. Indeed, the very same genomic insta-bility that leads to BRCA restoration might subsequently lead to an addi-tional mutation of restored BRCA that would re-establish the original HR deficiency. In addition, extrusion of PARPi due to increased expression of

P-glycoprotein should reduce their intracellular concentration and thereby restore PARP1-dependent suppression of P-glycoprotein expression. It is currently not known whether molecular mechanisms underlying PIR in cells or animal models are also responsible for PIR in cancer patients, which represents a further ambiguity around PIR. Functional restoration of BRCA in patients on cisplatin has been reported [26,27], but it is not known whether this also occurs in those receiving PARPi. Importantly, it is also not known whether PARPi promote error-prone DNA repair pathways or affect P-glycoprotein expression in cancer patients.

Additional events triggered by PARPi could contribute to PIR. For instance, PARP1 is a master epigenetic regulator of gene expression, and PARPi significantly affect gene expression profiles in different cells and tissues [36,37]. Hence, PARPi-dependent alterations in tumor progression gene transcription might contribute to PIR. In this regard, upregulation of RAD51 expression can occur in BRCA1-null breast cancer cells and is sufficient to restore efficient HR and cell viability [42]. In principle, therefore, acute and/or chronic treatment with PARPi might alter expression levels of RAD51 (or additional DNA repair proteins) in BRCA patients and thereby reduce sensitivity to the drugs (Figure 2). In addition, PARPi-dependent perturbation of miRNAs could also contribute to PIR development. According to a very recent study, BRCA1 transcripts are downregulated by miR-182, whereas BRCA1 expression increases and induces PIR when miR-182 levels are reduced [43].

In terms of strategies potentially able to circumvent PIR, preclinical evidence demonstrates that it can be overcome by P-glycoprotein inhibition or the use of 6-thioguanine [21,22,35]. These approaches (which have already been used to overcome chemoresistance of different tumors) might also be applied to overcome PIR in patients. Furthermore, development of PARPi that are not substrates of extrusion carriers might help in preventing PIR. It is also conceivable that PIR might not develop during treatment of malignancies without deficits in HR. As an alternative strategy, given the increased recombination frequency in cells null for PARP1 [31,32] or exposed to PARPi [30], PIR might also be counteracted by repression of recombination activity in tumor cells. Finally, evidence that a mild increase in temperature from 37° to 41°C induces BRCA2 degradation and sensitizes cancer cells to PARPi [44] suggests that mild hyperthermia might be harnessed to reduce levels of restored BRCA and thereby defeat PIR.

In the future, it will be necessary to avoid clinical PARPi testing in premature development stages. For instance, iniparib is a weak PARP1 inhibitor that has several targets in addition to PARP1 [45]. This lack of selectivity is of major concern when interpreting clinical data on iniparib, such as the failure of the drug to prolong overall survival in patients with metastatic triple-negative breast cancer enrolled in a recent Phase III trial [46]. Another key issue that must be considered for successful clinical PARPi development is accurate genetic profiling of BRCA1 and -2 mutations in patients. Lack of this information, such as in the above-mentioned trial with iniparib, renders interpretation of clinical results very difficult and hampers an understanding of the real therapeutic potential of PARPi. In addition, knowledge about PARP1 expression levels in neoplasms treated with PARPi could help to avoid false-negative results [47].

CONCLUDING REMARKS

Chemoresistance is a crucial issue for clinical PARPi development. However, this issue should not be allowed to obscure the anti-tumor potential of these drugs or to become a concern for pharmaceutical industries or regulatory agencies. It should be remembered that PARPi have unprecedented therapeutic potential for the treatment of various human disorders [48] and it would be disappointing if hasty decisions were made. An additional major question that must be answered in the near future is the safety profile of PARPi. Although they seem to be sufficiently well tolerated in clinical trials, because of genomic instability directly caused by impairment of DNA repair, their carcinogenic potential as mono- or add-on therapies is still unknown. A fundamental requirement is an understanding of the frequency of PIR in large cohort of patients receiving PARPi and whether it also occurs in neoplasms without HR defects.

In conclusion, PARPi represent an extraordinarily important new class of compounds that, as usual during drug development, is facing problems in clinical translation, including PIR. At this stage I would recommend that indications stemming from basic science and the clinic should be followed, rather than be confused by misleading financial or regulatory concerns.

ACKNOWLEDGMENTS

I thank Flavio Moroni, Guy Poirier, Robert Sobol, Zhao-Qi Wang and Mathias Ziegler for helpful discussion and comments.

REFERENCES

1 Janne, P.A. *et al*. (2009) Factors underlying sensitivity of cancers to small-molecule kinase inhibitors. *Nat. Rev. Drug Discov.* 8, 709–723

2 Force, T. and Kolaya, K.L. (2011) Cardiotoxicity of kinase inhibitors: the prediction and translation of preclinical models to clinical outcomes. *Nat. Rev. Drug Discov.* 10, 111–126

3 Chan, D.A. and Giaccia, A.J. (2011) Harnessing synthetic lethal interactions in anticancer drug discovery. *Nat. Rev. Drug Discov.* 10, 351–364

4   Kaelin, W.G., Jr (2005) The concept of synthetic lethality in the context of anticancer therapy. *Nat. Rev. Cancer* 5, 689–698

5   de Bono, J.S. *et al.* (2003) The future of cytotoxic therapy: selective cytotoxicity based on biology is the key. *Breast Cancer Res.* 5, 154–159

6   Yap, T.A. *et al.* (2011) Poly(ADP-ribose) polymerase (PARP) inhibitors: exploiting a synthetic lethal strategy in the clinic. *CA Cancer J. Clin.* 61, 31–49

7   Carey, L.A. and Sharpless, N.E. (2011) PARP and cancer – if it's broke, don't fix it. *N. Engl. J. Med.* 364, 277–279

8   Hottiger, M.O. *et al.* (2010) Toward a unified nomenclature for mammalian ADP-ribosyltransferases. *Trends Biochem. Sci.* 35, 208–219

9   Ame, J.C. *et al.* (2004) The PARP superfamily. *Bioessays* 26, 882–893

10 Schreiber, V. *et al.* (2006) Poly(ADP-ribose): novel functions for an old molecule. *Nat. Rev. Mol. Cell Biol.* 7, 517–528

11 Sancar, A. *et al.* (2004) Molecular mechanisms of mammalian DNA repair and the DNA damage checkpoints. *Annu. Rev. Biochem.* 73, 39–85

12 Rouleau, M. *et al.* (2010) PARP inhibition: PARP1 and beyond. *Nat. Rev. Cancer* 10, 293–301

13 Mangerich, A. and Burkle, A. (2011) How to kill tumor cells with inhibitors of poly(ADP-ribosyl)ation. *Int. J. Cancer* 128, 251–265

14 Helleday, T. (2011) The underlying mechanisms for the PARP and BRCA synthetic lethality: clearing up the misunderstandings. *Mol. Oncol.* 5, 387–393

15 Papeo, G. *et al.* (2009) Poly(ADP-ribose) polymerase inhibition in cancer therapy: are we close to maturity? *Expert Opin. Ther. Pat.* 19, 1377–1400

16 Calvert, H. and Azzariti, A. (2011) The clinical development of inhibitors of poly(ADP-ribose) polymerase. *Ann. Oncol.* 22(Suppl. 1), i53–i59

17 Bryant, H.E. *et al.* (2005) Specific killing of BRCA2-deficient tumours with inhibitors of poly(ADP-ribose) polymerase. *Nature* 434, 913–917

18 Farmer, H. *et al.* (2005) Targeting the DNA repair defect in BRCA mutant cells as a therapeutic strategy. *Nature* 434, 917–921

19 Edwards, S.L. *et al.* (2008) Resistance to therapy caused by intragenic deletion in *BRCA2*. *Nature* 451, 1111–1115

20 Sakai, W. *et al.* (2008) Secondary mutations as a mechanism of cisplatin resistance in BRCA2-mutated cancers. *Nature* 451, 1116–1120

21 Rottenberg, S. *et al.* (2008) High sensitivity of BRCA1-deficient mammary tumors to the PARP inhibitor AZD2281 alone and in combination with platinum drugs. *Proc. Natl. Acad. Sci. U.S.A.* 105, 17079–17084

22 Issaeva, N. *et al.* (2010) 6-thioguanine selectively kills BRCA2-defective tumors and overcomes PARP inhibitor resistance. *Cancer Res.* 70, 6268–6276

23 Tutt, A. *et al.* (2010) Oral poly(ADP-ribose) polymerase inhibitor olaparib in patients with *BRCA1* or *BRCA2* mutations and advanced breast cancer: a proof-of-concept trial. *Lancet* 376, 235–244

24 Audeh, M.W. *et al.* (2010) Oral poly(ADP-ribose) polymerase inhibitor olaparib in patients with *BRCA1* or *BRCA2* mutations and recurrent ovarian cancer: a proof-of-concept trial. *Lancet* 376, 245–251

25 O'Shaughnessy, J. *et al.* (2011) Iniparib plus chemotherapy in metastatic triple-negative breast cancer. *N. Engl. J. Med.* 364, 205–214

26 Swisher, E.M. *et al.* (2008) Secondary *BRCA1* mutations in BRCA1-mutated ovarian carcinomas with platinum resistance. *Cancer Res.* 68, 2581–2586

27 Sakai, W. *et al.* (2009) Functional restoration of BRCA2 protein by secondary *BRCA2* mutations in BRCA2-mutated ovarian carcinoma. *Cancer Res.* 69, 6381–6386

28 Hirschhorn, R. (2003) *In vivo* reversion to normal of inherited mutations in humans. *J. Med. Genet.* 40, 721–728

29 Tutt, A. *et al.* (2001) Mutation in *Brca2* stimulates error-prone homology-directed repair of DNA double-strand breaks occurring between repeated sequences. *EMBO J.* 20, 4704–4716

30 Schreiber, V. *et al.* (1995) A dominant-negative mutant of human poly(ADP-ribose) polymerase affects cell recovery, apoptosis, and sister chromatid exchange following DNA damage. *Proc. Natl. Acad. Sci. U.S.A.* 92, 4753–4757

31 Wang, Z.Q. *et al.* (1997) PARP is important for genomic stability but dispensable in apoptosis. *Genes Dev.* 11, 2347–2358

32 Morrison, C. *et al.* (2011) Genetic interaction between PARP and DNA-PK in V(D)J recombination and tumorigenesis. *Nat. Med.* 17, 479–482

33 Patel, A.G. *et al.* (2011) Nonhomologous end joining drives poly(ADP-ribose) polymerase (PARP) inhibitor lethality in homologous recombination-deficient cells. *Proc. Natl. Acad. Sci. U.S.A.* 108, 3406–3411

34 Lee, C.A. *et al.* (2010) P-glycoprotein related drug interactions: clinical importance and a consideration of disease states. *Expert Opin. Drug Metab. Toxicol.* 6, 603–619

35 Wurzer, G. *et al.* (2000) Increased resistance to anticancer therapy of mouse cell lacking the poly(ADP-ribose) polymerase attributable to up-regulation of the multidrug resistance gene product P-glycoprotein. *Cancer Res.* 60, 4238–4244

36 Krishnakumar, R. and Kraus, W.L. (2010) The PARP side of the nucleus: molecular actions, physiological outcomes, and clinical targets. *Mol. Cell* 39, 8–24

37 Kraus, W.L. (2008) Transcriptional control by PARP-1: chromatin modulation, enhancer-binding, coregulation, and insulation. *Curr. Opin. Cell Biol.* 20, 294–302

38 Dumitriu, I.E. *et al.* (2004) UV irradiation inhibits ABC transporters via generation of ADP-ribose by concerted action of poly(ADP-ribose) polymerase-1 and glycohydrolase. *Cell Death Differ.* 11, 314–320

39 Rossi, L. *et al.* (2002) Poly(ADP-ribose) degradation by post-nuclear extracts from human cells. *Biochimie* 84, 1229–1235

40 Formentini, L. *et al.* (2009) Poly(ADP-ribose) catabolism triggers AMP-dependent mitochondrial energy failure. *J. Biol. Chem.* 284, 17668–17676

41 McLennan, A.G. (2006) The Nudix hydrolase superfamily. *Cell Mol. Life Sci.* 63, 123–143

42 Martin, R.W. *et al.* (2007) RAD51 up-regulation bypasses BRCA1 function and is a common feature of BRCA1-deficient breast tumors. *Cancer Res.* 67, 9658–9665

43 Moskwa, P. *et al.* (2011) miR-182-mediated downregulation of BRCA1 impacts DNA repair and sensitivity to PARP inhibitors. *Mol. Cell* 41, 210–220

44 Krawczyk, P.M. *et al.* (2011) Mild hyperthermia inhibits homologous recombination, induces BRCA2 degradation, and sensitizes cancer cells to poly (ADP-ribose) polymerase-1 inhibition. *Proc. Natl. Acad. Sci. U.S.A.* 108, 9851–9856

45 Bauer, P.I. *et al.* (2002) Anti-cancer action of 4-iodo-3-nitrobenzamide in combination with buthionine sulfoximine: inactivation of poly(ADP-ribose) polymerase and tumor glycolysis and the appearance of a poly(ADP-ribose) polymerase protease. *Biochem. Pharmacol.* 63, 455–462

46 Wendling, P. (2011) Iniparib loses blockbuster image in triple-negative breast cancer. *Oncol. Rep.* 13–14

47 Domagala, P. *et al.* (2011) I PARP-1 expression in breast cancer including BRCA1-associated, triple negative and basal-like tumors: possible implications for PARP-1 inhibitor therapy. *Breast Cancer Res. Treat.* 127, 861–869

48 Virag, L. and Szabo, C. (2002) The therapeutic potential of poly(ADP-ribose) polymerase inhibitors. *Pharmacol. Rev.* 54, 375–429

Index

Note: Page numbers with "f" denote figures; "t" tables; "b" boxes.

Printed in the United States
By Bookmasters